F

ΔI

Robot Modelling
Control and Applications with Software

ROBOT MODELLING
CONTROL AND APPLICATIONS
WITH SOFTWARE

P. G. Ránky
C. Y. Ho

IFS (Publications) Ltd, UK
Springer-Verlag
Berlin · Heidelberg · New York · Tokyo
1985

British Library Cataloguing in Publication Data

Ránky, Paul G.
 Robot modelling: control and applications with software
 1. Robots, Industrial — Data processing
 I. Title   II, Ho, C. Y.
 629.8'92   TS191.8

ISBN 0-903608-72-3   IFS (Publications) Ltd
ISBN 3-540-15373-X Springer-Verlag Berlin Heidelberg New York Tokyo
ISBN 0-387-15373-X Springer-Verlag New York Heidelberg Berlin Tokyo

© 1985 **IFS (Publications) Ltd**, 35–39 High Street, Kempston,
   Bedford MK42 7BT, England
   and **Springer-Verlag** Berlin Heidelberg New York Tokyo

Phototypeset by Fleetlines Typesetters, Southend-on-Sea, Essex
Printed and bound by Short Run Press Ltd, Exeter

ACKNOWLEDGEMENTS

We would like to thank all our friends and colleagues on both sides of the Atlantic, whose kind help contributed in some way to this book.

It would be impossible to list here all names, companies and institutions, but we would like to use the available space to express our thanks in particular to Jim Corlett, Technical Librarian at Trent Polytechnic, and to Mike Innes, our Editor at IFS Publications for editing the manuscript and for useful suggestions.

Special thanks are also expressed to Professor Philip Francis at ITI in Ann Arbor, Professor Ferdinand Leimkuhler at Purdue University, Ian Taylor of Cincinnati Milacron and Robin Truman at Plessey Office Systems for their constant encouragement in the robotics and FMS research work.

We would also like to express our thanks to Trent Polytechnic and University of Missouri for their support and to all those institutions and companies which have provided us with useful information, photographs and experience through consultancy and part-time lecturing, or through employment, including: IBM Corporation, Plessey Office Systems Ltd, Csepel Machine Tool Company, University of Michigan, Industrial Technology Institution (ITI) in Ann Arbor, Michigan, Purdue University in West Lafayette, Sandvik Coromant, Hertel and the Lucas Group.

In addition, thanks are due to Paul Ránky's British students who helped in testing simulation algorithms and/or programs written for this book, in particular Vijay Chhatralia, Simon Jones, Mark Walker and Robert Walker.

Equally warm thanks are expressed to the following USA students for their help in organising the notes of Peter Ho's Robotic Systems course at Missouri for Chapters 2 to 7 and 9: David Burt, Alan Dragoo, William Eldridge, Robert Guehne, Louis Gentry, Starline Judkins, Edward Hammerand, B. V. Satish, K. Vishweshwara and Jen Sriwattanathamma.

Finally let us thank our darling wives annd families for their continuous support and hard work during weekends, early mornings and late nights when this text was written, as well as British Telecom and AT & T for the expensive but technically perfect telephone link between the two of us in Nottingham, Rolla and/or Boca Raton.

Paul Ránky and C. Y. (Peter) Ho

CONTENTS

PREFACE

The idea of this book was borne in my mind in the Wollaton Park in Nottingham on a beautiful sunny day early in 1983, when Peter came to visit me to discuss some technical aspects of Flexible Manufacturing and Assembly Systems. We both felt that there was no book on robotics which covered both theory and application, including programming, simulation of robotised manufacturing and assembly systems utilising not only one arm, but many robots and many different end-effectors.

Our basic aim was to explain why industrial robots are important for industry, how they work, how they are controlled and programmed, how their implementation should be prepared, what kind of robot tools are available and how to build up complex manufacturing systems of robots rather than utilising a single arm only.

The hard work of developing the project and finalising the structure of the book was done in 1984. The text includes several new fields of interests relating to FAS (Flexible Assembly Systems), robotised system simulation by means of the FMS Software Library programs (created by myself), robot positioning and orientation test, hand integrated sensors, automated robot hand changers (ARHC) and robot programming.

This book has been written for the engineering manager in industry as well as for lecturers and students at Universities and Polytechnics.

Part One gives technical insights on general purpose mathematical models including the link equations, kinematic and arm position control equations, differential relationships and speed control, dynamic control and model.

Part Two is more applications and systems orientated. Here we discuss robot programming languages with several examples, task planning and compliance, systematic planning of robot installations and robot projects, three-dimensional robot error testing and robotised manufacturing and assembly system design, and simulation by means of software.

Besides giving technical details, numerical examples, sample runs of simulation programs and sample robot programs, we also maintain our interest in providing the reader with an overview of each field discussed.

It is certainly our hope that these goals have been achieved, and that this book will contribute to the understanding and development of robotised manufacturing and assembly systems.

Paul G. Ránky
Nottingham, June 1985

Chapter One

INTRODUCTION TO INDUSTRIAL ROBOTS AND ROBOTISED ASSEMBLY SYSTEMS

IT IS instructive to analyse the way systems develop for industrial robots, manipulators and complex robotised manufacturing, assembly, painting, wire looming, welding, and sheet metal processing, etc. The most interesting trend is that machining centres, designed both for milling or turning, utilise integrated robot loaders or manipulators for part and tool changing (Fig. 1.1), and these features have been adapted for plastic moulding machines, sheet metal manufacturing equipment and other machines in a variety of different process industries.

The robot industry is also catching up with the flexible manufacturing cell concept and practice utilised in the typically machining orientated industries. This is proved by the fact that up-to-date robots are capable of working in a system, rather than as stand-alone machines, that they require sophisticated controllers and structured programming languages, that they offer data communication between robots and computers working in local area networks, and that they need hand-changing to make them more flexible and integrated sensors to make them more intelligent (Fig. 1.2).

There are several different devices used for loading and unloading machines for automated assembly, welding, painting, sheet metal manufacturing, laser cutting, etc., and many are called 'industrial robots', although only a few are really intelligent enough to satisfy the criteria of a 'robot'.

The definitions of industrial robots vary. The British Robot Association (BRA) defines a robot as "A reprogrammable device designed to both manipulate and transport parts, tools or specialised manufacturing implements through variable programmed motions for the performance of specific manufacturing tasks."

According to the Robot Institute of America (RIA) a robot is "a programmable, multifunction manipulator designed to move material, parts, tools, or specialised devices through variable programmed motions for the performance of a variety of tasks."

Fig. 1.1 Machine integrated part loading and unloading manipulator

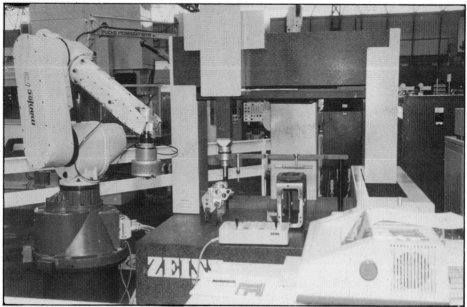

Fig. 1.2 Inspection machine and robot working in an integrated system

The Japanese Industrial Robot Association (JIRA) specifies five different levels of industrial robots:

● Manipulators, which are directly operated by human beings.
● Sequence robots, which fall into a further two categories, i.e. fixed and variable sequence robots.
● Playback robots, which execute taught fixed instructions.
● The NC robots, which execute numerically loaded information.
● Intelligent robots, having their own sensory based system which helps their programs to make decisions in real-time.

More generally, industrial robots are discussed in terms of generations. The first generation had no sensing or computing power, whereas the second has sensory feedback, a reasonably high level of computing power and an explicit-text high-level language. The third generation robots are those which will allow decision making and problem solving through sensory feedback processing based on AI techniques.

It is important to emphasise that one must always think in terms of a system, in which the robot itself is only one component. Such a system generally consists of the robot arm itself, with its controller, the necessary grippers and/or robot tools, and the part feeding, locating, orientating and other mechanical and sensory based devices which are interfaced with it. In flexible manufacturing and assembly systems a further important feature is the necessary provision of interfaces and communication facilities between the robot controller and an intelligent node of the computer network.

Robot applications and programming methods have expanded far beyond merely handling components since their controllers and sensors have evolved the capacity to communicate with other intelligent devices. Sensors particularly have been the focus of much development in recent years.

Robots are able to provide information not only about their arm position, but also about the part being handled, and about the gripper or more complex robot tool. They are capable of seeing, hearing, smelling, detecting and analysing force, torque, heat, pressure, colour and other environmental changes or conditions. Simple touch-triggered and complex non-contact sensors can provide information about the presence of a part in a buffer store, in the robot tool, or in the chuck of a lathe, and they can, for example, interact with the robot controller and modify a preprogrammed sequence in real-time.

It is important to realise that distributed, sensory feedback processing makes robots – and other devices – more intelligent, more reliable and more flexible, and these important trends are not to be ignored when designing robot integrated systems.

Industrial robots have a very wide range of potential applications, because they are reprogrammable, flexible devices themselves. The increasing power of their sensory feedback processing system allows them to work at a high level of intelligence, although we are far from being able to say that robots are capable of solving most material handling, assembly, and inspection tasks, etc. At present, a relatively low proportion of robots are capable of communicating at a high level with other robots and

Fig. 1.3 The Cincinatti Milacron T³ utilised as a machine loading/unloading robot

Fig. 1.4 Welding petrol tanks using an ASEA robot

computers, and their sensory systems are usually expensive compared to their performance. A considerable amount of R & D work is therefore still required before they reach the realistic limits of their potential industrial application.

Without aiming to provide a full list of applications in this rapidly changing field, the major application areas of industrial robots include:

● Pick-and-place type operations, in which case point-to-point controlled devices are usually sufficient: machine loading and unloading (Fig. 1.3), simple palletising, glass handling, brick manufacture, press work, packaging and package distributing, warehouse service, foundry work, heat treating, coating and electroplating, investment casting, die casting, forging, and metal coating.

● Welding robots (Fig. 1.4), which are the largest in terms of population since the automotive industry employs the greatest number (over 40% of the total robot population). Both point-to-point controlled and continuous-path controlled robots are widely applied for: spot welding, arc welding, seam welding, flame cutting, laser welding, and plasma cutting.

● Machining robots (Fig. 1.5) requiring a rigid body and arm structure with adequate positioning repeatability. They often have to change tools and/or hands to accommodate different jobs. Machining applications include: drilling, routing, sheet metal fabrication, and composite materials manufacturing.

● Cleaning and deburring applications with special purpose tools (Fig. 1.6). Such robot applications often need adaptive force feedback sensing as well as automated robot tool changing.

● Spray painting (Fig. 1.7).

● Automated assembly and inspection (Fig. 1.8), which is probably the most interesting and fastest growing area of applications. This is due to the potential of the rapidly growing market of the assembly industry and its complexity in programming, tooling, sensory feedback processing, interfacing with other devices and communicating with remote computers.

It has been emphasised in many publications and by different robot manufacturers that in the long run robots can do jobs faster than human operators, they are more consistent and more reliable than humans, and they can work in hazardous environments. In addition they cost less than employing human workers (for repetitive jobs – obviously there are tasks which only humans can solve, but unfortunately this requires a lot of training and hard work).

Even in mechanised assembly, where dedicated transfer lines dominated the industry for several decades (Figs. 1.9 and 1.10), robots started to compete with the faster, but less flexible, dedicated machines.

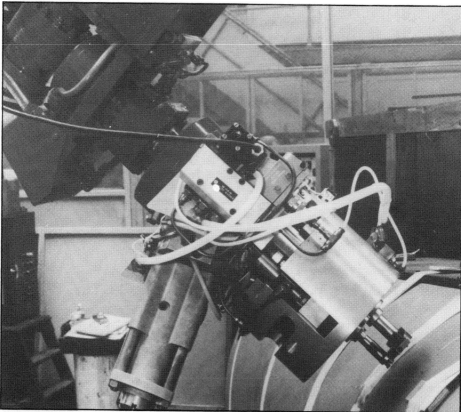

Fig. 1.5 Cincinatti Milacron robot with compliant head used for drilling operations

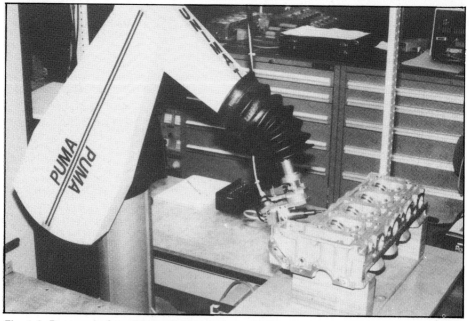

Fig. 1.6 Removal of machining looms using a PUMA industrial robot

Fig. 1.7 Robot spraying a boat hull

Fig. 1.8 KUKA robot mounting car wheels with the aid of a force/torque sensor

Fig. 1.9 Dedicated assembly machine

The reasons are clear:

- Very few manufacturers can ensure reliable orders of the same product for several years, thus it is often not possible to justify the large capital investment for a dedicated machine or line.
- Products change rapidly, thus alterations in the manufacturing process are unavoidable.
- Batch sizes also change, thus mixed production and assembly lines are required. Changes may even be as a result of seasonal requirements (e.g. fans are required mainly in the summer).
- The maintenance of flexible systems is much easier than that of dedicated systems, where maintenance usually upsets normal production.

It is important to remember that robotised assembly, or other system design processes, must start with the analysis and very often with the redesign of the parts to be assembled. Design for assembly and in general for manufacturing is of crucial importance. In this area CAD/CAM systems need to work with expert systems and should become increasingly intelligent.

The best solution would be to avoid assembly entirely, but in many cases this is not possible, because of functional constraints, because different materials must be used (e.g. rubber and steel), and because of maintenance, the requirement of mobility of components, or a combination of the above reasons.

If assembly cannot be avoided the part should be made suitable for flexible, robotised assembly. In general this can be assured by: shape, weight, material, mechanical interface (i.e. the way it can be gripped for manipulation purposes), manufacturing accuracy, possible part orientation, and other less important factors.

Fig. 1.10 Mechanised assembly system (courtesy of Plessey Office Systems Ltd)

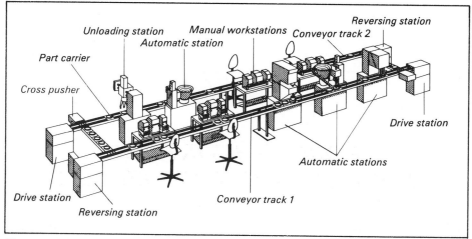

Fig. 1.11 Double-belt assembly conveyor system arranged in a rectangle. Note the way manual and automated workstations are arranged (courtesy Robert Bosch GmbH)

To summarise, up-to-date assembly systems require:

● A suitable modular product design process prior to assembly.
● High dimensional accuracy of the components to be assembled (usually in the region of ±0.005mm).
● Assembly without the need for adjustments and in-process inspection of certain dimensional values or performance characteristics.
● Full computer control and report generation of each assembled component.
● 100% quality control.
● Mixed batch assembly with flexible robotic cells, capable of changing hands (or indexing robot tools) and parts, and utilising sensory feedback processing.

Although robots are computers with hands and thus theoretically can do any work, it is often not yet economic to let them to do every task, thus assembly systems sometimes incorporate human operators. In such cases systems must be very carefully balanced and sufficient buffers must be designed to decrease the pressure generated by the cycle times of the robots working in the same system (see Fig. 1.11). (See also Chapter Thirteen, Section 13.3.)

Robots are not only coming but are here, and are often waiting to be integrated into more intelligent, more complex and more productive flexible manufacturing and assembly systems. But before discussing their programming and application possibilities, let us study some very important theoretical and mathematical modelling concepts relating to the control of the single robot arm.

PART I

ROBOT THEORY AND MATHEMATICAL MODELLING

Chapter Two

MATHEMATICAL PRELIMINARY OF ROBOT MODELLING

THE ROBOT arm consists of several rigid bodies, called links, connected sequentially by revolute or prismatic joints, to achieve the required rotational and/or translational motion. The study of robot manipulation comprises establishing the special relationship between the manipulator and the manipulated object. The link motion equation and the spatial geometry are described by the homogeneous transformation matrix. (The concept of homogeneous transformation was first introduced by Denavit and Hartenberg[1] and was later applied to robotics by Pieper[2] and Paul[3].)

Vector and matrix algebra are used to develop a systematic and generalised approach for representing the location and orientation of robot links with respect to a reference frame. Since each robot link can rotate and/or translate with respect to a reference coordinate frame, a body attached coordinate frame is established at each joint. The direct kinematic problem is then reduced to finding a transformation matrix that relates the body attached coordinate frame to the fixed reference coordinate frame.

The numerical examples in this chapter are extracted from Paul[3].

2.1 Position of a point

The position of a point is a vector from the origin of a specified reference coordinate system to the point. The position vector in terms of its components along the coordinate axes is given by

$$\bar{v} = a\bar{i} + b\bar{j} + c\bar{k} \tag{2.1}$$

where $\bar{i}$, $\bar{j}$ and $\bar{k}$ are unit vectors along X, Y and Z coordinate axes respectively, as shown in Fig. 2.1.

In addition, the direction of the position vector can be expressed by the direction cosines:

$$\cos\alpha = a/v; \cos\beta = b/v; \cos\gamma = c/v \tag{2.2}$$

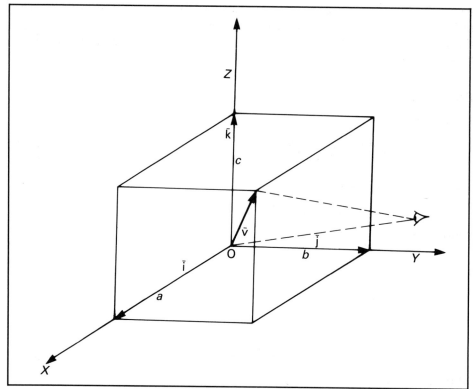

Fig. 2.1 Vector representation

where α, β and γ are, respectively, the angles measured from the positive coordinate axes to the vector $\bar{v}$, which has length $v = (a^2+b^2+c^2)^{1/2}$. The vector can then be represented in homogeneous coordinates as a column matrix:

$$\bar{v} = \begin{bmatrix} x \\ y \\ z \\ w \end{bmatrix} \tag{2.3}$$

where $a = x/w$, $b = y/w$, and $c = z/w$, where w is a scale factor.
   Let vector $\bar{v} = 3\bar{i}+4\bar{j}+8\bar{k}$. Thus,

$$\bar{v} = \begin{bmatrix} 3 \\ 4 \\ 8 \\ 1 \end{bmatrix} \tag{2.4}$$

in an homogeneous coordinate system.
   Also,

$$\bar{v} = \begin{bmatrix} 30 \\ 40 \\ 80 \\ 10 \end{bmatrix} = \begin{bmatrix} -6 \\ -8 \\ -16 \\ -2 \end{bmatrix} = [3,4,8,1]^{\mathrm{T}} \tag{2.5}$$

where $[3,4,8,1]^T$ represents the transpose of the matrix in Eqn.(2.4). The null vector is $[0,0,0,n]^T$, where n is any non-zero scale factor. The directional vector is given by $[a,b,c,0]^T$, which points direction with infinite length.

Dot product

Let $\bar{a}$ and $\bar{b}$ represent two vectors, i.e.

$$\bar{a} = a_x\bar{i}+a_y\bar{j}+a_z\bar{k} \qquad (2.6)$$
$$\bar{b} = b_x\bar{i}+b_y\bar{j}+b_z\bar{k} \qquad (2.7)$$

The vector dot product is

$$\bar{a}.\bar{b} = a_xb_x+a_yb_y+a_zb_z \qquad (2.8)$$

which is scalar.

Cross product

The vector cross product of $\bar{a}$ and $\bar{b}$ is

$$\bar{a} \times \bar{b} = \begin{bmatrix} \bar{i} & \bar{j} & \bar{k} \\ a_x & a_y & a_z \\ b_x & b_y & b_z \end{bmatrix} \qquad (2.9)$$

which is a vector.

2.2 Planes

A plane can be represented as a row matrix

$$P = [a,b,c,d] \qquad (2.10)$$

A point V is said to lie in the plane if PV $= 0$, or

$$[a,b,c,d] \begin{bmatrix} x \\ y \\ z \\ w \end{bmatrix} = 0 \qquad (2.11)$$

or $\qquad\qquad ax + by + cz + dw = 0 \qquad (2.12)$

If PV > 0, then the point lies above the plane. If PV < 0, the point lies below the plane.

2.3 Homogeneous transformation

The homogeneous transformation is used to describe the position and orientations of coordinate frames in space. A homogeneous transformation matrix is a 4×4 matrix that maps an object defined in a homogeneous coordinate system. A homogeneous transformation can be thought of as two submatrices, i.e. a translation matrix and a rotational matrix.

$$T = \begin{bmatrix} & : \text{ origin} \\ \text{orientation} & : \text{ position} \\ \text{matrix} & : \text{ column} \\ & : \text{ vector} \\ 4 \times 3 & 4 \times 1 \end{bmatrix}_{4 \times 4}$$ (2.13)

As an example, let XYZ be a fixed coordinate system, let $X'Y'Z'$ be a body attached coordinate system, and let the axes of both coordinate systems be parallel at the instant. The origin O of a body attached coordinate system with respect to a fixed coordinate system is given by a vector $a\bar{i} + b\bar{j} + c\bar{k}$, where $\bar{i}$, $\bar{j}$ and $\bar{k}$ are unit vectors in the XYZ directions, respectively.

The process describing such displacements (Fig. 2.2) is called translation transformation and is given by

$$\text{Trans}(a,b,c) = \begin{bmatrix} 1 & 0 & 0 & a \\ 0 & 1 & 0 & b \\ 0 & 0 & 1 & c \\ 0 & 0 & 0 & 1 \end{bmatrix}$$ (2.14)

From this relation it can be seen that the directional vectors are the same as in XYZ and there is no rotational motion.

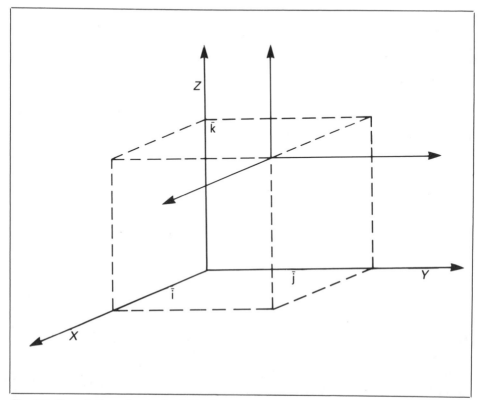

Fig. 2.2 Translation of coordinate system

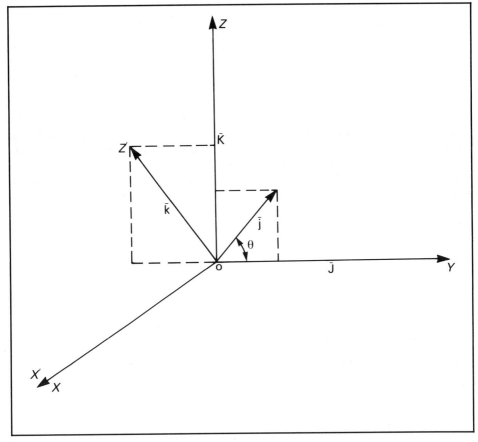

Fig. 2.3 Effect of pure rotation

Given a position vector $\bar{u} = [x, y, z, w]^T$ in the $X'Y'Z'$ system, the transformed vector $\bar{v}$ is given by

$$\bar{v} = \begin{bmatrix} 1 & 0 & 0 & a \\ 0 & 1 & 0 & b \\ 0 & 0 & 1 & c \\ 0 & 0 & 0 & 1 \end{bmatrix} \begin{bmatrix} x \\ y \\ z \\ w \end{bmatrix} = \begin{bmatrix} x + aw \\ y + bw \\ z + cw \\ w \end{bmatrix} \tag{2.15}$$

or

$$\bar{v} = \begin{bmatrix} x/w + a \\ y/w + b \\ z/w + c \\ 1 \end{bmatrix} \tag{2.16}$$

This result can be interpreted as addition of two vectors, $x/w\,\bar{i} + y/w\,\bar{j} + z/w\,\bar{k}$ and $a\bar{i} + b\bar{j} + c\bar{k}$, both expressed in the XYZ system.

Now, let us investigate the effect of pure rotation. Let $X'Y'Z'$ be rotated about the X axis by an angle θ and assume that the origins of the XYZ and $X'Y'Z'$ systems coincide. Fig. 2.3 shows the pure rotation about the X axis.

The direction of the X axis is not changed, since we are rotating about that axis. Thus the directional vector remains the same, i.e.

$$[1\ 0\ 0\ 0]^T \tag{2.17}$$

The directional vectors of the Y' and Z' axes can be determined as

$$\bar{j} = \cos\theta\,\bar{J} + \sin\theta\bar{K}$$

$$\bar{k} = \cos\theta\bar{K} - \sin\theta\bar{J} \tag{2.18}$$

or

$$\begin{bmatrix} 0 \\ \cos\theta \\ \sin\theta \\ 0 \end{bmatrix} \text{ and } \begin{bmatrix} 0 \\ -\sin\theta \\ \cos\theta \\ 0 \end{bmatrix} \tag{2.19}$$

Combining the above results, the homogeneous transformation matrix is given as

$$\text{Rot}(X,\theta) = \begin{bmatrix} 1 & 0 & 0 & 0 \\ 0 & \cos\theta & -\sin\theta & 0 \\ 0 & \sin\theta & \cos\theta & 0 \\ 0 & 0 & 0 & 1 \end{bmatrix} \tag{2.20}$$

Similarly, the homogeneous transformation matrix for rotation about the Y and Z axes by angle θ are given by

$$\text{Rot}(Y,\theta) = \begin{bmatrix} \cos\theta & 0 & \sin\theta & 0 \\ 0 & 1 & 0 & 0 \\ -\sin\theta & 0 & \cos\theta & 0 \\ 0 & 0 & 0 & 1 \end{bmatrix} \tag{2.21}$$

$$\text{Rot}(Z,\theta) = \begin{bmatrix} \cos\theta & -\sin\theta & 0 & 0 \\ \sin\theta & \cos\theta & 0 & 0 \\ 0 & 0 & 1 & 0 \\ 0 & 0 & 0 & 1 \end{bmatrix} \tag{2.22}$$

Let us consider an example. A vector $\bar{u} = 7\bar{i} + 3\bar{j} + 2\bar{k}$ is rotated about the Z axis by 90°. The new position of the vector $\bar{u}$ is given by vector $\bar{v}$ where

$$\bar{v} = \text{Rot}(Z,90°)\bar{u} = \begin{bmatrix} \cos\theta & -\sin & 0 & 0 \\ \sin\theta & \cos\theta & 0 & 0 \\ 0 & 0 & 1 & 0 \\ 0 & 0 & 0 & 1 \end{bmatrix} \tag{2.23}$$

Substituting $\cos 90° = 0$ and $\sin 90° = 1$, we have

$$\bar{v} = \begin{bmatrix} 0 & -1 & 0 & 0 \\ 1 & 0 & 0 & 0 \\ 0 & 0 & 1 & 0 \\ 0 & 0 & 0 & 1 \end{bmatrix} \begin{bmatrix} 7 \\ 3 \\ 2 \\ 1 \end{bmatrix} = \begin{bmatrix} -3 \\ 7 \\ 2 \\ 1 \end{bmatrix} \tag{2.24}$$

If this vector is further rotated about the Y axis by 90°, the new position vector is given by vector $\bar{w}$,

$$\bar{w} = \text{Rot}(Y,90°)\bar{v} \;=\; \begin{bmatrix} 0 & 0 & 1 & 0 \\ 0 & 1 & 0 & 0 \\ -1 & 0 & 0 & 0 \\ 0 & 0 & 0 & 1 \end{bmatrix} \begin{bmatrix} -3 \\ 7 \\ 2 \\ 1 \end{bmatrix} = \begin{bmatrix} 2 \\ 7 \\ 3 \\ 1 \end{bmatrix} \quad (2.25)$$

or $\qquad \bar{w} = 2\bar{i} + 7\bar{j} + 3\bar{k}$

The same result can be obtained by

$$\bar{w} = \text{Rot}(Y,90°)\,\text{Rot}\,(Z,90°)\bar{u} \;=\; \begin{bmatrix} 0 & 0 & 1 & 0 \\ 0 & 1 & 0 & 0 \\ -1 & 0 & 0 & 0 \\ 0 & 0 & 0 & 1 \end{bmatrix} \begin{bmatrix} 0 & -1 & 0 & 0 \\ 1 & 0 & 0 & 0 \\ 0 & 0 & 1 & 0 \\ 0 & 0 & 0 & 1 \end{bmatrix} \begin{bmatrix} 7 \\ 3 \\ 2 \\ 1 \end{bmatrix}$$

$$= \begin{bmatrix} 0 & 0 & 1 & 0 \\ 1 & 0 & 0 & 0 \\ 0 & 1 & 0 & 0 \\ 0 & 0 & 0 & 1 \end{bmatrix} \begin{bmatrix} 7 \\ 3 \\ 2 \\ 1 \end{bmatrix}$$

$$= \begin{bmatrix} 2 \\ 7 \\ 3 \\ 1 \end{bmatrix} \quad (2.26)$$

If the vector $\bar{w}$ is further translated by the distance given by the vector

$$\bar{d} = 4\bar{i} - 3\bar{j} + 7\bar{k} \quad (2.27)$$

then the new position of vector $\bar{w}$ is given by vector $\bar{x}$,

$$\begin{aligned} \bar{x} &= \text{Trans}(4,-3,7)\bar{w} \\ &= \text{Trans}(4,-3,7)\text{Rot}(Y,90°)\text{Rot}(X,90°)\bar{u} \end{aligned} \quad (2.28)$$

$$= \begin{bmatrix} 1 & 0 & 0 & 4 \\ 0 & 1 & 0 & -3 \\ 0 & 0 & 1 & 7 \\ 0 & 0 & 0 & 1 \end{bmatrix} \begin{bmatrix} 2 \\ 7 \\ 3 \\ 1 \end{bmatrix} = \begin{bmatrix} 6 \\ 4 \\ 10 \\ 1 \end{bmatrix} \quad (2.29)$$

or $\qquad \bar{x} = 6\bar{i} + 4\bar{j} + 10\bar{k}$

2.4 Orthonormal property of orientation submatrix

The vector $\bar{a}$ is in the direction of approaching an object and hence is called the approach vector. The unit vector $\bar{o}$ which specifies orientation of link is called the orientation vector. The unit vector $\bar{n}$ is called the normal vector and is defined in such a way that $\bar{n} = \bar{o} \times \bar{a}$. Fig. 2.4 shows the relationship of these three vectors.

Thus the transformation T has the following elements:

$$T = \begin{bmatrix} n_x & o_x & a_x & p_x \\ n_y & o_y & a_y & p_y \\ n_z & o_z & a_z & p_z \\ 0 & 0 & 0 & 1 \end{bmatrix} \quad (2.30)$$

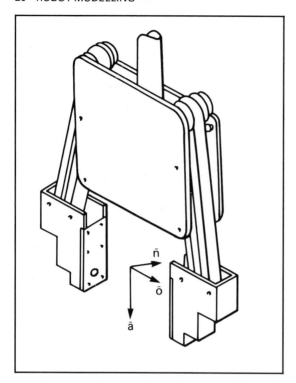

Fig. 2.4 Relationship of orientation vectors n̄, ō, and ā for a manipulator wrist

The first three columns represent the new XYZ axes direction with respect to the old reference frame.

The orientation submatrix

$$\begin{bmatrix} n_x\, o_x\, a_x \\ n_y\, o_y\, a_y \\ n_z\, o_z\, a_z \end{bmatrix} \tag{2.31}$$

is an orthonormal matrix, and has the following properties:

- The scalar product of two different columns is zero.
- The scalar product of the same columns is unity.
- The cross product of two different columns results in the third column in a cyclic manner.
- The determinant of the matrix is unity $(+1)$ if the three axes form the right-handed triad rule (i.e. right-hand coordinate system).

Thus,

$$\bar{n}.\bar{o} = \bar{o}.\bar{a} = \bar{a}.\bar{n} = 0 \tag{2.32}$$
$$\bar{n}.\bar{n} = \bar{o}.\bar{o} = \bar{a}.\bar{a} = 1 \tag{2.33}$$
$$\bar{n} = \bar{o} \times \bar{a} \; ; \bar{o} = \bar{a} \times \bar{n} \; ; \bar{a} = \bar{n} \times \bar{o} \tag{2.34}$$

and

$$\begin{vmatrix} n_x\, o_x\, a_x \\ n_y\, o_y\, a_y \\ n_z\, o_z\, a_z \end{vmatrix} = +1 \tag{2.35}$$

It is interesting to note that the value of the determinant will be -1 for a left-hand coordinate system.

2.5 Geometric interpretation of relative transformation

The rotations and translations described so far have all been made with respect to the fixed base coordinate frame. Thus, in the earlier example:

$$\text{Trans}(4,-3,7)\,\text{Rot}(Y,90°)\,\text{Rot}(Z,90°) = \begin{bmatrix} 0 & 0 & 1 & 4 \\ 1 & 0 & 0 & -3 \\ 0 & 1 & 0 & 4 \\ 0 & 0 & 0 & 1 \end{bmatrix} \qquad (2.36)$$

The sequence of transformation is, proceeding from right to left, rotation of the frame around the Z axis of the base frame by 90°, followed by rotation around the Y axis of the base frame by 90°, and finally translation of the origin by $4\bar{i} - 3\bar{j} + 7\bar{k}$. The same final configuration may also be achieved in a different manner, by proceeding from left to right, i.e. first translate the origin by $4\bar{i} - 3\bar{j} + 7\bar{k}$, then rotate 90° around the Y axis of the current frame axes, and finally rotate 90° around the Z axis of the current frame axes.

This concept of postmultiplying a transform plays an important role in a robotic link relationship which indicates the translation and/or rotation of the present link with respect to the immediately preceding link frame. This transform is reiterated in Chapter Three.

2.6 Inverse transformation

This transformation is simply the description of the reference coordinate frame with respect to the transformed frame. Given a transform with elements:

$$T = \begin{bmatrix} n_x & o_x & a_x & p_x \\ n_y & o_y & a_y & p_y \\ n_z & o_z & a_z & p_z \\ 0 & 0 & 0 & 1 \end{bmatrix} \qquad (2.37)$$

then the inverse is

$$T^{-1} = \begin{bmatrix} n_x & n_y & n_z & -\bar{p}.\bar{n} \\ o_x & o_y & o_z & -\bar{p}.\bar{o} \\ a_x & a_y & a_z & -\bar{p}.\bar{a} \\ 0 & 0 & 0 & 1 \end{bmatrix} \qquad (2.38)$$

It can be easily seen that $T\,T^{-1} = I$, where

$$I = \begin{bmatrix} 1 & 0 & 0 & 0 \\ 0 & 1 & 0 & 0 \\ 0 & 0 & 1 & 0 \\ 0 & 0 & 0 & 1 \end{bmatrix} \qquad (2.39)$$

References

[1] Denavit, J. and Hartenberg, R. S. 1955. A kinematic notation for lower pair mechanisms based on matrices. *ASME J. Applied Mechanics*, June: 215–221.

[2] Pieper, D. L. 1968. *The Kinematics of Manipulators Under Computer Control.* Stanford Artificial Intelligence Laboratory, Stanford University, CA, USA.

[3] Paul, R. P. 1983 *Robot Manipulators – Mathematics, Programming and Control.* MIT Press, Cambridge, MA, USA.

Chapter Three

LINK EQUATIONS AND RELATIONSHIPS

MECHANICALLY, a robot manipulator is composed of an arm and a wrist attached to a tool fixture. The arm typically has three degrees-of-freedom, which accomplish the major positioning of the robot arm and place the wrist unit at the workpiece. The wrist consists of up to three rotary motions which help to attain an appropriate orientation of the tool toward the object. Thus, the arm is the major positioning mechanism, while the wrist is the minor positioning and orientation mechanism.

In this chapter homogeneous transformations are developed to represent the spatial configuration between neighbouring links. Here the notations of Paul[3], Denavit and Hartenberg[1], and Pieper[2] are employed.

An A_1 transformation matrix is simply the transformation describing the relative translation and rotation between the link i coordinate system and the link $i-1$ coordinate system. Thus, given a six degree-of-freedom robot manipulator,

$$T_6 = A_1 A_2 A_3 A_4 A_5 A_6 \tag{3.1}$$

which completely describes the kinematic motion of a manipulator.

3.1 Specification of 'A' matrices

Commercially available robot arms nominally have a simple structure. The joints are either revolute or prismatic. The links are straight, and the twist direction between links is either parallel or perpendicular. It is therefore necessary to introduce only a simple relationship to describe the link transformations.

The conventions[4] for determining the coordinate frame of the robot arm (see Fig. 3.1) are:

● Align the X axes of all of the joint frames in the same direction as the fixed base frame.

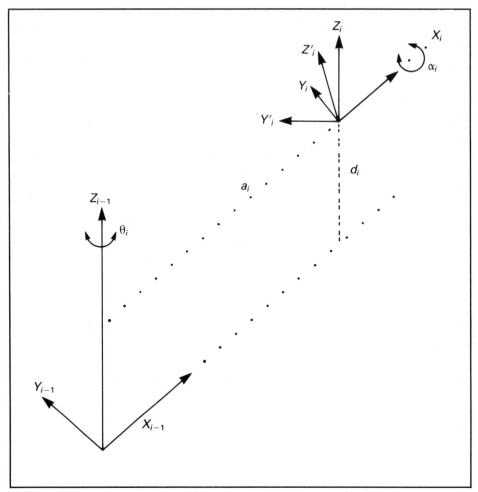

Fig. 3.1 The relationship between link i−1 and link i

● Make the revolute joints rotate about their respective Z axes.
● Make the prismatic joints travel along their respective Z axes.

Having assigned the coordinate frames according to the prescribed rules, the successive relationship between link i and link $i-1$, is described by the following set of parameters:

θ_i – rotating an angle θ_i about the Z_{i-1} axis.
d_i – translating a distance d_i along the Z_{i-1} axis.
a_i – translating a length a_i along the X_{i-1} axis.
α_i – rotating about the X_i axis clockwise by a twist angle α_1, between the Z_1 and the Z_{i-1} axes.

It is common practice that the geometrical configuration of link is represented by the fixed parameters, the link length a_i and the twist angle α_i, and that the moving variables are represented by the rotational variable θ_i

and the translational variable d_i. However, this rule is somewhat modified to accommodate exceptional cases such as the RPY wrist and Cartesian arm discussed in this chapter.

Thus the homogeneous transformation matrix A_1 between link i and link $i-1$ is:

$$A_i = \text{Rot}(Z,\theta_i)\,\text{Trans}(0,0,d_i)\,\text{Trans}(a_i,0,0)\,\text{Rot}(X,\alpha_i) \tag{3.2}$$

The order of the transformation matrix here is from left to right due to the relationship between successive links.

Fig. 3.1 shows the relationship between successive links and the parameter measurements. Expanding Eqn. (3.2) and ignoring the sub-scripts results in the following transformation matrix:

$$A_1 = \begin{bmatrix} \cos\theta & -\sin\theta & 0 & 0 \\ \sin\theta & \cos\theta & 0 & 0 \\ 0 & 0 & 1 & 0 \\ 0 & 0 & 0 & 1 \end{bmatrix} \begin{bmatrix} 1 & 0 & 0 & 0 \\ 0 & 1 & 0 & 0 \\ 0 & 0 & 1 & d \\ 0 & 0 & 0 & 1 \end{bmatrix} \begin{bmatrix} 1 & 0 & 0 & a \\ 0 & 1 & 0 & 0 \\ 0 & 0 & 1 & 0 \\ 0 & 0 & 0 & 1 \end{bmatrix} \begin{bmatrix} 1 & 0 & 0 & 0 \\ 0 & \cos\alpha & -\sin\alpha & 0 \\ 0 & \sin\alpha & \cos\alpha & 0 \\ 0 & 0 & 0 & 1 \end{bmatrix}$$

Thus,

$$A_i = \begin{bmatrix} \cos\theta & -\sin\theta\cos\alpha & \sin\theta\sin\alpha & a\cos\theta \\ \sin\theta & \cos\theta\cos\alpha & -\cos\theta\sin\alpha & a\sin\theta \\ 0 & \sin\alpha & \cos\alpha & d \\ 0 & 0 & 0 & 1 \end{bmatrix} \tag{3.3}$$

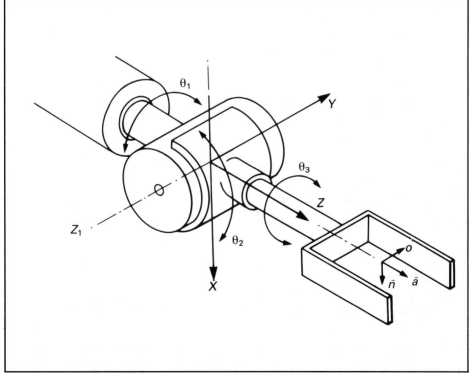

Fig. 3.2 Euler angles

3.2 Wrist kinematic control equations

The orientation of a robot manipulator is determined by the motion of the wrist. Orientation is frequently specified by a sequence of rotations about the X, Y, and Z axes. Two types of angles are used to describe the possible orientation of the wrist motion. They are called the Euler angles and the roll–pitch–yaw (RPY) angles (Figs. 3.2 and 3.3).

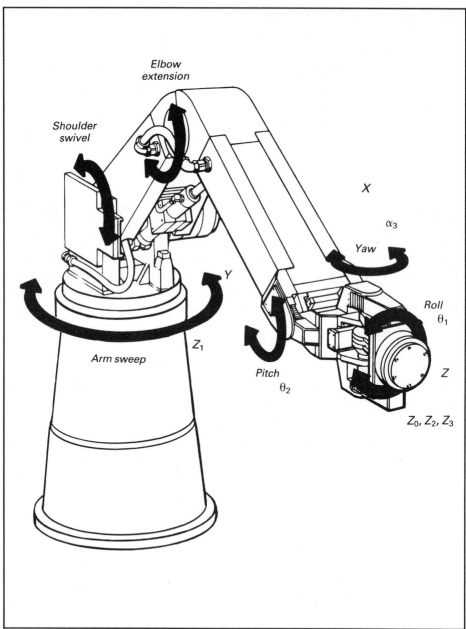

Fig. 3.3 Robot axes arrangement showing roll–pitch–yaw (RPY) angles

Euler angles

With Euler angles any possible orientation is described in terms of a rotation θ_1 about the Z axis, then a rotation θ_2 about the new rotated Y axis and finally a rotation θ_3 about the rotated Z axis once again. The PUMA–560 robot manipulator utilises a Euler type wrist (Fig. 3.2). The conventional approach to writing the Euler angles transform equation is as follows:

$$T_E = \text{Rot}(Z,\theta_1)\,\text{Rot}(Y,\theta_2)\,\text{Rot}(Z,\theta_3)$$

$$= \begin{bmatrix} C\theta_1 & -S\theta_1 & 0 & 0 \\ S\theta_1 & C\theta_1 & 0 & 0 \\ 0 & 0 & 1 & 0 \\ 0 & 0 & 0 & 1 \end{bmatrix} \begin{bmatrix} C\theta_2 & 0 & S\theta_2 & 0 \\ 0 & 1 & 0 & 0 \\ -S\theta_2 & 0 & C\theta_2 & 0 \\ 0 & 0 & 0 & 1 \end{bmatrix} \begin{bmatrix} C\theta_3 & -S\theta_3 & 0 & 0 \\ S\theta_3 & C\theta_3 & 0 & 0 \\ 0 & 0 & 1 & 0 \\ 0 & 0 & 0 & 1 \end{bmatrix}$$

$$= \begin{bmatrix} C\theta_1 C\theta_2 C\theta_3 - S\theta_1 S\theta_3 & -C\theta_1 C\theta_2 S\theta_3 - S\theta_1 C\theta_3 & C\theta_1 S\theta_2 & 0 \\ S\theta_1 C\theta_2 C\theta_3 + C\theta_1 S\theta_3 & -S\theta_1 C\theta_2 S\theta_3 + C\theta_1 C\theta_3 & S\theta_1 S\theta_2 & 0 \\ -S\theta_2 C\theta_3 & S\theta_2 S\theta_3 & C\theta_2 & 0 \\ 0 & 0 & 0 & 1 \end{bmatrix} \quad (3.4)$$

To conform to the transformation notation, it is necessary to write the transformation in terms of the 'A' matrix. The XYZ axes joints are assigned according to the rules outlined earlier. The specified axes alignments result in a positioning of the Euler wrist as shown in Fig. 3.2. The link parameter table of the three rotational joints is then stated in the following form:

Link	Variable	α	a	d	$\cos\alpha$	$\sin\alpha$
1	θ_1	$-90°$	0	0	0	-1
2	θ_2	$90°$	0	0	0	1
3	θ_3	$0°$	0	0	1	0

The A transformations for the individual Euler angles and the total transformation matrix are as follows:

$$A_1 = \begin{bmatrix} \cos\theta_1 & 0 & -\sin\theta_1 & 0 \\ \sin\theta_1 & 0 & \cos\theta_1 & 0 \\ 0 & -1 & 0 & 0 \\ 0 & 0 & 0 & 1 \end{bmatrix} \quad (3.5)$$

$$A_2 = \begin{bmatrix} \cos\theta_2 & 0 & \sin\theta_2 & 0 \\ \sin\theta_2 & 0 & -\cos\theta_2 & 0 \\ 0 & 1 & 0 & 0 \\ 0 & 0 & 0 & 1 \end{bmatrix} \quad (3.6)$$

$$A_3 = \begin{bmatrix} \cos\theta_3 & -\sin\theta_3 & 0 & 0 \\ \sin\theta_3 & \cos\theta_3 & 0 & 0 \\ 0 & 0 & 1 & 0 \\ 0 & 0 & 0 & 1 \end{bmatrix} \quad (3.7)$$

Thus,

$$T_E = A_1 A_2 A_3$$

$$= \begin{bmatrix} C\theta_1 C\theta_2 C\theta_3 - S\theta_1 S\theta_3 & -C\theta_1 C\theta_2 S\theta_3 - S\theta_1 C\theta_3 & C\theta_1 S\theta_2 & 0 \\ S\theta_1 C\theta_2 C\theta_3 + C\theta_1 S\theta_3 & -S\theta_1 C\theta_2 S\theta_3 + C\theta_1 C\theta_3 & S\theta_1 S\theta_2 & 0 \\ -S\theta_2 C\theta_3 & S\theta_2 S\theta_3 & C\theta_2 & 0 \\ 0 & 0 & 0 & 1 \end{bmatrix} \quad (3.8)$$

The result of the total transformation matrix is completely identical to that derived in Eqn. (3.4).

Roll–pitch–yaw angles

Fig. 3.3 shows the rotation angles of roll, pitch and yaw. The first two angles, roll and pitch, are identical to the first two Euler angles. The final angular movement, which deviates from that of the final Euler angle movement, consists of an angle of rotation α_3 about its current X axis. The advantage of having this type of rotation is that the degeneracy of angular movement will not occur in the normal working space of the robot motion. (Angular degeneracy is discussed in Chapter Four.) The conventional approach to writing the transform equation is:

$$T_{RPY} = Rot(Z,\theta_1)\, Rot(Y,\theta_2)\, Rot(X,\alpha_3)$$

$$= \begin{bmatrix} C\theta_1 & -S\theta_1 & 0 & 0 \\ S\theta_1 & C\theta_1 & 0 & 0 \\ 0 & 0 & 1 & 0 \\ 0 & 0 & 0 & 1 \end{bmatrix} \begin{bmatrix} C\theta_2 & 0 & S\theta_2 & 0 \\ 0 & 1 & 0 & 0 \\ -S\theta_2 & 0 & C\theta_2 & 0 \\ 0 & 0 & 0 & 1 \end{bmatrix} \begin{bmatrix} 1 & 0 & 0 & 0 \\ 0 & C\alpha_3 & -S\alpha_3 & 0 \\ 0 & S\alpha_3 & C\alpha_3 & 0 \\ 0 & 0 & 0 & 1 \end{bmatrix}$$

$$= \begin{bmatrix} C\theta_1 C\theta_2 & C\theta_1 S\theta_2 S\alpha_3 - S\theta_1 C\alpha_3 & C\theta_1 S\theta_2 C\alpha_3 + S\theta_1 S\alpha_3 & 0 \\ S\theta_1 C\theta_2 & S\theta_1 S\theta_2 S\alpha_3 + C\theta_1 C\alpha_3 & S\theta_1 S\theta_2 C\alpha_3 - C\theta_1 S\alpha_3 & 0 \\ -S\theta_2 & C\theta_2 S\alpha_3 & C\theta_2 C\alpha_3 & 0 \\ 0 & 0 & 0 & 1 \end{bmatrix} \quad (3.9)$$

The link parameter table for expressing the transform A is:

Link	Variable	α	a	d	$\cos\alpha$	$\sin\alpha$
1	θ_1	$-90°$	0	0	0	-1
2	θ_2	$90°$	0	0	0	1
3	α_3	α_3	0	0	$\cos\alpha_3$	$\sin\alpha_3$

$$A_1 = \begin{bmatrix} \cos\theta_1 & 0 & -\sin\theta_1 & 0 \\ \sin\theta_1 & 0 & \cos\theta_1 & 0 \\ 0 & -1 & 0 & 0 \\ 0 & 0 & 0 & 1 \end{bmatrix} \quad (3.10)$$

$$A_2 = \begin{bmatrix} \cos\theta_2 & 0 & \sin\theta_2 & 0 \\ \sin\theta_2 & 0 & -\cos\theta_2 & 0 \\ 0 & 1 & 0 & 0 \\ 0 & 0 & 0 & 1 \end{bmatrix} \quad (3.11)$$

$$A_3 = \begin{bmatrix} 1 & 0 & 0 & 0 \\ 0 & \cos\alpha_3 & -\sin\alpha_3 & 0 \\ 0 & \sin\alpha_3 & \cos\alpha_3 & 0 \\ 0 & 0 & 0 & 1 \end{bmatrix} \qquad (3.12)$$

Thus,

$$T_{RPY} = A_1 A_2 A_3$$

$$= \begin{bmatrix} C\theta_1 C\theta_2 & C\theta_1 S\theta_2 S\alpha_3 - S\theta_1 C\alpha_3 & C\theta_1 S\theta_2 C\alpha_3 + S\theta_1 S\alpha_3 & 0 \\ S\theta_1 C\theta_2 & S\theta_1 S\theta_2 S\alpha_3 + C\theta_1 C\alpha_3 & S\theta_1 S\theta_2 C\alpha_3 - C\theta_1 S\alpha_3 & 0 \\ -S\theta_2 & C\theta_2 S\alpha_3 & C\theta_2 C\alpha_3 & 0 \\ 0 & 0 & 0 & 1 \end{bmatrix} \qquad (3.13)$$

Again, the resulting transformation matrix is identical when approaching the transformation by a series of rotations about the *X, Y,* and *Z* axes (Eqn. (3.9)) or by utilising the 'A' matrix solution (Eqn. (3.13)).

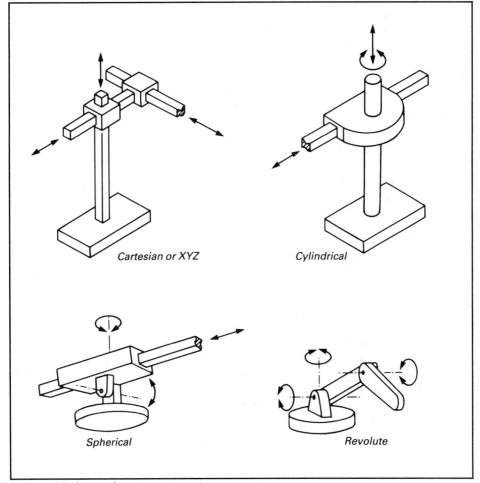

Cartesian or XYZ

Cylindrical

Spherical

Revolute

Fig. 3.4 Various robot arm categories

3.3 Basic motion of the robotic arm

Many commercially available industrial robots can be broken down into four basic groups according to their characteristics of arm motion and geometrical appearance. Fig. 3.4 shows these four groups and their characteristic design.

The four designs can be classified as follows:

- Cartesian coordinate (PPP) – three prismatic linear axes, e.g. IBM RS–7565.
- Cylindrical coordinate (PPR) – two linear axes and one rotary axis, e.g. Fanuc M1.
- Spherical or polar coordinate (PRR) – one linear axis and two rotary axes, e.g. Unimate 4000B.
- Revolute or articulated coordinate (RRR) – three rotary axes, e.g. PUMA–560.

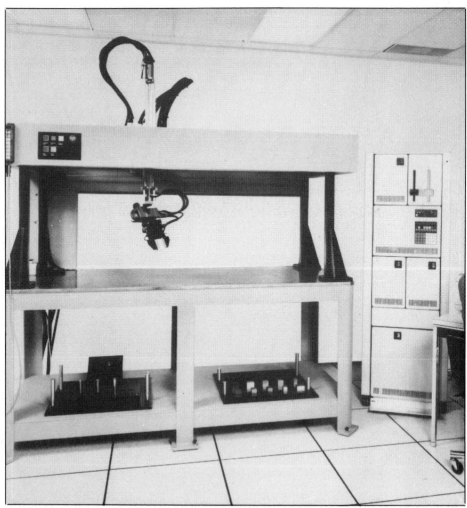

Fig. 3.5 The IBM RS–7565 gantry frame robot

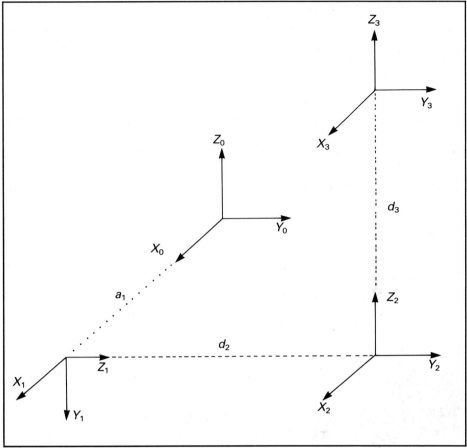

Fig. 3.6 The coordinate frame assignments for the Cartesian coordinate arm

3.4 Arm kinematic control equations

In the preceding section the wrist kinematic control equations were developed. We will now develop the arm kinematic control equations according to the four conventional types of manipulator arms mentioned above.

Cartesian coordinate arm (PPP)

The IBM RS–7565 gantry frame robot is shown in Fig. 3.5 with its three prismatic joints and respective travel axes. The coordinate frame assignment of the RS–7565's three linear translations are shown in Fig. 3.6. The link parameter table for the Cartesian arm is then:

Link	Variable	α	a	d	$cos\alpha$	$sin\alpha$
1	a_1	$-90°$	a_1	0	0	-1
2	d_2	$90°$	0	d_2	0	1
3	d_3	0	0	d_3	1	0

The corresponding 'A' matrices for the three prismatic joints are:

$$A_1 = \begin{bmatrix} 1 & 0 & 0 & a_1 \\ 0 & 0 & 1 & 0 \\ 0 & -1 & 0 & 0 \\ 0 & 0 & 0 & 1 \end{bmatrix} \tag{3.14}$$

$$A_2 = \begin{bmatrix} 1 & 0 & 0 & 0 \\ 0 & 0 & -1 & 0 \\ 0 & 1 & 0 & d_2 \\ 0 & 0 & 0 & 1 \end{bmatrix} \tag{3.15}$$

$$A_3 = \begin{bmatrix} 1 & 0 & 0 & 0 \\ 0 & 1 & 0 & 0 \\ 0 & 0 & 1 & d_3 \\ 0 & 0 & 0 & 1 \end{bmatrix} \tag{3.16}$$

Thus,

$$T_{PPP} = A_1 A_2 A_3$$
$$= \begin{bmatrix} 1 & 0 & 0 & a_1 \\ 0 & 1 & 0 & d_2 \\ 0 & 0 & 1 & d_3 \\ 0 & 0 & 0 & 1 \end{bmatrix} \tag{3.17}$$

Eqn. (3.17) agrees completely with the conventional expression for three linear XYZ motions.

Cylindrical coordinate arm (PRP)

The 600 Fanuc M1 robot can be utilised as a representation of a cylindrical coordinate arm. The unit is shown in Fig. 3.7. Fig. 3.8 shows the coordinate frame assignment where the two linear motions move along the Z_0 and Z_2 axes, respectively. The link parameter table is as follows:

Link	Variable	α	a	d	$\cos\alpha$	$\sin\alpha$
1	d_1	0°	0	d_1	1	0
2	θ_2	−90°	a_2	0	0	−1
3	d_3	0	0	d_3	1	0

The corresponding 'A' matrices for the two prismatic and one rotational joints are:

$$A_1 = \begin{bmatrix} 1 & 0 & 0 & 0 \\ 0 & 1 & 0 & 0 \\ 0 & 0 & 1 & d_1 \\ 0 & 0 & 0 & 1 \end{bmatrix} \tag{3.18}$$

$$A_2 = \begin{bmatrix} \cos\theta_2 & 0 & -\sin\theta_2 & a_2\cos\theta_2 \\ \sin\theta_2 & 0 & \cos\theta_2 & a_2\sin\theta_2 \\ 0 & -1 & 0 & 0 \\ 0 & 0 & 0 & 1 \end{bmatrix} \tag{3.19}$$

$$A_3 = \begin{bmatrix} 1 & 0 & 0 & 0 \\ 0 & 1 & 0 & 0 \\ 0 & 0 & 1 & d_3 \\ 0 & 0 & 0 & 1 \end{bmatrix} \tag{3.20}$$

Thus,

$$T_{PRP} = A_1 A_2 A_3$$
$$= \begin{bmatrix} \cos\theta_2 & 0 & -\sin\theta_2 & -d_3\sin\theta_2 + a_2\cos\theta_2 \\ \sin\theta_2 & 0 & \cos\theta_2 & d_3\cos\theta_2 + a_2\sin\theta_2 \\ 0 & -1 & 0 & d_1 \\ 0 & 0 & 0 & 1 \end{bmatrix} \tag{3.21}$$

Fig. 3.7. The Fanuc M1 robot

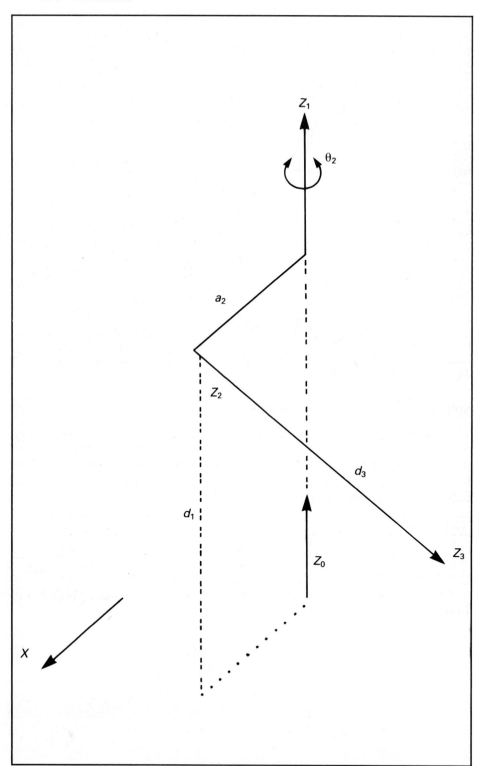

Fig. 3.8 The coordinate frame assignments for the cylindrical coordinate arm

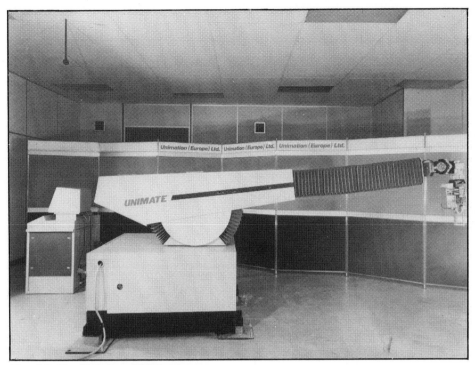

Fig. 3.9 The Unimate 4000B robot

Spherical or polar coordinate arm (RRP)

The Unimate 4000B is an example of an industrial spherical robot (Fig. 3.9). The boom has two rotary motions that rotate left or right or pivot up or down. The third axis is prismatic and allows the boom to move in and out in an axial motion. The coordinate frame assignments are shown in Fig. 3.10. The link parameter table is given below:

Link	Variable	α	a	d	$\cos\alpha$	$\sin\alpha$
1	θ_1	$-90°$	0	0_1	0	-1
2	θ_2	$90°$	0	0	0	0
3	d_3	$0°$	0	d_3	1	0

The corresponding 'A' matrices for the two rotational and one prismatic joints are:

$$A_1 = \begin{bmatrix} \cos\theta_1 & 0 & -\sin\theta_1 & 0 \\ \sin\theta_1 & 0 & \cos\theta_1 & 0 \\ 0 & -1 & 0 & 0 \\ 0 & 0 & 0 & 1 \end{bmatrix} \qquad (3.22)$$

$$A_2 = \begin{bmatrix} \cos\theta_2 & 0 & \sin\theta_2 & 0 \\ \sin\theta_2 & 0 & -\cos\theta_2 & 0 \\ 0 & 1 & 0 & 0 \\ 0 & 0 & 0 & 1 \end{bmatrix} \qquad (3.23)$$

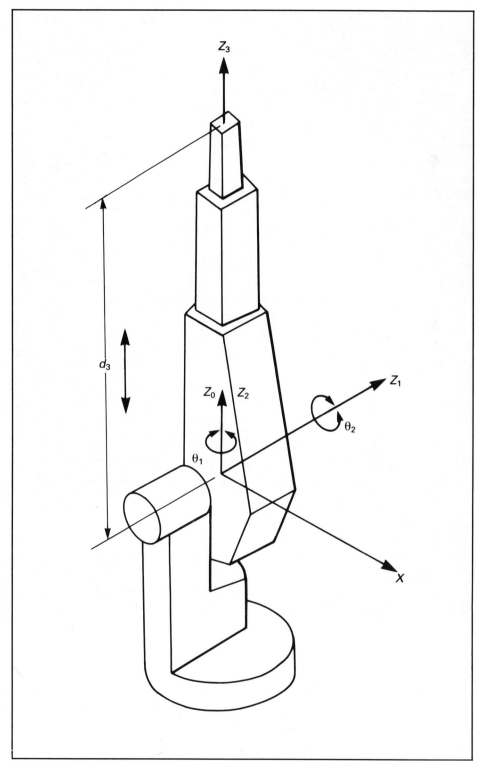

Fig. 3.10 The coordinate frame assignments for the spherical coordinate arm

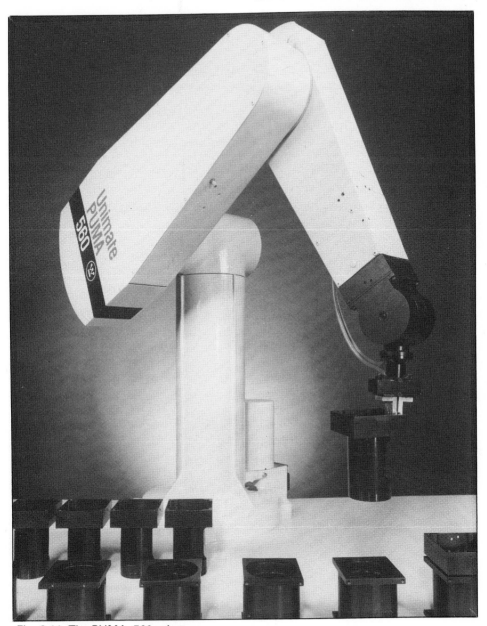

Fig. 3.11 The PUMA–560 robot

$$A_3 = \begin{bmatrix} 1 & 0 & 0 & 0 \\ 0 & 1 & 0 & 0 \\ 0 & 0 & 1 & d_3 \\ 0 & 0 & 0 & 1 \end{bmatrix} \qquad (3.24)$$

Thus,

$$T_{RRP} = A_1 A_2 A_3$$

$$
= \begin{bmatrix}
\cos\theta_1\cos\theta_2 & -\sin\theta_1 & \cos\theta_1\sin\theta_2 & d_3\cos\theta_1\sin\theta_2 \\
\sin\theta_1\cos\theta_2 & \cos\theta_2 & \sin\theta_1\sin\theta_2 & d_3\sin\theta_1\sin\theta_2 \\
\sin\theta_2 & 0 & 0 & d_3\cos\theta_2 \\
0 & 0 & 0 & 1
\end{bmatrix} \tag{3.25}
$$

Revolute or articulated coordinate arm (RRR)

This type of robot arm is analogous to an anthropomorphic arm. The body trunk, shoulder and elbow represent three rotational joints. Figs. 3.11 and 3.12 show the PUMA–560 and the coordinate frame assignments for an RRR arm without joint offset. The revolute arm link parameters are as follows:

Link	Variable	α	a	d	$\cos\alpha$	$\sin\alpha$
1	θ_1	90°	0	0	0	1
2	θ_2	0°	a_2	0	1	0
3	θ_3	0°	a_3	0	1	0

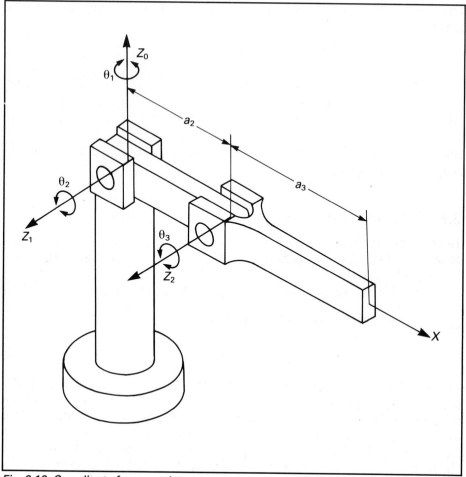

Fig. 3.12 *Coordinate frame assignments for the revolute arm*

The corresponding 'A' matrices for the three rotational joints are:

$$A_1 = \begin{bmatrix} \cos\theta_1 & 0 & \sin\theta_1 & 0 \\ \sin\theta_1 & 0 & -\cos\theta_1 & 0 \\ 0 & 1 & 0 & 0 \\ 0 & 0 & 0 & 1 \end{bmatrix} \tag{3.26}$$

$$A_2 = \begin{bmatrix} \cos\theta_2 & -\sin\theta_2 & 0 & a_2\cos\theta_2 \\ \sin\theta_2 & \cos\theta_2 & 0 & a_2\sin\theta_2 \\ 0 & 0 & 1 & 0 \\ 0 & 0 & 0 & 1 \end{bmatrix} \tag{3.27}$$

$$A_3 = \begin{bmatrix} \cos\theta_3 & -\sin\theta_3 & 0 & a_3\cos\theta_3 \\ \sin\theta_3 & \cos\theta_2 & 0 & a_3\sin\theta_3 \\ 0 & 0 & 1 & 0 \\ 0 & 0 & 0 & 1 \end{bmatrix} \tag{3.28}$$

Thus, the transformation matrix for the final position and orientation of the arm is:

$$T_{RRR} = A_1 A_2 A_3$$

$$= \begin{bmatrix} C_1C_{23} & -C_1C_{23} & S_1 & C_1(a_3C_{23}+a_2C_2) \\ S_1C_{23} & -S_2S_{23} & -C_1 & S_1(a_3C_{23}+a_2C_2) \\ S_{23} & C_{23} & 0 & a_2C_{23}+a_2C_2 \\ 0 & 0 & 0 & 1 \end{bmatrix} \tag{3.29}$$

where S_1, C_1, S_{23} and C_{23} refer to $\sin\theta_1$, $\cos\theta_1$, $\sin(\theta_2 + \theta_3)$ and $\cos(\theta_2 + \theta_3)$, respectively.

3.5 Kinematic control equation for robot manipulators

In any robot manipulator, the total motion is comprised of the arm motion and the wrist motion. The general equation can be written in the form:

$$[\text{Total motion}] = [\text{Arm motion}]\ [\text{Wrist motion}] \tag{3.30}$$

or
$$^0T = {}^0T_{arm}\ {}^aT_{wrist}$$

$$= [A_1A_2A_3]\ [A_4A_5A_6] \tag{3.31}$$

where T describes the end-effector position and orientation with respect to the base coordinate frame, and the superscripts denote the referenced coordinate frame. Eqn. (3.31) represents a product of six 'A' matrices. Due to the complexity involved in the matrix multiplication it is not advisable to expand Eqn. (3.31) into lengthy individual elements. In the following chapter it will be shown that the solution of the kinematic equation can be readily obtained through a segmented equation of the arm and wrist.

References

[1] Denavit, J. and Hartenberg, R. S. 1955. A kinematic notation for lower-pair mechanisms based on matrices. *J. Applied Mechanics,* June: 215–221.

[2] Pieper, D. J. 1968. *The Kinematics of a Manipulator Under Computer Control.* Stanford Artificial Intelligence Laboratory, Stanford University, CA, USA.

[3] Paul, R. P. 1983. *Robot Manipulators – Mathematics, Programming and Control.* MIT Press, Cambridge, MA, USA.

[4] Ho, C. Y. and Copeland, K. W. 1982. Solution of kinematic equations for robot manipulators. *Digital Systems for Industrial Automation,* 1 (4): 335–352.

Chapter Four

KINEMATIC AND POSITION CONTROL

IN THE previous chapter, the forms of wrist motion and arm motion were categorised according to their geometrical configurations. There are two types of angular motions to represent the wrist movement, these being the Euler and roll–pitch–yaw motions. The arm motions are represented by Cartesian (PPP), cylindrical (PRP), spherical (RRP), and articulated (RRR) arms, where P and R stand for prismatic (translational) and revolute (rotational) joints, respectively. The majority of all existing commercial robot manipulators are the combination of one of these four types of arms attached to one of the two types of wrists. This chapter will examine the position control aspect of their manipulation.

The method of specifying the position-control equations is well established[1], whereby 4 × 4 homogeneous transform 'A' matrices are used to describe the linkage relationships. The kinematic equations are then the result of the products of the 'A' matrices linking the base coordinate frame to the frame of the end-effector. Many papers have been published to solve the kinematic control equations in terms of the joint control variables, given the position and orientation of the end-effector[1–4]. This accomplished, the control system can actuate the joint control variables to attain and maintain the appropriate values obtained from the solution of the kinematic control equations.

The technique employed by Paul, Shimano and Meyer[2] was to obtain the solution for the joint control variables in a sequential manner. Each successive variable was isolated from the end-effector by premultiplication of successive inverse 'A' matrices. The complexity and frequency of matrix multiplications in this method are error-prone, often discouraging researchers from extensive study. The method proposed by Lee[3], Ho and Copeland[4], and Siegler[5] was to obtain the joint control coordinates by breaking the overall problem into two smaller parts: arm and wrist. The solutions for the arm and wrist are then combined into the total manipulator motion solution. This method has reduced the tedium of the formation of the solution and expedited the computation effort.

4.1 Solution for wrist equations

In this section, the Euler and RPY wrist variables are determined using the wrist configurations of Chapter Three. Determination of the 'A' matrices used for the wrist solutions in this section and the arm solutions in the next section was also accomplished in that chapter.

Euler wrist

The wrist equation is expressed by Euler angles as a product of the three 'A' matrices in the following equation (where C_i and S_i are the cosine and sine of angle θ_i, respectively):

$$T_E = A_1 A_2 A_3 \tag{4.1}$$

Substituting for A_1, A_2 and A_3 from Chapter Three gives:

$$T_E = \begin{bmatrix} C_1 & 0 & -S_1 & 0 \\ S_1 & 0 & C_1 & 0 \\ 0 & -1 & 0 & 0 \\ 0 & 0 & 0 & 1 \end{bmatrix} \begin{bmatrix} C_2 & 0 & S_2 & 0 \\ S_2 & 0 & -C_2 & 0 \\ 0 & 1 & 0 & 0 \\ 0 & 0 & 0 & 1 \end{bmatrix} \begin{bmatrix} C_3 & -S_3 & 0 & 0 \\ S_3 & C_3 & 0 & 0 \\ 0 & 0 & 1 & 0 \\ 0 & 0 & 0 & 1 \end{bmatrix} \tag{4.2}$$

or

$$T_E = \begin{bmatrix} C_1 C_2 C_3 - S_1 S_3 & -C_1 C_2 S_3 - S_1 C_3 & C_1 S_2 & 0 \\ S_1 C_2 C_3 + C_1 S_3 & -S_1 C_2 S_3 + C_1 C_3 & S_1 S_2 & 0 \\ -S_2 C_3 & S_2 S_3 & C_2 & 0 \\ 0 & 0 & 0 & 1 \end{bmatrix} \tag{4.3}$$

Premultiplying both sides of Eqn. (4.1) by A_1^{-1} yields:

$$A_1^{-1} T_E = A_2 A_3 \tag{4.4}$$

The orientation of the wrist is given by

$$T_E = \begin{bmatrix} n_x & o_x & a_x & 0 \\ n_y & o_y & a_y & 0 \\ n_z & o_z & a_z & 0 \\ 0 & 0 & 0 & 1 \end{bmatrix} \tag{4.5}$$

where the vectors $\bar{n}$, $\bar{o}$, and $\bar{a}$ are as shown in Fig. 2.4.

It should be noted that for a matrix of the form given in Eqn. (4.5), its inverse is simply its transposed form. This is due to the orthonormality of its component column vectors. Obtaining A_1^{-1} from A_1^T and expanding Eqn. (4.4) yields:

$$\begin{bmatrix} C_1 & S_1 & 0 & 0 \\ 0 & 0 & -1 & 0 \\ -S_1 & C_1 & 0 & 0 \\ 0 & 0 & 0 & 1 \end{bmatrix} \begin{bmatrix} n_x & o_x & a_x & 0 \\ n_y & o_y & a_y & 0 \\ n_z & o_z & a_z & 0 \\ 0 & 0 & 0 & 1 \end{bmatrix}$$

$$= \begin{bmatrix} C_2 & 0 & S_2 & 0 \\ S_2 & 0 & -C_2 & 0 \\ 0 & 1 & 0 & 0 \\ 0 & 0 & 0 & 1 \end{bmatrix} \begin{bmatrix} C_3 & -S_3 & 0 & 0 \\ S_3 & C_3 & 0 & 0 \\ 0 & 0 & 1 & 0 \\ 0 & 0 & 0 & 1 \end{bmatrix} \qquad (4.6)$$

or

$$\begin{bmatrix} C_1n_x+S_1n_y & C_1o_x+S_1o_y & C_1a_x+S_1a_y & 0 \\ -n_z & -o_z & -a_z & 0 \\ -S_1n_x+C_1n_y & -S_1o_x+C_1o_y & -S_1a_x+C_1a_y & 0 \\ 0 & 0 & 0 & 1 \end{bmatrix}$$

$$= \begin{bmatrix} C_2C_3 & -C_2S_3 & S_2 & 0 \\ S_2C_3 & -S_2S_3 & -C_2 & 0 \\ S_3 & C_3 & 0 & 0 \\ 0 & 0 & 0 & 1 \end{bmatrix} \qquad (4.7)$$

In determining the value of an angle, it is always advisable to use $\tan^{-1}y/x$, since it will encompass the entire range of the angle from $0°$ to $360°$. This is because the signs of the numerator y and denominator x determine the quadrant of the angle.

With this in mind, elements (3,3) from both sides of Eqn. (4.7) are equated as:

$$-S_1a_x + C_1a_y = 0 \qquad (4.8)$$

Thus

$$\frac{S_1}{C_1} = \frac{a_y}{a_x} \qquad (4.9)$$

$$\tan\theta_1 = \frac{a_y}{a_x} \qquad (4.10)$$

or

$$\tan\theta_1 = \frac{-a_y}{-a_x} \qquad (4.11)$$

Then

$$\theta_1 = \tan^{-1}\frac{a_y}{a_x} \qquad (4.12)$$

or

$$\theta_1 = \theta_1 + 180° \qquad (4.13)$$

which takes into account the fact that $\tan\phi = \tan(\phi + 180°)$.

Equating elements (1,3) and (2,3) of Eqn. (4.7) gives:

$$S_2 = C_1 a_x + S_1 a_y \qquad (4.14)$$

$$-C_2 = -a_z \qquad (4.15)$$

or

$$\frac{S_2}{C_2} = \frac{C_1 a_x + S_1 a_y}{a_z} \qquad (4.16)$$

and

$$\theta_2 = \tan^{-1} \frac{C_1 a_x + S_1 a_y}{a_z} \qquad (4.17)$$

Finally, by equating elements (3,1) and (3,2) of Eqn. (4.7), we have:

$$S_3 = -S_1 n_x + C_1 n_y \qquad (4.18)$$

$$C_3 = -S_1 o_x + C_1 o_y \qquad (4.19)$$

so that

$$\theta_3 = \tan^{-1} \frac{-S_1 n_x + C_1 n_y}{-S_1 o_x + C_1 o_y} \qquad (4.20)$$

Eqns. (4.10) and (4.17) are further examined to obtain a different equation for θ_2. Since

$$\tan\theta_1 = \frac{a_y}{a_x} \qquad (4.21)$$

it follows that

$$\sin\theta_1 = \frac{a_y}{\pm\sqrt{a_x^2 + a_y^2}} \qquad (4.22)$$

and

$$\cos\theta_1 = \frac{a_x}{\pm\sqrt{a_x^2 + a_y^2}} \qquad (4.23)$$

Making these substitutions in Eqn. (4.17),

$$\theta_2 = \tan^{-1} \frac{\pm\sqrt{a_x^2 + a_y^2}}{a_z} \qquad (4.24)$$

Due to the normality of vector $\bar{a}$, the condition $a_z = \pm 1$ results in $a_x = a_y = 0$; Eqn. (4.24) shows this to be the condition where $\theta_2 = 0°$ or $180°$, indicating that the end-effector is pointing straight up or down. From Eqn. (4.12), this condition further implies that θ_1 is not defined. This is known as a degenerate condition. At this particular position/orientation configuration, one degree of freedom has been lost. When this condition arises, arbitrarily setting $\theta_1 = 0$ or the current value of θ_1 then allows $\theta_1 + \theta_3$ to be the total angle required to align the orientation vector of the end-effector.

Roll–pitch–yaw wrist

Alternatively, roll–pitch–yaw angles may be used to express wrist motion. Here, the product of the three 'A' matrices is:

$$T_{RPY} = A_1 A_2 A_3 \tag{4.25}$$

Substituting for A_1, A_2 and A_3 from Chapter Three gives:

$$T_{RPY} = \begin{bmatrix} C_1 & 0 & -S_1 & 0 \\ S_1 & 0 & C_1 & 0 \\ 0 & -1 & 0 & 0 \\ 0 & 0 & 0 & 1 \end{bmatrix} \begin{bmatrix} C_2 & 0 & S_2 & 0 \\ S_2 & 0 & -C_2 & 0 \\ 0 & 1 & 0 & 0 \\ 0 & 0 & 0 & 1 \end{bmatrix} \begin{bmatrix} 1 & 0 & 0 & 0 \\ 0 & C\alpha_3 & -S\alpha_3 & 0 \\ 0 & S\alpha_3 & C\alpha_3 & 0 \\ 0 & 0 & 0 & 1 \end{bmatrix} \tag{4.26}$$

or

$$T_{RPY} = \begin{bmatrix} C_1 C_2 & C_1 S_2 S\alpha_3 - S_1 C\alpha_3 & C_1 S_2 C\alpha_3 + S_1 S\alpha_3 & 0 \\ S_1 C_2 & S_1 S_2 S\alpha_3 + C_1 C\alpha_3 & S_1 S_2 C\alpha_3 - C_1 S\alpha_3 & 0 \\ -S_2 & C_2 S\alpha_3 & C_2 C\alpha_3 & 0 \\ 0 & 0 & 0 & 1 \end{bmatrix} \tag{4.27}$$

The reader should note that the sine and cosine of the third angle above are denoted by $S\alpha_3$ and $C\alpha_3$, respectively, rather than S_3 and C_3. This is because the third angle of rotation in an RPY wrist is a rotation about an X axis, rather than a Z axis. In keeping with the notation established previously, it is labelled α_3 instead of θ_3.

Premultiplying both sides of Eqn. (4.25) by A_1^{-1} yields:

$$A_1^{-1} T_{RPY} = A_2 A_3 \tag{4.28}$$

The orientation of the wrist is

$$T_{RPY} = \begin{bmatrix} n_x & o_x & a_x & 0 \\ n_y & o_y & a_y & 0 \\ n_z & o_z & a_z & 0 \\ 0 & 0 & 0 & 1 \end{bmatrix} \tag{4.29}$$

(See Fig. 2.4 for the geometrical configuration of the column vectors of T_{RPY}.)

Obtaining A_1^{-1} from A_1^T and expanding Eqn. (4.28):

$$
\begin{bmatrix}
C_1 & S_1 & 0 & 0 \\
0 & 0 & -1 & 0 \\
-S_1 & C_1 & 0 & 0 \\
0 & 0 & 0 & 1
\end{bmatrix}
\begin{bmatrix}
n_x & o_x & a_x & 0 \\
n_y & o_y & a_y & 0 \\
n_z & o_z & a_z & 0 \\
0 & 0 & 0 & 1
\end{bmatrix}
$$

$$
=
\begin{bmatrix}
C_2 & 0 & S_2 & 0 \\
S_2 & 0 & -C_2 & 0 \\
0 & 1 & 0 & 0 \\
0 & 0 & 0 & 1
\end{bmatrix}
\begin{bmatrix}
1 & 0 & 0 & 0 \\
0 & C\alpha_3 & -S\alpha_3 & 0 \\
0 & S\alpha_3 & C\alpha_3 & 0 \\
0 & 0 & 0 & 1
\end{bmatrix}
\tag{4.30}
$$

or

$$
\begin{bmatrix}
C_1n_x+S_1n_y & C_1o_x+S_1o_y & C_1a_x+S_1a_y & 0 \\
-n_z & -o_z & -a_z & 0 \\
-S_1n_x+C_1n_y & -S_1o_x+C_1o_y & -S_1a_x+C_1a_y & 0 \\
0 & 0 & 0 & 1
\end{bmatrix}
$$

$$
=
\begin{bmatrix}
C_2 & S_2S\alpha_3 & S_2C\alpha_3 & 0 \\
S_2 & -C_2S\alpha_3 & -C_2C\alpha_3 & 0 \\
0 & C\alpha_3 & -S\alpha_3 & 0 \\
0 & 0 & 0 & 1
\end{bmatrix}
\tag{4.31}
$$

Equating elements (3,1) from both sides of Eqn. (4.31):

$$
-S_1n_x + C_1n_y = 0 \tag{4.32}
$$

$$
\frac{S_1}{C_1} = \frac{n_y}{n_x} \tag{4.33}
$$

$$
\tan\theta_1 = \frac{n_y}{n_x} \tag{4.34}
$$

or

$$
\tan\theta_1 = \frac{-n_y}{-n_x} \tag{4.35}
$$

Thus

$$
\theta_1 = \tan^{-1}\frac{n_y}{n_x} \tag{4.36}
$$

or

$$
\theta_1 = \theta_1 + 180° \tag{4.37}
$$

Equating elements (2,1) and (1,1) of Eqn. (4.31),

$$
S_2 = -n_z \tag{4.38}
$$

$$C_2 = C_1 n_x + S_1 n_y \tag{4.39}$$

So

$$\theta_2 = \tan^{-1} \frac{-n_z}{C_1 n_x + S_1 n_y} \tag{4.40}$$

Further, equating elements (3,3) and (3,2) of Eqn. (4.31),

$$-S\alpha_3 = -S_1 a_x + C_1 a_y \tag{4.41}$$

$$C\alpha_3 = -S_1 o_x + C_1 o_y \tag{4.42}$$

So

$$\alpha_3 = \tan^{-1} \frac{S_1 a_x - C_1 a_y}{-S_1 o_x + C_1 o_y} \tag{4.43}$$

As was done for the Euler wrist, the condition of degeneracy is now examined for the RPY wrist. Eqn. (4.34) states that:

$$\tan\theta_1 = \frac{n_y}{n_x} \tag{4.44}$$

Thus, it follows that:

$$\sin\theta_1 = \frac{n_y}{\pm\sqrt{n_x^2 + n_y^2}} \tag{4.45}$$

and

$$\cos\theta_1 = \frac{n_x}{\pm\sqrt{n_x^2 + n_y^2}} \tag{4.46}$$

Substituting for S_1 and C_1 in Eqn. (4.40), it is seen that θ_2 may be rewritten as:

$$\theta_2 = \tan^{-1} \frac{-n_z}{\pm\sqrt{n_x^2 + n_y^2}} \tag{4.47}$$

When the condition $n_z = \pm 1$ arises, the result is $n_x = n_y = 0$, due to the normality of vector $\bar{n}$. Eqn. (4.47) shows θ_2 to be either 90° or 270° in this situation. From Eqn. (4.36), it is clear that angle θ_1 is not defined. When this situation occurs, θ_1 is arbitrarily set to 0.

The solutions for the angles of the Euler and roll–pitch–yaw wrists given the orientation of the wrist are summarised in Table 4.1.

Table 4.1 Solution for wrist given orientation

Euler wrist	RPY wrist
$\theta_1 = \tan^{-1} \dfrac{a_y}{a_x}$ or $\theta_1 + 180°$	$\theta_1 = \tan^{-1} \dfrac{n_y}{n_x}$ or $\theta_1 + 180°$
$\theta_2 = \tan^{-1} \dfrac{C_1 a_x + S_1 a_y}{a_z}$	$\theta_2 = \tan^{-1} \dfrac{-n_z}{C_1 n_x + S_1 n_y}$
$\theta_3 = \tan^{-1} \dfrac{-S_1 n_x + C_1 n_y}{-S_1 o_x + C_1 o_y}$	$\alpha_3 = \tan^{-1} \dfrac{S_1 a_x - C_1 a_y}{-S_1 o_x + C_1 o_y}$
$\theta_2 = 0°$ or $180°$	$\theta_2 = 90°$ or $270°$
(degeneracy)	(degeneracy)

4.2 Solution for arm equations

In this section, the solutions for the control variables of each of the four types of arms as discussed in Chapter Three are determined. It should be noted that the solutions presented are for the simplest possible cases. It is possible that additional offsets may be involved in a specific arm, but the methods of determining the solutions are the same.

Cartesian arm

The Cartesian arm has three prismatic joints, denoted PPP. The arm equation is the product of the three 'A' matrices from Chapter Three as follows:

$$T_{PPP} = A_1 A_2 A_3 \tag{4.48}$$

$$= \begin{bmatrix} 1 & 0 & 0 & a_1 \\ 0 & 0 & 1 & 0 \\ 0 & -1 & 0 & 0 \\ 0 & 0 & 0 & 1 \end{bmatrix} \begin{bmatrix} 1 & 0 & 0 & 0 \\ 0 & 0 & -1 & 0 \\ 0 & 1 & 0 & d_2 \\ 0 & 0 & 0 & 1 \end{bmatrix} \begin{bmatrix} 1 & 0 & 0 & 0 \\ 0 & 1 & 0 & 0 \\ 0 & 0 & 1 & d_3 \\ 0 & 0 & 0 & 1 \end{bmatrix} \tag{4.49}$$

$$= \begin{bmatrix} 1 & 0 & 0 & a_1 \\ 0 & 1 & 0 & d_2 \\ 0 & 0 & 1 & d_3 \\ 0 & 0 & 0 & 1 \end{bmatrix} \tag{4.50}$$

The orientation/position of the arm is given by

$$T_{PPP} = \begin{bmatrix} n_x & o_x & a_x & p_x \\ n_y & o_y & a_y & p_y \\ n_z & o_z & a_z & p_z \\ 0 & 0 & 0 & 1 \end{bmatrix} \tag{4.51}$$

where the position of the tip of the arm is given by

$$\bar{p} = \begin{bmatrix} p_x \\ p_y \\ p_z \end{bmatrix} \qquad (4.52)$$

The solution of the control variables can be expressed very simply in terms of the given position:

$$a_1 = p_x \qquad (4.53)$$
$$d_2 = p_y \qquad (4.54)$$
$$d_3 = p_z \qquad (4.55)$$

Cylindrical arm

The cylindrical arm combines linkage of prismatic, revolute, and prismatic joints (PRP). The transform equation is:

$$T_{PRP} = A_1 A_2 A_3 \qquad (4.56)$$

$$= \begin{bmatrix} 1 & 0 & 0 & 0 \\ 0 & 1 & 0 & 0 \\ 0 & 0 & 1 & d_1 \\ 0 & 0 & 0 & 1 \end{bmatrix} \begin{bmatrix} C_2 & 0 & -S_2 & a_2C_2 \\ S_2 & 0 & C_2 & a_2S_2 \\ 0 & -1 & 0 & 0 \\ 0 & 0 & 0 & 1 \end{bmatrix} \begin{bmatrix} 1 & 0 & 0 & 0 \\ 0 & 1 & 0 & 0 \\ 0 & 0 & 1 & d_3 \\ 0 & 0 & 0 & 1 \end{bmatrix} \qquad (4.57)$$

$$= \begin{bmatrix} C_2 & 0 & -S_2 & -d_3S_2 + a_2C_2 \\ S_2 & 0 & C_2 & d_3C_2 + a_2S_2 \\ 0 & -1 & 0 & d_1 \\ 0 & 0 & 0 & 1 \end{bmatrix} \qquad (4.58)$$

Equating the translation vector components of T_{PRP} to the given position vector components p_x, p_y and p_z of Eqn. (4.52) gives:

$$-d_3S_2 + a_2C_2 = p_x \qquad (4.59)$$
$$d_3C_2 + a_2S_2 = p_y \qquad (4.60)$$
$$d_1 = p_z \qquad (4.61)$$

Variable d_1 is immediately given by Eqn. (4.61). Eqn. (4.59) is an equation of one variable, θ_2. For an equation of the form

$$-A \sin\theta + B \cos\theta = D \qquad (4.62)$$

the solution may be obtained as follows:

Let

$$A = r \cos\phi \qquad (4.63)$$

and

$$B = r \sin\phi \qquad (4.64)$$

then

$$r = \sqrt{A^2 + B^2} \tag{4.65}$$

and

$$\phi = \tan^{-1} \frac{B}{A} \tag{4.66}$$

Fig. 4.1 shows the trigonometric relationships.

   Substituting for A and B in Eqn. (4.62),

$$-r \cos\phi \sin\theta + r \sin\phi \cos\theta = D \tag{4.67}$$

Dividing by r and rearranging terms,

$$\sin\phi \cos\theta - \cos\phi \sin\theta = D/r \tag{4.68}$$

Eqn. (4.68) is the expression for the sine of the difference of two angles, so

$$\sin(\phi-\theta) = D/r \tag{4.69}$$

Since $\sin^2\alpha + \cos^2\alpha = 1$,

$$\cos(\phi-\theta) = \pm\sqrt{1 - D^2/r^2} \tag{4.70}$$

and hence

$$\tan(\phi-\theta) = \frac{D/r}{\pm\sqrt{1 - D^2/r^2}} \tag{4.71}$$

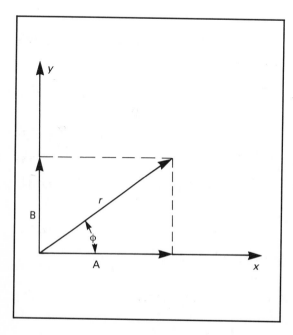

Fig. 4.1 *Trigonometric relationships between* A, B, *r and* ϕ

or

$$\tan(\phi-\theta) = \frac{\pm D}{\sqrt{r^2 - D^2}} \tag{4.72}$$

Substituting for r from Eqn. (4.65),

$$\tan(\phi-\theta) = \frac{\pm D}{\sqrt{A^2 + B^2 - D^2}} \tag{4.73}$$

and

$$\phi - \theta = \tan^{-1} \frac{\pm D}{\sqrt{A^2 + B^2 - D^2}} \tag{4.74}$$

Substituting for ϕ from Eqn. (4.66) and rearranging terms gives:

$$\theta = \tan^{-1} \frac{B}{A} - \tan^{-1} \frac{\pm D}{\sqrt{A^2 + B^2 - D^2}} \tag{4.75}$$

Further, from Eqn. (4.69), if D is positive, then

$$0 < \sin(\phi-\theta) \leq 1 \tag{4.76}$$

which implies that:

$$0 < \phi - \theta < \pi \tag{4.77}$$

$$0 < \tan^{-1} B/A - \theta < \pi \tag{4.78}$$

$$-\tan^{-1} B/A < -\theta < \pi - \tan^{-1} B/A \tag{4.79}$$

and

$$\tan^{-1} B/A - \pi < \theta < \tan^{-1} B/A \tag{4.80}$$

If, on the other hand, D is negative, then

$$-1 \leq \sin(\phi-\theta) < 0 \tag{4.81}$$

so

$$-\pi < \phi - \theta < 0 \tag{4.82}$$

$$-\pi < \tan^{-1} B/A - \theta < 0 \tag{4.83}$$

$$-\pi - \tan^{-1} B/A < -\theta < -\tan^{-1} B/A \tag{4.84}$$

and

$$\tan^{-1} B/A < \theta < \pi + \tan^{-1} B/A \tag{4.85}$$

For Eqn. (4.59) the solution from Eqn. (4.75) is:

$$\theta_2 = \tan^{-1} \frac{a_2}{d_3} - \tan^{-1} \frac{\pm p_x}{\sqrt{d_3^2 + a_2^2 - p_x^2}} \tag{4.86}$$

with the restriction that if p_x is positive,

$$\tan^{-1}\frac{a_2}{d_3} - \pi < \theta_2 < \tan^{-1}\frac{a_2}{d_3} \qquad (4.87)$$

or if p_x is negative,

$$\tan^{-1}\frac{a_2}{d_3} < \theta_2 < \pi + \tan^{-1}\frac{a_2}{d_3} \qquad (4.88)$$

To derive a relationship for d_3, Eqns. (4.59) and (4.60) are each squared and then added as follows:

$$(-d_3S_2 + a_2C_2)^2 = p_x^2 \qquad (4.89)$$

$$(d_3C_2 + a_2S_2)^2 = p_y^2 \qquad (4.90)$$

$$(d_3^2S_2^2 - 2d_3a_2S_2C_2 + a_2^2C_2^2) = p_x^2 \qquad (4.91)$$

$$(d_3^2C_2^2 + 2d_3a_2S_2C_2 + a_2^2S_2^2) = p_y^2 \qquad (4.92)$$

and

$$d_3^2 + a_2^2 = p_x^2 + p_y^2 \qquad (4.93)$$

so

$$d_3 = \sqrt{p_x^2 + p_y^2 - a_2^2} \qquad (4.94)$$

Spherical arm

The spherical arm combines two revolute joints and one prismatic joint (RRP). The transform equation is:

$$T_{RRP} = A_1A_2A_3 \qquad (4.95)$$

$$= \begin{bmatrix} C_1 & 0 & -S_1 & 0 \\ S_1 & 0 & C_1 & 0 \\ 0 & -1 & 0 & 0 \\ 0 & 0 & 0 & 1 \end{bmatrix} \begin{bmatrix} C_2 & 0 & S_2 & 0 \\ S_2 & 0 & -C_2 & 0 \\ 0 & 1 & 0 & 0 \\ 0 & 0 & 0 & 1 \end{bmatrix} \begin{bmatrix} 1 & 0 & 0 & 0 \\ 0 & 1 & 0 & 0 \\ 0 & 0 & 1 & d_3 \\ 0 & 0 & 0 & 1 \end{bmatrix} \qquad (4.96)$$

$$= \begin{bmatrix} C_1C_2 & -S_1 & C_1S_2 & C_1S_2d_3 \\ S_1C_2 & C_1 & S_1S_2 & S_1S_2d_3 \\ -S_2 & 0 & C_2 & C_2d_3 \\ 0 & 0 & 0 & 1 \end{bmatrix} \qquad (4.97)$$

Equating the translation vector components of T_{RRP} to p_x, p_y and p_z of Eqn. (4.52) gives:

$$C_1S_2d_3 = p_x \qquad (4.98)$$

$$S_1S_2d_3 = p_y \qquad (4.99)$$

$$C_2 d_3 = p_z \tag{4.100}$$

Dividing Eqn. (4.99) by (4.98),

$$\frac{S_1 S_2 d_3}{C_1 S_2 d_3} = \frac{p_y}{p_x} \tag{4.101}$$

or

$$\theta_1 = \tan^{-1} \frac{p_y}{p_x} \tag{4.102}$$

Squaring Eqns. (4.98) and (4.99) and adding,

$$C_1^2 S_2^2 d_3^2 = p_x^2 \tag{4.103}$$

$$S_1^2 S_2^2 d_3^2 = p_y^2 \tag{4.104}$$

$$S_2^2 d_3^2 = p_x^2 + p_y^2 \tag{4.105}$$

and

$$S_2 d_3 = \pm \sqrt{p_x^2 + p_y^2} \tag{4.106}$$

Dividing by Eqn. (4.100),

$$\frac{S_2 d_3}{C_2 d_3} = \frac{\pm \sqrt{p_x^2 + p_y^2}}{p_z} \tag{4.107}$$

or

$$\theta_2 = \tan^{-1} \frac{\pm \sqrt{p_x^2 + p_y^2}}{p_z} \tag{4.108}$$

Now squaring Eqn. (4.100) and adding it to Eqn. (4.105),

$$C_2^2 d_3^2 = p_z^2 \tag{4.109}$$

$$S_2^2 d_3^2 = p_x^2 + p_y^2 \tag{4.110}$$

$$d_3^2 = p_x^2 + p_y^2 + p_z^2 \tag{4.111}$$

so

$$d_3 = \sqrt{p_x^2 + p_y^2 + p_z^2} \tag{4.112}$$

Revolute arm

The revolute arm imitates the motion of the human arm, i.e. it has motion corresponding to that of the human trunk, shoulder, and elbow. Each motion is revolute (RRR). The transform equation is:

$$T_{RRR} = A_1 A_2 A_3 \tag{4.113}$$

$$= \begin{bmatrix} C_1 & 0 & S_1 & 0 \\ S_1 & 0 & -C_1 & 0 \\ 0 & 1 & 0 & 0 \\ 0 & 0 & 0 & 1 \end{bmatrix} \begin{bmatrix} C_2 & -S_2 & 0 & a_2C_2 \\ S_2 & C_2 & 0 & a_2S_2 \\ 0 & 0 & 1 & 0 \\ 0 & 0 & 0 & 1 \end{bmatrix} \begin{bmatrix} C_3 & -S_3 & 0 & a_3C_3 \\ S_3 & C_3 & 0 & a_3S_3 \\ 0 & 0 & 1 & 0 \\ 0 & 0 & 0 & 1 \end{bmatrix}$$

$$\text{(4.114)}$$

$$= \begin{bmatrix} C_1C_{23} & -C_1S_{23} & S_1 & C_1(a_3C_{23}+a_2C_2) \\ S_1C_{23} & -S_1S_{23} & -C_1 & S_1(a_3C_{23}+a_2C_2) \\ S_{23} & C_{23} & 0 & a_2S_{23}+a_2S_2 \\ 0 & 0 & 0 & 1 \end{bmatrix} \quad \text{(4.115)}$$

where S_{23} and C_{23} represent $\sin(\theta_2 + \theta_3)$ and $\cos(\theta_2 + \theta_3)$, respectively.

Equating the translation vector to the given position p_x, p_y and p_z of Eqn. (4.52),

$$C_1(a_3C_{23} + a_2C_2) = p_x \quad \text{(4.116)}$$

$$S_1(a_3C_{23} + a_2C_2) = p_y \quad \text{(4.117)}$$

$$a_3S_{23} + a_2S_2 = p_z \quad \text{(4.118)}$$

Dividing Eqn. (4.117) by (4.116),

$$\frac{S_1(a_3C_{23} + a_2C_2)}{C_1(a_3C_{23} + a_2C_2)} = \frac{p_y}{p_x} \quad \text{(4.119)}$$

or

$$\theta_1 = \tan^{-1} \frac{p_y}{p_x} \quad \text{(4.120)}$$

Squaring Eqns. (4.116) and (4.117) and then adding,

$$C_1{}^2(a_3C_{23} + a_2C_2)^2 = p_x{}^2 \quad \text{(4.121)}$$

$$S_1{}^2(a_3C_{23} + a_2C_2)^2 = p_y{}^2 \quad \text{(4.122)}$$

$$(a_3C_{23} + a_2C_2)^2 = p_x{}^2 + p_y{}^2 \quad \text{(4.123)}$$

$$a_3C_{23} + a_2C_2 = \sqrt{p_x{}^2 + p_y{}^2} \quad \text{(4.124)}$$

To simplify the expression, let $\alpha = \sqrt{p_x{}^2 + p_y{}^2}$

Then,

$$C_{23} = \frac{\alpha - a_2C_2}{a_3} \quad \text{(4.125)}$$

Isolating S_{23} in Eqn. (4.118) and employing the identity $S_{23}{}^2 + C_{23}{}^2 = 1$,

$$S_{23} = \frac{P_z - a_2S_2}{a_3} \quad \text{(4.126)}$$

$$\left(\frac{p_z - a_2S_2}{a_3}\right)^2 + \left(\frac{\alpha - a_2C_2}{a_3}\right)^2 = 1 \tag{4.127}$$

$$p_z^2 - 2a_2p_zS_2 + a_2^2S_2^2 + \alpha^2 - 2\alpha a_2C_2 + a_2^2C_2^2 = a_3^2 \tag{4.128}$$

Regrouping,

$$-2a_2p_zS_2 - 2\alpha a_2C_2 = a_3^2 - (a_2^2S_2^2 + a_2^2C_2^2) - p_z^2 - \alpha^2 \tag{4.129}$$

$$p_zS_2 + \alpha C_2 = \frac{\alpha^2 + p_z^2 + a_2^2 - a_3^2}{2a_2} \tag{4.130}$$

Again, to simplify, let $\beta = \dfrac{\alpha^2 + p_z^2 + a_2^2 - a_3^2}{2a_2}$

Thus,

$$p_zS_2 + \alpha C_2 = \beta \tag{4.131}$$

This equation is solved using the same technique as that employed for Eqn. (4.59) of the spherical arm.

Let

$$p_z = r\cos\phi \tag{4.132}$$
$$\alpha = r\sin\phi \tag{4.133}$$

Then,

$$r = \sqrt{p_z^2 + \alpha^2} \tag{4.134}$$

and

$$\phi = \tan^{-1}\frac{\alpha}{p_z} \tag{4.135}$$

Substituting for p_z and α in Eqn. (4.131),

$$r\cos\phi\,\sin\theta_2 + r\sin\phi\,\cos\theta_2 = \beta \tag{4.136}$$

or

$$\sin(\theta_2 + \phi) = \beta/r \tag{4.137}$$

then

$$\cos(\theta_2 + \phi) = \pm\sqrt{1 - \beta^2/r^2} \tag{4.138}$$

and

$$\tan(\theta_2 + \phi) = \frac{\pm\beta}{\sqrt{r^2 - \beta^2}} \tag{4.139}$$

$$\theta_2 + \phi = \tan^{-1}\frac{\pm\beta}{\sqrt{r^2 - \beta^2}} \qquad (4.140)$$

Substituting for ϕ and r,

$$\theta_2 = \tan^{-1}\frac{\pm\beta}{\sqrt{p_z^2 + \alpha^2 - \beta^2}} - \tan^{-1} \qquad (4.141)$$

Also, from Eqn. (4.137), if β is positive,

$$0 < \sin(\theta_2 + \phi) \leqslant 1 \qquad (4.142)$$

$$0 < \theta_2 + \phi < \pi \qquad (4.143)$$

and

$$-\tan^{-1}\frac{\alpha}{p_z} < \theta_2 < \pi - \tan^{-1}\frac{\alpha}{p_z} \qquad (4.144)$$

If β is negative,

$$-\pi < \theta_2 + \phi < 0 \qquad (4.145)$$

and

$$-\pi - \tan^{-1}\frac{\alpha}{p_z} < \theta_2 < -\tan^{-1}\frac{\alpha}{p_z} \qquad (4.146)$$

To obtain a solution for θ_3, divide Eqn. (4.126) by Eqn. (4.125),

$$\frac{S_{23}}{C_{23}} = \frac{\dfrac{p_z - a_2S_2}{a_3}}{\dfrac{\alpha - a_2C_2}{a_3}} \qquad (4.147)$$

$$\tan(\theta_2 + \theta_3) = \frac{p_z - a_2S_2}{\alpha - a_2C_2} \qquad (4.148)$$

and

$$\theta_3 = \tan^{-1}\frac{p_z - a_2S_2}{\alpha - a_2C_2} - \theta_2 \qquad (4.149)$$

The solutions for the four types of arms as determined in this section are summarised in Table 4.2.

4.3 Solution for the robot manipulator

The solution for a robot manipulator at first appears to be a difficult task, but when solved as two parts, it turns out to be relatively simple. With a six-degree-of-freedom robot manipulator, the total equation is comprised

Table 4.2 *Solution for arm given position*

Cartesian arm (PPP)	Cylindrical arm (PRP)

$$a_1 = p_x$$
$$d_2 = p_y$$
$$d_3 = p_z$$

$$d_1 = p_z$$
$$\theta_2 = \tan^{-1} \frac{a_2}{d_3} - \tan^{-1} \frac{\pm p_x}{p_y}$$
$$d_3 = \sqrt{p_x^2 + p_y^2 - a_2^2}$$

Spherical arm (RRP)	Revolute arm (RRR)

$$\theta_1 = \tan^{-1} \frac{p_y}{p_z}$$

$$\theta_2 = \tan^{-1} \frac{\pm \sqrt{p_x^2 + p_y^2}}{p_z}$$

$$d_3 = \sqrt{p_x^2 + p_y^2 + p_z^2}$$

$$\theta_1 = \tan^{-1} \frac{p_y}{p_x}$$

$$\theta_2 = \tan^{-1} \frac{\pm \beta}{\sqrt{p_z^2 + \alpha^2 - \beta^2}} - \tan^{-1} \frac{\alpha}{p_z}$$

$$\theta_3 = \tan^{-1} \frac{p_z - a_2 S_2}{\alpha - a_2 C_2} - \theta_2$$

$$\alpha = \sqrt{p_x^2 + p_y^2}$$

$$\beta = \frac{\alpha^2 + p_z^2 + a_2^2 - a_3^2}{2a_2}$$

of the arm motion of one of the four previously mentioned possible arms and the wrist motion comprises either Euler angles or roll–pitch–yaw angles. The general equation can be written as:

$$[\text{total motion}] = [\text{arm motion}] [\text{wrist motion}] \qquad (4.150)$$

or

$$^0T_6 = {}^0T_a \, {}^aT_w \qquad (4.151)$$

where $^0T_6$ describes the end-effector position and orientation with respect to the base coordinate frame. The superscripts designate the reference frame; zero represents the base frame and a represents the tip of the arm. The subscripts designate the new frame reached by the transform; a is again the tip of the arm and w is the position centred on by the wrist.

Using Denavit and Hartenberg's[5] method, the end-effector position and orientation can be written as

$$^0T_6 = A_1 A_2 A_3 A_4 A_5 A_6 \qquad (4.152)$$

By the associative law the product of matrices can be regrouped into two subsets which represent the arm and the wrist respectively:

$$^0T_6 = (A_1 A_2 A_3) (A_4 A_5 A_6) \qquad (4.153)$$

where

$$A_1 A_2 A_3 = {}^0T_3 = {}^0T_a \tag{4.154}$$

and

$$A_4 A_5 A_6 = {}^3T_6 = {}^aT_w \tag{4.155}$$

The ${}^0T_6$ given for the end-effector can be written as a 4×4 homogeneous matrix composed of an orientation submatrix R and a position vector $\bar{P}$:

$$
{}^0T_6 =
\begin{bmatrix}
N_x & O_x & A_x & P_x \\
N_y & O_y & A_y & P_y \\
N_z & O_z & A_z & P_z \\
0 & 0 & 0 & 1
\end{bmatrix}
\tag{4.156}
$$

$$
=
\begin{bmatrix}
\begin{bmatrix}
\text{orientation} \\
\text{matrix} \\
3 \times 3 \\
0 \quad 0 \quad 0
\end{bmatrix}
&
\begin{bmatrix}
\text{position} \\
\text{vector} \\
3 \times 1 \\
1
\end{bmatrix}
\end{bmatrix}
\tag{4.157}
$$

$$
=
\begin{bmatrix}
R & \bar{P} \\
\hline
0 & 1
\end{bmatrix}
\tag{4.158}
$$

Likewise, the orientation and position of the arm with respect to the base coordinate frame can be written:

$$
{}^0T_3 = {}^0T_a =
\begin{bmatrix}
n_{ax} & o_{ax} & a_{ax} & P_{ax} \\
n_{ay} & o_{ay} & a_{ay} & P_{ay} \\
n_{az} & o_{az} & a_{az} & P_{az} \\
0 & 0 & 0 & 1
\end{bmatrix}
\tag{4.159}
$$

$$
{}^0T_3 =
\begin{bmatrix}
{}^0R_3 & {}^0\bar{P}_a \\
\hline
0 & 1
\end{bmatrix}
\tag{4.160}
$$

The orientation and position of the centre of the grip of the wrist with respect to the third link frame is:

$$
{}^3T_6 = {}^aT_w =
\begin{bmatrix}
n_{wx} & o_{wx} & a_{wx} & P_{wx} \\
n_{wy} & o_{wy} & a_{wy} & P_{wy} \\
n_{wz} & o_{wz} & a_{wz} & P_{wz} \\
0 & 0 & 0 & 1
\end{bmatrix}
\tag{4.161}
$$

$$
{}^3T_6 =
\begin{bmatrix}
{}^3R_6 & {}^3\bar{P}_w \\
\hline
0 & 1
\end{bmatrix}
\tag{4.162}
$$

From Eqns. (4.151), (4.160) and (4.162),

$$
{}^0T_6 = {}^0T_a {}^aT_w \tag{4.163}
$$

$$= \left[\begin{array}{c|c} {}^{0}R_3 & {}^{0}\bar{P}_a \\ \hline 0 & 1 \end{array} \right] \left[\begin{array}{c|c} {}^{3}R_6 & {}^{3}\bar{P}_w \\ \hline 0 & 1 \end{array} \right] \tag{4.164}$$

$$= \left[\begin{array}{c|c} {}^{0}R_3 & {}^{0}R_3{}^{3}\bar{P}_w + {}^{0}\bar{P}_a \\ \hline 0 & 1 \end{array} \right] \tag{4.165}$$

Therefore, the rotation part is:

$$R = {}^{0}R_3{}^{3}R_6 \tag{4.166}$$

and the translation part is

$$\bar{P} = {}^{0}R_3{}^{3}\bar{P}_w + {}^{0}\bar{P}_a \tag{4.167}$$

The matrix product ${}^{0}R_3{}^{3}\bar{P}_w$ of Eqn. (4.167) may be thought of as follows. ${}^{0}R_3$ is an orientation matrix serving as a connection between coordinate frames 0 and 3. When postmultiplied by a set of coordinates of frame 3 (such as ${}^{3}\bar{P}_w$), it converts them to their equivalent with reference to frame 0. Thus, the product ${}^{0}R_3{}^{3}\bar{P}_w$ is simply ${}^{0}\bar{P}_w$, i.e. the position of the centre of the grip of the wrist measured from the tip of the arm, all with reference to frame 0. Eqn. (4.167) may then be written:

$$\bar{P} = {}^{0}\bar{P}_a + {}^{0}\bar{P}_w \tag{4.168}$$

This states that the total translation of the manipulator is the sum of the translation from the base to the tip of the arm plus the translation from the tip of the arm to the centre of the grip of the wrist.

In order to exemplify the method presented, a well–known robot arm has been chosen for solution. It is then easy for the reader to compare the technique presented with other existing methods. The PUMA–560, a six degree-of-freedom articulated manipulator (Fig. 4.2), is the working example to be demonstrated. It is a combination of a revolute arm and a Euler wrist.

The axes of its joint coordinate frames are assigned according to the rules established by Denavit and Hartenberg[5] and explained in Chapter Three. The link parameter table is given in Table 4.3.

The coordinate frame links are completely described by Table 4.3. The D–H transformation matrix gives A_i, where link i is the transformation with respect to the preceding link $i-1$. A_i may be written:

Table 4.3 Link parameter table for PUMA–560

Link	Variable	α_i(degrees)	a_i(mm)	d_i(mm)	Range (degrees)
1	θ_1	−90	0	0	±160
2	θ_2	0	a_2=432	d_2=149.5	−225 to 45
3	θ_3	90	0	0	−45 to 225
4	θ_4	−90	0	d_4=432	−110 to 170
5	θ_5	90	0	0	±100
6	θ_6	0	0	d_6=56.5	±266

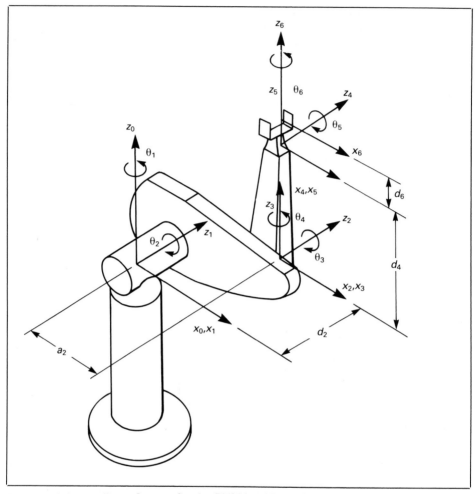

Fig. 4.2 Link coordinate frames for the PUMA–560 manipulator

$$A_i = \begin{bmatrix} \cos\theta_i & -\sin\theta_i\cos\alpha_i & \sin\theta_i\sin\alpha_i & a\cos\theta_i \\ \sin\theta_i & \cos\theta_i\cos\alpha_i & -\cos\theta_i\sin\alpha_i & a\sin\theta_i \\ 0 & \sin\alpha_i & \cos\alpha_i & d_i \\ 0 & 0 & 0 & 1 \end{bmatrix} \qquad (4.169)$$

Thus,

$$A_1 = \begin{bmatrix} C_1 & 0 & -S_1 & 0 \\ S_1 & 0 & C_1 & 0 \\ 0 & -1 & 0 & 0 \\ 0 & 0 & 0 & 1 \end{bmatrix} \qquad (4.170)$$

$$A_2 = \begin{bmatrix} C_2 & -S_2 & 0 & a_2C_2 \\ S_2 & C_2 & 0 & a_2S_2 \\ 0 & 0 & 1 & d_2 \\ 0 & 0 & 0 & 1 \end{bmatrix} \qquad (4.171)$$

$$A_3 = \begin{bmatrix} C_3 & 0 & S_3 & 0 \\ S_3 & 0 & -C_3 & 0 \\ 0 & 1 & 0 & 0 \\ 0 & 0 & 0 & 1 \end{bmatrix} \tag{4.172}$$

$$A_4 = \begin{bmatrix} C_4 & 0 & -S_4 & 0 \\ S_4 & 0 & C_4 & 0 \\ 0 & -1 & 0 & d_4 \\ 0 & 0 & 0 & 1 \end{bmatrix} \tag{4.173}$$

$$A_5 = \begin{bmatrix} C_5 & 0 & S_5 & 0 \\ S_5 & 0 & -C_5 & 0 \\ 0 & 1 & 0 & 0 \\ 0 & 0 & 0 & 1 \end{bmatrix} \tag{4.174}$$

$$A_6 = \begin{bmatrix} C_6 & -S_6 & 0 & 0 \\ S_6 & C_6 & 0 & 0 \\ 0 & 0 & 1 & d_6 \\ 0 & 0 & 0 & 1 \end{bmatrix} \tag{4.175}$$

Furthermore,

$$^0T_a = {}^0T_3 = A_1 A_2 A_3 \tag{4.176}$$

$$^0T_a = \begin{bmatrix} C_1 C_{23} & -S_1 & C_1 S_{23} & a_2 C_1 C_2 - d_2 S_1 \\ S_1 C_{23} & C_1 & S_1 S_{23} & a_2 S_1 C_2 + d_2 C_1 \\ -S_{23} & 0 & C_{23} & -a_2 S_2 \\ 0 & 0 & 0 & 1 \end{bmatrix} \tag{4.177}$$

and

$$^aT_w = {}^3T_6 = A_4 A_5 A_6 \tag{4.178}$$

$$^aT_w = \begin{bmatrix} C_4 C_5 C_6 - S_4 S_6 & -C_4 C_5 S_6 - S_4 C_6 & C_4 S_5 & d_6 C_4 S_5 \\ S_4 C_5 C_6 + C_4 S_6 & -S_4 C_5 S_6 + C_4 C_6 & S_4 S_5 & d_6 S_4 S_5 \\ -S_5 C_6 & S_5 S_6 & C_5 & d_6 C_5 + d_4 \\ 0 & 0 & 0 & 1 \end{bmatrix}$$

$$\tag{4.179}$$

Solution for arm control angles θ_1, θ_2 and θ_3

Figs. 4.3 and 4.4 show the relationship between the arm and wrist vectors. $\bar{P}$ is the position vector from the base coordinate frame to the centre of the wrist grip; vector $\bar{A}$ represents the approach direction of the end-effector, as in Fig. 2.4. (Vectors $\bar{N}$, $\bar{O}$, $\bar{A}$, and $\bar{P}$ are given as columns of $^0T_6$.) $\bar{P}_a$ is the arm vector measured from the origin to the connecting point of the arm and wrist, while $\bar{P}_w$ is the wrist vector having the same direction as the $\bar{A}$ vector and length measured from the connection point of the arm and wrist to the wrist grip centre. All quantities are in terms of the base coordinate system.

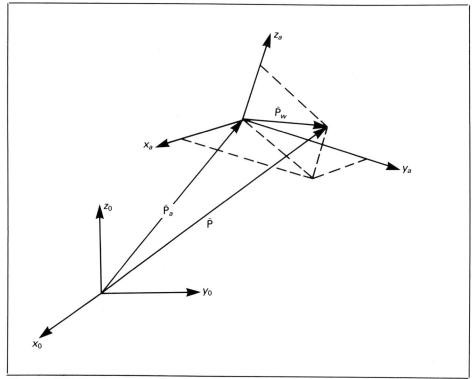

Fig. 4.3 Relationship between vectors $\bar{P}$, $\bar{P}_a$ *and* $\bar{P}_w$

In examining the translation and rotation of the robot manipulator, the first step is to analyse the translation part. This results in the determination of the arm control angles θ_1, θ_2 and θ_3. The translation is:

$$[\text{total translation}] = [\text{arm translation}] + [\text{wrist translation}] \qquad (4.180)$$

Eqn. (4.180) is identical to Eqn. (4.168), which stated:

$$\bar{P} = \bar{P}_a + \bar{P}_w \qquad (4.181)$$

or

$$\begin{bmatrix} P_x \\ P_y \\ P_z \end{bmatrix} = \begin{bmatrix} P_{ax} \\ P_{ay} \\ P_{az} \end{bmatrix} + \begin{bmatrix} P_{wx} \\ P_{wy} \\ P_{wz} \end{bmatrix} \qquad (4.182)$$

Fig. 4.3 represents the relationship between vectors $\bar{P}$, $\bar{P}_a$ and $\bar{P}_w$. The three vector components of $\bar{P}_w$ can be expressed in terms of the direction of the approach vector $\bar{A}$ and the magnitude of $\bar{P}_w$ (see Fig. 4.4). The azimuthal angle ϕ and polar angle θ of vector A with respect to the base frame can be determined as:

$$\phi = \tan^{-1} \frac{A_y}{A_x} \qquad (4.183)$$

and

$$\theta = \tan^{-1} \frac{\sqrt{A_x^2 + A_y^2}}{A_z} \tag{4.184}$$

Furthermore, examination of Fig. 4.4 shows the components of $\bar{P}_w$ to be:

$$P_{wx} = |\bar{P}_w| \sin\theta \cos\phi \tag{4.185}$$

$$P_{wy} = |\bar{P}_w| \sin\theta \sin\phi \tag{4.186}$$

$$P_{wz} = |\bar{P}_w| \cos\theta \tag{4.187}$$

From Eqn. (4.181), the translation vector of the arm is:

$$\bar{P}_a = \bar{P} - \bar{P}_w \tag{4.188}$$

or

$$\begin{bmatrix} P_{ax} \\ P_{ay} \\ P_{az} \end{bmatrix} = \begin{bmatrix} P_x \\ P_y \\ P_z \end{bmatrix} - \begin{bmatrix} |\bar{P}_w| \sin\theta \cos\phi \\ |\bar{P}_w| \sin\theta \sin\phi \\ |\bar{P}_w| \cos\theta \end{bmatrix} \tag{4.189}$$

Vector $\bar{P}_a$ is measured from the origin to the connecting point of the arm and wrist. The PUMA–560 configuration given in Fig. 4.2 shows that the transformation ^0T_a does not have a translation to the tip of the arm; an additional offset d_4 along the z_3 axis must be considered. Thus,

$$\bar{P}_a = {^0T_a} \begin{bmatrix} 0 \\ 0 \\ d_4 \\ 1 \end{bmatrix} \tag{4.190}$$

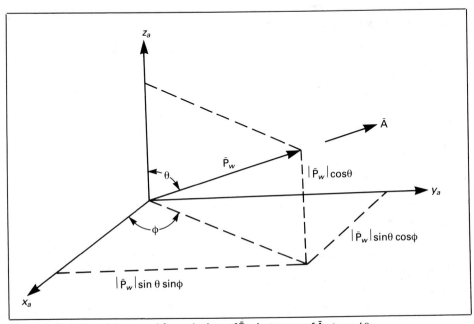

Fig. 4.4 Relationships used for solution of $\bar{P}_w$ in terms of $\bar{A}$, ϕ and θ

or

$$\bar{P}_a = \begin{bmatrix} C_1 S_{23} d_4 + C_1 C_2 a_2 - S_1 d_2 \\ S_1 S_{23} d_4 + S_1 C_2 a_2 + C_1 d_2 \\ C_{23} d_4 - a_2 S_2 \\ 1 \end{bmatrix} \tag{4.191}$$

Equating the vector components of $\bar{P}_a$ with the components of Eqn. (4.191),

$$C_1 S_{23} d_4 + C_1 C_2 a_2 - S_1 d_2 = P_{ax} \tag{4.192}$$

$$S_1 S_{23} d_4 + S_1 C_2 a_2 + C_1 d_2 = P_{ay} \tag{4.193}$$

$$C_{23} d_4 - a_2 S_2 = P_{az} \tag{4.194}$$

To solve for the control angles, begin by multiplying Eqns. (4.192) and (4.193) by $-S_1$ and C_1, respectively, and then adding:

$$-S_1 C_1 S_{23} d_4 - S_1 C_1 C_2 a_2 + S_1^2 d_2 = -S_1 P_{ax} \tag{4.195}$$

$$C_1 S_1 S_{23} d_4 + S_1 C_1 C_2 a_2 + C_1^2 d_2 = C_1 P_{ay} \tag{4.196}$$

and

$$-P_{ax} S_1 + P_{ay} C_1 = d_2 \tag{4.197}$$

which is of the form observed in Eqn. (4.62):

$$-A\sin\theta + B\cos\theta = D \tag{4.198}$$

for which the solution was derived earlier (see the section on the spherical arm in Sect. 4.2). Its solution is:

$$\theta = \tan^{-1} \frac{B}{A} - \tan^{-1} \frac{\pm D}{\sqrt{A^2 + B^2 - D^2}} \tag{4.199}$$

so

$$\theta_1 = \tan^{-1} \frac{P_{ay}}{P_{ax}} - \tan^{-1} \frac{\pm d_2}{\sqrt{P_{ax}^2 + P_{ay}^2 - d_2^2}} \tag{4.200}$$

The restriction on θ_1 (since d_2 is measured as a positive offset) is:

$$\tan^{-1} \frac{P_{ay}}{P_{ax}} - \pi < \theta_1 < \tan^{-1} \frac{P_{ay}}{P_{ax}} \tag{4.201}$$

To obtain a solution for θ_2, begin by multiplying Eqns. (4.192) and (4.193) by C_1 and S_1, respectively:

$$C_1^2 S_{23} d_4 + C_1^2 C_2 a_2 - S_1 C_1 d_2 = P_{ax} C_1 \tag{4.202}$$

$$S_1^2 S_{23} d_4 + S_1^2 C_2 a_2 + S_1 C_1 d_2 = P_{ay} S_1 \tag{4.203}$$

Adding the above equations results in:

$$d_4 S_{23} + a_2 C_2 = P_{ax} C_1 + P_{ay} S_1 \qquad (4.204)$$

To simplify, let $\alpha = P_{ax} C_1 + P_{ay} S_1$. Isolating S_{23} yields

$$S_{23} = \frac{\alpha - a_2 C_2}{d_4} \qquad (4.205)$$

Isolating C_{23} in Eqn. (4.194),

$$C_{23} = \frac{P_{az} + a_2 S_2}{d_4} \qquad (4.206)$$

Applying the identity $S_{23}{}^2 + C_{23}{}^2 = 1$,

$$\left(\frac{\alpha - a_2 C_2}{d_4} \right)^2 + \left(\frac{P_{az} + a_2 S_2}{d_4} \right)^2 = 1 \qquad (4.207)$$

$$\alpha^2 - 2\alpha a_2 C_2 + a_2{}^2 C_2{}^2 + P_{az}{}^2 + 2a_2 P_{az} S_2 + a_2{}^2 S_2{}^2 = d_4{}^2 \qquad (4.208)$$

$$-2a_2 \alpha C_2 + 2a_2 P_{az} S_2 = d_4{}^2 - \alpha^2 - a_2{}^2 - P_{az}{}^2 \qquad (4.209)$$

$$\alpha C_2 - P_{az} S_2 = \frac{\alpha^2 + a_2{}^2 + P_{az}{}^2 - d_4{}^2}{2a_2} \qquad (4.210)$$

To simplify further, let $\beta = \dfrac{\alpha^2 + a_2{}^2 + P_{az}{}^2 - d_4{}^2}{2a_2}$. The result is then

$$-P_{az} S_2 + \alpha C_2 = \beta \qquad (4.211)$$

Again apply the solution for an equation of this form,

$$\theta_2 = \tan^{-1} \frac{\alpha}{P_{az}} - \tan^{-1} \frac{\pm \beta}{\sqrt{\alpha^2 + P_{az}{}^2 - \beta^2}} \qquad (4.212)$$

Assuming β is positive, the restriction on θ_2 would be:

$$\tan^{-1} \frac{\alpha}{P_{az}} - \pi < \theta_2 < \tan^{-1} \frac{\alpha}{P_{az}} \qquad (4.213)$$

Returning to the equations for S_{23} and C_{23}, a relationship for θ_3 may be found by dividing S_{23} by C_{23}; thus,

$$\frac{S_{23}}{C_{23}} = \frac{\dfrac{\alpha - a2C_2}{d_4}}{\dfrac{P_{az} + a_2 S_2}{d_4}} \qquad (4.214)$$

$$\tan(\theta_2 + \theta_3) = \frac{\alpha - a_2 C_2}{P_{az} + a_2 S_2} \tag{4.215}$$

and

$$\theta_3 = \tan^{-1} \frac{\alpha - a_2 C_2}{P_{az} + a_2 S_2} - \theta_2 \tag{4.216}$$

Solution for wrist control angles θ_4, θ_5 and θ_6

After the arm control angles θ_1, θ_2 and θ_3 are determined, the next step is to determine the wrist control variables θ_4, θ_5 and θ_6. From Eqn. (4.166), the total rotation of the end-effector is the product of two sub-rotations, i.e. the arm rotation and the wrist rotation. Thus,

$$[\text{total rotation}] = [\text{arm rotation}] \, [\text{wrist rotation}] \tag{4.217}$$

$$R = {}^0 R_3 {}^3 R_6 \tag{4.218}$$

The wrist rotation may then be expressed as:

$${}^3 R_6 = {}^0 R_3{}^{-1} R \tag{4.219}$$

The matrix ${}^3 R_6$ is the wrist rotation in terms of the three Euler angles.
From Eqns. (4.161) and (4.162),

$${}^3 R_6 = \begin{bmatrix} n_{wx} & o_{wx} & a_{wx} \\ n_{wy} & o_{wy} & a_{wy} \\ n_{wz} & o_{wz} & a_{wz} \end{bmatrix} \tag{4.220}$$

and from Eqns. (4.179) and (4.220),

$$\begin{bmatrix} n_{wx} & o_{wx} & a_{wx} \\ n_{wy} & o_{wy} & a_{wy} \\ n_{wz} & o_{wz} & a_{wz} \end{bmatrix} = \begin{bmatrix} C_4 C_5 C_6 - S_4 S_6 & -C_4 C_5 S_6 - S_4 C_6 & C_4 S_5 \\ S_4 C_5 C_6 + C_4 S_6 & -S_4 C_5 S_6 + C_4 C_6 & S_4 S_5 \\ -S_5 C_6 & S_5 S_6 & C_5 \end{bmatrix} \tag{4.221}$$

${}^0 R_3$ may be obtained from Eqn. (4.177). Since its columns are orthonormal, its inverse is its transpose, or

$${}^0 R_3{}^{-1} = \begin{bmatrix} C_1 C_{23} & S_1 C_{23} & -S_{23} \\ -S_1 & C_1 & 0 \\ C_1 S_{23} & S_1 S_{23} & C_{23} \end{bmatrix} \tag{4.222}$$

Substituting from Eqns. (4.220) and (4.222) into Eqn. (4.219) yields:

$$\begin{bmatrix} n_{wx} & o_{wx} & a_{wx} \\ n_{wy} & o_{wy} & a_{wy} \\ n_{wz} & o_{wz} & a_{wz} \end{bmatrix} = \begin{bmatrix} C_1 C_{23} & S_1 C_{23} & -S_{23} \\ -S_1 & C_1 & 0 \\ C_1 S_{23} & S_1 S_{23} & C_{23} \end{bmatrix} \begin{bmatrix} N_x & O_x & A_x \\ N_y & O_y & A_y \\ N_z & O_z & A_z \end{bmatrix} \tag{4.223}$$

With the orientation matrix now solved, Table 4.1 may be used to determine the Euler angles of Eqn. (4.223):

$$\theta_4 = \tan^{-1} \frac{a_{wy}}{a_{wx}} \tag{4.224}$$

or

$$\theta_4 = \theta_4 + 180° \tag{4.225}$$

$$\theta_5 = \tan^{-1} \frac{C_4 a_{wx} + S_4 a_{wy}}{a_{wz}} \tag{4.226}$$

$$\theta_6 = \tan^{-1} \frac{-S_4 n_{wx} + C_4 n_{wy}}{-S_4 o_{wx} + C_4 o_{wy}} \tag{4.227}$$

Thus, the wrist control angles are established. The six angular control variables of the PUMA–560 are now completely determined for any given position and orientation of the end-effector.

4.4 Numerical example

The given position and orientation of the end-effector are expressed in terms of the base coordinate system as follows:

$$^0T_6 = \begin{bmatrix} -0.0474 & -0.7892 & -0.6124 & 300 \\ 0.6597 & 0.4356 & -0.6124 & 200 \\ 0.7500 & -0.4330 & 0.5000 & 150 \\ 0 & 0 & 0 & 1 \end{bmatrix} \tag{4.228}$$

The example of the PUMA–560 will be solved in a step-by-step manner given the $^0T_6$ of Eqn. (4.228).

The first step is to determine the polar angle θ and the azimuthal angle ϕ of the approach vector $\bar{A}$ of the end-effector from Eqn. (4.183):

$$\phi = \tan^{-1} \frac{A_y}{A_x} \tag{4.229}$$

$$= \tan^{-1} \frac{-0.6124}{-0.6124} \tag{4.230}$$

$$= 45° \tag{4.231}$$

or

$$\phi = 45° + 180° \tag{4.232}$$

Since both A_x and A_y are negative values, $\theta = 180° + 45°$, or $225°$, is chosen. From Eqn. (4.184),

$$\theta = \tan^{-1} \frac{\sqrt{A_x{}^2 + A_y{}^2}}{A_z} \tag{4.233}$$

$$= \tan^{-1} \frac{\sqrt{(-0.1624)^2 + (-0.6124)^2}}{0.500} = 60° \tag{4.234}$$
$$\tag{4.235}$$

Secondly, determine the components of the arm vector $\bar{P}_a$. From Eqn. (4.189),

$$\begin{bmatrix} P_{ax} \\ P_{ay} \\ P_{az} \end{bmatrix} = \begin{bmatrix} P_x \\ P_y \\ P_z \end{bmatrix} - \begin{bmatrix} |\bar{P}_w| \sin\theta \, \cos\phi \\ |\bar{P}_w| \sin\theta \, \sin\phi \\ |\bar{P}_w| \cos\theta \end{bmatrix} \tag{4.236}$$

The magnitude of $\bar{P}_w$ is the distance from the point where the arm and wrist meet to the centre of the wrist grip, or d_6 in Fig. 4.2. The value is given in Table 4.3. Thus,

$$\begin{bmatrix} P_{ax} \\ P_{ay} \\ P_{az} \end{bmatrix} = \begin{bmatrix} 300 \\ 200 \\ 150 \end{bmatrix} - \begin{bmatrix} 56.5 \sin 60° \, \cos 225° \\ 56.5 \sin 60° \, \sin 225° \\ 56.5 \cos 60° \end{bmatrix} \tag{4.237}$$

and

$$\begin{bmatrix} P_{ax} \\ P_{ay} \\ P_{az} \end{bmatrix} = \begin{bmatrix} 334.60 \\ 234.60 \\ 121.75 \end{bmatrix} \tag{4.238}$$

The third step is to evaluate the three arm control angles θ_1, θ_2, and θ_3. From Eqn. (4.191),

$$\begin{bmatrix} P_{ax} \\ P_{ay} \\ P_{az} \end{bmatrix} = \begin{bmatrix} C_1(S_{23}d_4 + C_2a_2) - S_1d_2 \\ S_1(S_{23}d_4 + C_2a_2) + C_1d_2 \\ C_{23}d_4 - S_2a_2 \end{bmatrix} \tag{4.239}$$

Combining this with Eqn. (4.238),

$$\begin{bmatrix} C_1(S_{23}d_4 + C_2a_2) - S_1d_2 \\ S_1(S_{23}d_4 + C_2a_2) + C_1d_2 \\ C_{23}d_4 - S_2a_2 \end{bmatrix} = \begin{bmatrix} 334.60 \\ 234.60 \\ 121.75 \end{bmatrix} \tag{4.240}$$

The solutions for the control variables are as follows. From Eqn. (4.200),

$$\theta_1 = \tan^{-1} \frac{P_{ay}}{P_{ax}} - \tan^{-1} \frac{\pm d_2}{\sqrt{P_{ax}^2 + P_{ay}^2 - d_2^2}} \tag{4.241}$$

$$= \tan^{-1} \frac{234.60}{334.60} - \tan^{-1} \frac{\pm 149.5}{\sqrt{334.60^2 + 234.60^2 - 149.5^2}} \tag{4.242}$$

$$= 35.036° \pm 21.459° \tag{4.243}$$

$$= 13.576° \text{ or } 56.495° \tag{4.244}$$

From Eqn. (4.201), the restriction on the values for θ_1 is

$$\tan^{-1} \frac{P_{ay}}{P_{ax}} - \pi < \theta_1 < \tan^{-1} \frac{P_{ay}}{P_{ax}} \tag{4.245}$$

$$\tan^{-1}\frac{234.60}{334.60} - 180° < \theta_1 < \tan^{-1}\frac{234.60}{334.60} \tag{4.246}$$

$$-144.964° < \theta_1 < 35.036° \tag{4.247}$$

Only one of the calculated values for θ_1 lies in this range. Thus,

$$\theta_1 = 13.576° \tag{4.248}$$

From Eqn. (4.212),

$$\theta_2 = \tan^{-1}\frac{\alpha}{P_{az}} - \tan^{-1}\frac{\pm\beta}{\sqrt{\alpha^2 + P_{az}^2 - \beta^2}} \tag{4.249}$$

where

$$\alpha = P_{ax}C_1 + P_{ay}S_1 \tag{4.250}$$

$$= 334.60\cos13.576° + 234.60\sin13.576° \tag{4.251}$$

$$= 380.321 \tag{4.252}$$

and

$$\beta = \frac{\alpha^2 + a_2^2 + P_{az}^2 - d_4^2}{2a_2} \tag{4.253}$$

$$= \frac{380.321^2 + 432^2 + 121.75^2 - 432^2}{2(432)} \tag{4.254}$$

$$= 184.568 \tag{4.255}$$

Thus,

$$\theta_2 = \tan^{-1}\frac{380.321}{121.75} - \tan^{-1}\frac{\pm 184.568}{\sqrt{380.321^2 + 121.75^2 - 184.568^2}} \tag{4.256}$$

$$= 72.249° \pm 27.529° \tag{4.257}$$
$$= 99.777° \text{ or } 44.720° \tag{4.258}$$

Since β is positive, the restriction on the values for θ_2 from Eqn. (4.213) is

$$\tan^{-1}\frac{\alpha}{P_{az}} - \pi < \theta_2 < \tan^{-1}\frac{\alpha}{P_{az}} \tag{4.259}$$

$$\tan^{-1}\frac{380.321}{121.75} - 180° < \theta_2 < \tan^{-1}\frac{380.321}{121.75} \tag{4.260}$$

$$-107.751° < \theta_2 < 72.249° \tag{4.261}$$

Only one of the calculated values for θ_2 lies in this range. Thus,

$$\theta_2 = 44.720° \tag{4.262}$$

From Eqn. (4.216),

$$\theta_3 = \tan^{-1} \frac{\alpha - a_2 C_2}{P_{az} + a_2 S_2} - \theta_2 \tag{4.263}$$

$$= \tan^{-1} \frac{380.321 - 432 \cos 44.720°}{121.75 + 432 \sin 44.720°} - 44.720° \tag{4.264}$$

$$= 9.777° - 44.720° \tag{4.265}$$

$$= -34.943° \tag{4.266}$$

Thus, the only possible set of control angles is:

$$\theta_1 = 13.576°, \ \theta_2 = 44.720°, \ \theta_3 = -34.943° \tag{4.267}$$

To verify that this set of angles will indeed reach the position specified by the vector $\bar{P}_{ax}$, the values are substituted into Eqn. (4.240), which does indeed hold true for this set. In the event that there is more than one possible solution set, Eqn. (4.240) will verify those correct.

Finally, the three wrist control angles are evaluated. The orientation of the wrist can now be evaluated in terms of known angles. From Eqns. (4.219) and (4.223),

$$^3R_6 = \, ^0R_3^{-1}R \tag{4.268}$$

$$= \begin{bmatrix} C_1 C_{23} & S_1 C_{23} & -S_{23} \\ -S_1 & C_1 & 0 \\ C_1 S_{23} & S_1 S_{23} & C_{23} \end{bmatrix} R \tag{4.269}$$

$$= \begin{bmatrix} 0.9579 & 0.2314 & -0.1699 \\ -0.2347 & 0.9720 & 0 \\ 0.1651 & 0.0399 & 0.9855 \end{bmatrix} \begin{bmatrix} -0.0474 & -0.7892 & -0.6124 \\ 0.6597 & 0.4356 & -0.6124 \\ 0.7500 & -0.4330 & 0.5000 \end{bmatrix} \tag{4.270}$$

$$\begin{bmatrix} n_{wx} & o_{wx} & a_{wx} \\ n_{wy} & o_{wy} & a_{wy} \\ n_{wz} & o_{wz} & a_{wz} \end{bmatrix} = \begin{bmatrix} -0.0202 & -0.5817 & -0.8132 \\ 0.6524 & 0.6087 & -0.4516 \\ 0.7576 & -0.5396 & 0.3672 \end{bmatrix} \tag{4.271}$$

From Eqns. (4.224) and (4.225),

$$\theta_4 = \tan^{-1} \frac{a_{wy}}{a_{wx}} \tag{4.272}$$

$$= \tan^{-1} \frac{-0.4516}{-0.8132} \tag{4.273}$$

$$= 29.045° \tag{4.274}$$

or

$$\theta_4 = 29.045° + 180° \tag{4.275}$$

$$= 209.045° \tag{4.276}$$

However, this second possible value for θ_4 is out of range for the variable according to the link parameter table for the PUMA–560. From Eqns. (4.226) and (4.227),

$$\theta_5 = \tan^{-1}\frac{a_{wx}C_4 + a_{wy}S_4}{a_{wz}} \tag{4.277}$$

$$= \tan^{-1}\frac{-0.8132\cos29.045° - 0.4516\sin29.045°}{0.3672} \tag{4.278}$$

$$= -68.458° \tag{4.279}$$

and

$$\theta_6 = \tan^{-1}\frac{-n_{wx}S_4 + n_{wy}C_4}{-o_{wx}S_4 + o_{wy}C_4} \tag{4.280}$$

$$= \tan^{-1}\frac{0.0202\sin29.045° + 0.6524\cos29.045°}{0.5817\sin29.045° + 0.6087\cos29.045°} \tag{4.281}$$

$$= 35.460° \tag{4.282}$$

Due to the limits of the control variables of the PUMA–560 given in Table 4.3, the only possible solution set is:

$$\theta_4 = 29.045°, \theta_5 = -68.458°, \theta_6 = 35.460° \tag{4.283}$$

Substitution of these values into Eqn. (4.221) verifies them as correct.

Thus, to attain the position and orientation given by $^0T_6$ at the beginning of this example, the values for the control variables are:

$$\theta_1 = 13.576° \tag{4.284}$$

$$\theta_2 = 44.720° \tag{4.285}$$

$$\theta_3 = -34.943° \tag{4.286}$$

$$\theta_4 = 29.045° \tag{4.287}$$

$$\theta_5 = -68.458° \tag{4.288}$$

$$\theta_6 = 35.460° \tag{4.289}$$

4.5 Concluding remarks

This chapter has dealt with position and orientation control of four types of robot arms, two wrists, and the PUMA–560. The control variables were determined in each case in a methodical manner. For the wrists, where orientation was the key, an inverse multiplication was performed to enable isolation of the control variables. For the arms, use of the given position of the tip was made to determine the variables in each case. For the

PUMA–560, the total manipulator solution was accomplished by use of the arm–wrist separation method presented. The general steps were as follows:

(1) The polar and azimuthal angles θ and ϕ of the end-effector's approach vector were determined.
(2) The components of the arm vector $\bar{P}_a$ were found, using θ and ϕ from step 1.
(3) Arm control angles θ_1, θ_2 and θ_3 were evaluated using the components of vector $\bar{P}_a$ from step 2.
(4) Wrist control variables θ_4, θ_5 and θ_6 were solved using θ_1, θ_2 and θ_3 from step 3.

The number of computations in this method are kept to a minimum by reducing the overall problem into separate steps, which in turn lowers the likelihood of errors and helps to reduce the tediousness of the work.

The step-by-step solution of the position control problem presented in this chapter has fulfilled the goal: given the orientation and position required for the end-effector of the manipulator, determine the values of the control variables for the specified situation. Position control is the first problem that must be considered when dealing with a robot manipulator. Trajectory control, speed control, and dynamic control problems are investigated in the following chapters, but position control stands alone in that it must be accomplished or no work can be done.

References

[1] Paul, R., Shimano, B. and Meyer, G. 1981. Kinematic control equation for a simple manipulator. *IEEE Trans. System, Man and Cybernetics*, SMC–11 June: 449–455.
[2] Lee, C. S. G. 1982. Robot arm kinetics, dynamics and control. *IEEE Computer*, December: 62–80.
[3] Ho, C. Y. and Copeland, K. W. 1982. Solution of kinematic equation for robot manipulators. *Digital Systems for Industrial Automation*, 1(4): 335–352.
[4] Siegler, A. 1979. *Kinematics and Microcomputer Control of a Six Degree-of-Freedom Manipulator*. University of Cambridge CUED/F – CAMS/TRi85.
[5] Denavit, J. and Hartenberg, R. S. 1955. A kinematic notation for lower pair mechanism based on matrices. *J. Applied Mechanics*, June: 215–221.

Chapter Five

DIFFERENTIAL RELATIONSHIPS AND SPEED CONTROL

THE GENERAL solutions of kinematic equations for robot manipulators have been developed and demonstrated in Chapter Four for most commercially available manipulators (i.e. PUMA, Unimate, Cincinnati Milacron, etc.). The kinematic equations have homogeneous transformations representing the position and orientation of the end-effector as the independent variables[1–3]. The method used in Chapter Four has simplified the procedures for obtaining the solutions.

In this chapter, a simplified method for obtaining the Jacobian of robot manipulators is developed. This set of equations has the differential change in position and orientation of the end-effector as dependent variables, and the differential change in the joint coordinates as independent variables[4]. The transformation matrix that will relate the differential changes in the joint coordinates to the differential changes in the base frame angular and linear velocities is the Jacobian matrix[5]. Finding this Jacobian matrix and relating it to speed control is the subject of this chapter.

5.1 Relative motion between frames

Considering the fixed and moving coordinate frames shown in Fig. 5.1, by vector addition it can be seen that:

$$(\bar{X})_F = (\bar{X}_0)_F + (\bar{P})_F \tag{5.1}$$

Differentiating this expression with respect to time yields the expression for the velocity:

$$(\bar{V})_F = (\bar{V}_0)_F + (\bar{P}')_F \tag{5.2}$$

Now let R be defined as the 3×3 orientation portion of the transformation matrix A from the fixed frame to the moving frame.

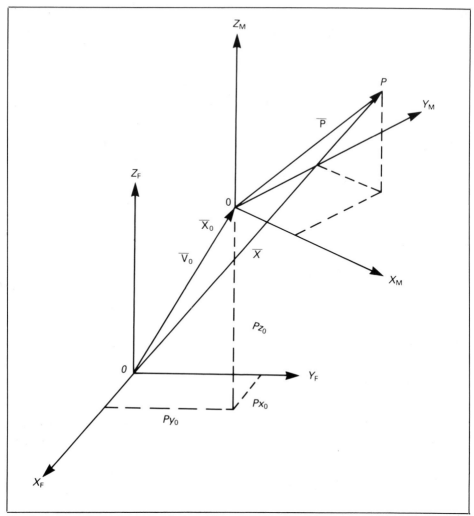

Fig. 5.1   Coordinate frames relationships: $(X,Y,Z)_F$, fixed frame; and $(X, Y, Z)_M$, moving frame

Let

$$(\bar{P})_F = (R)_F(\bar{P})_M \tag{5.3}$$

Differentiating both sides with respect to time yields:

$$(\bar{P}')_F = (R')_F(\bar{P})_M + (R)_F(\bar{P}')_M \tag{5.4}$$

Utilising the property that the R matrix is orthogonal, that is:

$$R^{-1} = R^T \tag{5.5}$$

Eqn. (5.4) can be expanded to:

$$(\bar{P}')_F = (R'R^T)_F(R)_F(\bar{P})_M + (R)_F(\bar{P}')_M \tag{5.6}$$

If point P is fixed with respect to the moving frame, then $(\bar{P}')_M = 0$.

So,

$$(\bar{P}')_F = (R'R^T)_F(R)_F(\bar{P})_M \tag{5.7}$$

The R matrix is defined[6] to be:

$$R = \begin{bmatrix} C\theta & -S\theta C\alpha & S\theta S\alpha \\ S\theta & C\theta C\alpha & -C\theta S\alpha \\ 0 & S\alpha & C\alpha \end{bmatrix} \tag{5.8}$$

Differentiating R with respect to time yields:

$$R' = \begin{bmatrix} -S\theta & -C\theta C\alpha & C\theta S\alpha \\ C\theta & -S\theta C\alpha & S\theta S\alpha \\ 0 & 0 & 0 \end{bmatrix} \theta' \tag{5.9}$$

Multiplying R' by R^T:

$$R'R^T = \begin{bmatrix} -S\theta & -C\theta C\alpha & C\theta S\alpha \\ C\theta & -S\theta C\alpha & S\theta S\alpha \\ 0 & 0 & 0 \end{bmatrix} \begin{bmatrix} C\theta & S\theta & 0 \\ -S\theta C\alpha & C\theta C\alpha & S\alpha \\ S\theta S\alpha & -C\theta S\alpha & C\alpha \end{bmatrix} \theta' \tag{5.10}$$

This can be simplified to:

$$R'R^T = \begin{bmatrix} 0 & -1 & 0 \\ 1 & 0 & 0 \\ 0 & 0 & 0 \end{bmatrix} \theta' \tag{5.11}$$

The quantity $R'R^T$ is defined to be Ω, the angular velocity matrix of the moving frame with respect to the fixed frame. Substituting this into Eqn. (5.7),

$$(\bar{P})_F = (\Omega)_F(R)_F(\bar{P})_M \tag{5.12}$$

And substituting Eqn. (5.12) into Eqn. (5.2) yields:

$$(\bar{V})_F = (\bar{V}_0)_F + (\Omega)_F(R)_F(\bar{P})_M \tag{5.13}$$

Since,

$$(\bar{X}_0)_F = \begin{bmatrix} aC\theta \\ aS\theta \\ d \end{bmatrix} \quad \text{then,} \quad (\bar{V}_0)_F = \begin{bmatrix} -aS\theta\ \theta' \\ aC\theta\ \theta' \\ d' \end{bmatrix} \tag{5.14}$$

it represents the translational velocity of the moving frame with respect to the fixed frame.

Now expressing Eqn. (5.13) in matrix form gives:

$$(\bar{V})_F = \begin{bmatrix} -aS\theta\ \theta' \\ aC\theta\ \theta' \\ d' \end{bmatrix} + \begin{bmatrix} 0 & -1 & 0 \\ 1 & 0 & 0 \\ 0 & 0 & 0 \end{bmatrix} \theta' \begin{bmatrix} C\theta & -S\theta C\alpha & S\theta S\alpha \\ S\theta & C\theta C\alpha & -C\theta\ S\alpha \\ 0 & S\alpha & C\alpha \end{bmatrix} \begin{bmatrix} \bar{P} \end{bmatrix}_M \tag{5.15}$$

The prismatic joint

Consider the prismatic joint shown in Fig. 5.2 with translation along the Z_F axis. For a prismatic joint whose displacement lies along the Z axis:

$$\theta' = \Omega = 0 \quad \text{and} \quad (\bar{V})_F = \begin{bmatrix} 0 \\ 0 \\ d' \end{bmatrix} \tag{5.16}$$

In order to find the Jacobian element of a prismatic joint, Eqn. (5.13) is used. Let F be a coordinate frame $(i-1)$, so that:

$$^{i-1}\bar{V}_i = {}^{i-1}\begin{bmatrix} 0 \\ 0 \\ d'_i \end{bmatrix} \tag{5.17}$$

where d'_i is the translation velocity of prismatic link i measured with respect to frame $(i-1)$. Expressing the linear changes in velocity in terms of the independent variable d'_i yields:

$$^{i-1}\begin{bmatrix} V_x \\ V_y \\ V_z \end{bmatrix}_i = {}^{i-1}\begin{bmatrix} 0 \\ 0 \\ d' \end{bmatrix}_i = {}^{i-1}\begin{bmatrix} 0 \\ 0 \\ 1 \end{bmatrix} d'_i \tag{5.18}$$

Expressing the velocity of link i with respect to the base frame yields:

$$^{0}\begin{bmatrix} V_x \\ V_y \\ V_z \end{bmatrix}_i = ({}^{0}A_1 {}^{1}A_2 {}^{2}A_3 ... {}^{i-2}A_{i-1}) \begin{bmatrix} 0 \\ 0 \\ 1 \end{bmatrix} d'_i \tag{5.19}$$

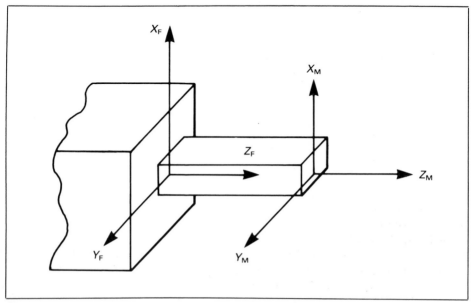

Fig. 5.2   A prismatic joint

For the transformation of velocity vector, we consider only the orientation portion of the A matrices. This simplifies to:

$$
{}^0\begin{bmatrix} V_x \\ V_y \\ V_z \end{bmatrix}_i = {}^0\begin{bmatrix} N_x & O_x & A_x \\ N_y & O_y & A_y \\ N_z & O_z & A_z \end{bmatrix}_{i-1} \begin{bmatrix} 0 \\ 0 \\ 1 \end{bmatrix} d'_i \tag{5.20}
$$

Matrix multiplication yields:

$$
{}^0\begin{bmatrix} X' \\ Y' \\ Z' \end{bmatrix}_i = {}^0\begin{bmatrix} A_x \\ A_y \\ A_z \end{bmatrix}_{i-1} d'_i \tag{5.21}
$$

It can be shown for prismatic joints whose motion lies along the X coordinate axis that:

$$
{}^0\begin{bmatrix} X' \\ Y' \\ Z' \end{bmatrix}_i = {}^0\begin{bmatrix} N_x \\ N_y \\ N_z \end{bmatrix}_{i-1} a'_i \tag{5.22}
$$

where the displacement along the X axis is represented by the variable a.

Since, $\theta' = 0$, the angular velocity with respect to the fixed coordinate frame is zero. Expressing the velocity relationships in matrix form yields:

$$
{}^0\begin{bmatrix} X' \\ Y' \\ Z' \\ \theta'_x \\ \theta'_y \\ \theta'_z \end{bmatrix}_i = {}^0\begin{bmatrix} A_x \\ A_y \\ A_z \\ 0 \\ 0 \\ 0 \end{bmatrix}_{i-1} d'_i \tag{5.23}
$$

For a motion along the X axis of a_i:

$$
{}^0\begin{bmatrix} X' \\ Y' \\ Z' \\ \theta'_x \\ \theta'_y \\ \theta'_z \end{bmatrix}_i = {}^0\begin{bmatrix} N_x \\ N_y \\ N_z \\ 0 \\ 0 \\ 0 \end{bmatrix}_{i-1} a'_i \tag{5.24}
$$

Revolute joint

Next consider the revolute joint (Fig. 5.3) with rotation about the Z_F axis. For a revolute joint without translation,

$$
(\bar{V}_0)_F = 0 \tag{5.25}
$$

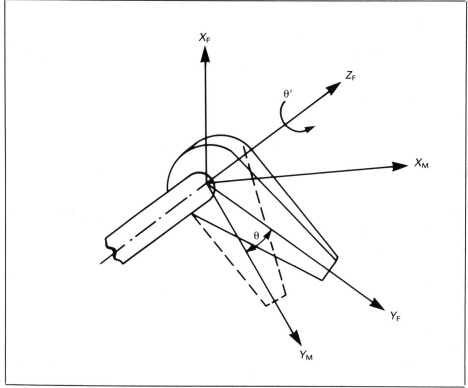

Fig. 5.3   A revolute joint

From Eqn. (5.13),

$$(\bar{V})_F = (\Omega)_F (R)_F (\bar{P})_M \tag{5.26}$$

Let F be a coordinate frame $(i-1)$, which yields:

$$^{i-1}(\bar{V})_i = {}^{i-1}(\Omega)_i {}^{i-1}(R)_i (\bar{P})_i \tag{5.27}$$

Substituting Eqn. (5.27) into the above yields:

$$^{i-1}(\bar{V})_i = \begin{bmatrix} 0 & -1 & 0 \\ 1 & 0 & 0 \\ 0 & 0 & 0 \end{bmatrix} {}^{i-1}\theta'_i \begin{bmatrix} C\theta & -S\theta C\alpha & S\theta S\alpha \\ S\theta & C\theta C\alpha & -C\theta S\alpha \\ 0 & S\alpha & C\alpha \end{bmatrix}_i \begin{bmatrix} \bar{P} \end{bmatrix}_i \tag{5.28}$$

Consider $(\bar{P})_i$ as the origin point of the end-effector frame n, measured with respect to frame i, as in Fig. 5.4. Eqn. (5.28) can be written as follows:

$$^{i-1}(\bar{V})_i = \begin{bmatrix} 0 & -1 & 0 \\ 1 & 0 & 0 \\ 0 & 0 & 0 \end{bmatrix} \theta'_i \, ({}^{i-1}R_i) \begin{bmatrix} P_x \\ P_y \\ P_z \end{bmatrix}_n^{i} \tag{5.29}$$

where $^{i-1}R_i$ is the 3×3 orientation matrix. It can be written as:

$$^{i-1}(\bar{V})_i = \begin{bmatrix} 0 & -1 & 0 \\ 1 & 0 & 0 \\ 0 & 0 & 0 \end{bmatrix}^{i-1} \begin{bmatrix} P_x \\ P_y \\ P_z \end{bmatrix}_n \tag{5.30}$$

This simplifies the above equation to:

$$^{i-1}(\bar{V})_i = \begin{bmatrix} -P_y \\ P_x \\ 0 \end{bmatrix}^{i-1}_n \theta'_i \tag{5.31}$$

In order to express this velocity in terms of the base coordinate frame, multiply the above equation by the orientation portion of the A_1 to A_{i-1} transformation matrices as shown:

$$^0(\bar{V})_i = (^0R_1\ ^1R_2 \ldots \,^{i-2}R_{i-1}) \begin{bmatrix} -P_y \\ P_x \\ 0 \end{bmatrix}^{i-1}_n \theta'_i \tag{5.32}$$

and

$$^0R_1\,^1R_2 \ldots\,^{i-2}R_{i-1} = \begin{bmatrix} N_x & O_x & A_x \\ N_y & O_y & A_y \\ N_z & O_z & A_z \end{bmatrix}^0_{i-1} \tag{5.33}$$

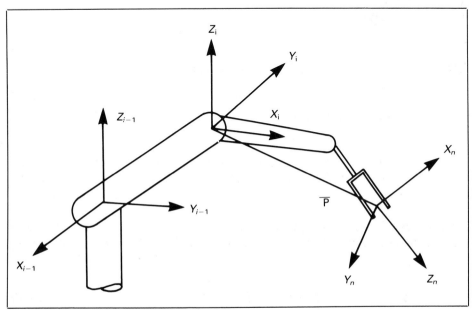

Fig. 5.4   Revolute link relationships

Substituting in the general form of the orientation transformation yields:

$$^0(\bar{V})_i = \,^0\begin{bmatrix} N_x & O_x & A_x \\ N_y & O_y & A_y \\ N_z & O_z & A_z \end{bmatrix}^{i-1}_{i-1} \begin{bmatrix} -P_y \\ P_x \\ 0 \end{bmatrix}_n \theta'_i \qquad (5.34)$$

Matrix multiplication results in:

$$^0\begin{bmatrix} X' \\ Y' \\ Z' \end{bmatrix}_i = \begin{bmatrix} -\,^0\begin{bmatrix} N_x \\ N_y \\ N_z \end{bmatrix}_{i-1}^{i-1}(P_y)_n + \,^0\begin{bmatrix} O_x \\ O_y \\ O_z \end{bmatrix}_{i-1}^{i-1}(P_x)_n \end{bmatrix} \theta'_i \qquad (5.35)$$

From this equation, the linear velocity along the base coordinate axes can be found.

It is now desired to find the rotational velocity around the base coordinate axes. The joint variable θ was defined to be a rotation about the Z axis of the previous link coordinate frame. With this in mind it can be seen that:

$$\bar{\theta}_i = 0\bar{i} + 0\bar{j} + \theta_i\bar{k} = \begin{bmatrix} 0 \\ 0 \\ 1 \end{bmatrix}\theta_i \qquad (5.36)$$

Differentiating both sides with respect to time yields:

$$\dot{\bar{\theta}}_i = \begin{bmatrix} 0 \\ 0 \\ 1 \end{bmatrix}\dot{\theta}_i \qquad (5.37)$$

As before, this velocity vector must be transformed back to the base coordinate frame by multiplying by the orientation portion of the A_1 through the A_{i-1} matrices. Multiplying yields:

$$^0\begin{bmatrix} \theta'_x \\ \theta'_y \\ \theta'_z \end{bmatrix}_i = \,^0\begin{bmatrix} N_x & O_x & A_x \\ N_y & O_y & A_y \\ N_z & O_z & A_z \end{bmatrix}_{i-1} \begin{bmatrix} 0 \\ 0 \\ 1 \end{bmatrix}\theta'_i = \,^0\begin{bmatrix} A_x \\ A_y \\ A_z \end{bmatrix}_{i-1}\theta'_i \qquad (5.38)$$

Combining Eqns. (5.35) and (5.38) to form the total velocity response yields:

$$^0\begin{bmatrix} X' \\ Y' \\ Z' \\ \theta'_x \\ \theta'_y \\ \theta'_z \end{bmatrix}_i = \begin{bmatrix} -\,^0\begin{bmatrix} N_x \\ N_y \\ N_z \end{bmatrix}_{i-1}^{i-1}(P_y)_n + \,^0\begin{bmatrix} O_x \\ O_y \\ O_z \end{bmatrix}_{i-1}^{i-1}(P_x)_n \\ \\ ^0\begin{bmatrix} A_x \\ A_y \\ A_z \end{bmatrix}_{i-1} \end{bmatrix} \theta'_i \qquad (5.39)$$

It can be shown from the preceding development that for a rotation about the X axis by an angle α, the following equation is used to find the relationship:

$$
{}^0\begin{bmatrix} X' \\ Y' \\ Z' \\ \theta'_x \\ \theta'_y \\ \theta'_z \end{bmatrix}_i = \left(-{}^0\begin{bmatrix} N_x \\ N_y \\ N_z \end{bmatrix}_{i-1} {}^{i-1}(P_y)_n + {}^0\begin{bmatrix} O_x \\ O_y \\ O_z \end{bmatrix}_{i-1} {}^{i-1}(P_x)_n \\ {}^0\begin{bmatrix} N_x \\ N_y \\ N_z \end{bmatrix}_{i-1} \right) \alpha'i \quad (5.40)
$$

5.2 Formulating the Jacobian

The velocity relationship between frames can be converted into the differential form. The differential change in position and orientation of the end-effector as a function of all n joint coordinates is written as a $6 \times n$ matrix consisting of the differential translation and rotation vector elements, and is known as the Jacobian matrix. Each column of the Jacobian matrix consists of the differential translation and rotation vector corresponding to the differential changes of each of the joint coordinates. These columns are calculated using Eqns. (5.23) and (5.38) for a prismatic and a revolute joint, respectively, except for the cases noted. In matrix form the Jacobian relationship is:

$$
{}^0\begin{bmatrix} \Delta X \\ \Delta Y \\ \Delta Z \\ \Delta \theta_x \\ \Delta \theta_y \\ \Delta \theta_z \end{bmatrix}_n = \begin{bmatrix} X_1 & X_2 & X_3 & \ldots & X_n \\ Y_1 & Y_2 & Y_3 & \ldots & Y_n \\ Z_1 & Z_2 & Z_3 & \ldots & Z_n \\ \theta_{1x} & \theta_{2x} & \theta_{3x} & \ldots & \theta_{nx} \\ \theta_{1y} & \theta_{2y} & \theta_{3y} & \ldots & \theta_{ny} \\ \theta_{1z} & \theta_{2z} & \theta_{3z} & \ldots & \theta_{nz} \end{bmatrix} \begin{bmatrix} \Delta q_1 \\ \Delta q_2 \\ \Delta q_3 \\ - \\ - \\ \Delta q_n \end{bmatrix} \quad (5.41)
$$

where $q_i = \theta_i$ for a revolute joint, $q_i = d_i$ for a prismatic joint, $x_i = \delta X/\delta q_i$ and $\theta_{ix} = \delta\theta_x/\delta q_i$.

5.3 Examples of formulating the Jacobian matrix

This section is devoted to examples illustrating the formulation of the Jacobian matrix. The examples to be included are the same systems referred to in Chapter Four.

Solution for wrist Jacobian

Euler wrist. In order to calculate the Jacobian matrix, each column is computed separately and then combined to form the total Jacobian matrix.

Since column 1 refers to joint 1, which is a revolute joint, the column is calculated using Eqn. (5.39). Since there are no prismatic joints or joint offsets in the Euler wrist, all of the $^{i-1}(\bar{P})_n$ elements are zero. This will also be true for the RPY wrist. Since $i = 1$, in this case the component $^0T_0$ is equal to the identity matrix. From these relationships column 1 is found to be:

$$
^0\begin{bmatrix} X' \\ Y' \\ Z' \\ \theta'_x \\ \theta'_y \\ \theta'_z \end{bmatrix}_1 = \left[-\begin{bmatrix} 1 \\ 0 \\ 0 \end{bmatrix}(0) + \begin{bmatrix} 0 \\ 1 \\ 0 \end{bmatrix}(0) \\ \begin{bmatrix} 0 \\ 0 \\ 1 \end{bmatrix} \right]\theta'_1
\tag{5.42}
$$

And simplification yields:

$$
^0\begin{bmatrix} X' \\ Y' \\ Z' \\ \theta'_x \\ \theta'_y \\ \theta'_z \end{bmatrix}_1 = \begin{bmatrix} 0 \\ 0 \\ 0 \\ 0 \\ 0 \\ 1 \end{bmatrix}\theta'_1
\tag{5.43}
$$

Column 2 is found by a similar method, except the $^0T_1$ elements are equal to the orientation portion of the A_1 matrix. Thus, column 2 is found to be:

$$
^0\begin{bmatrix} X' \\ Y' \\ Z' \\ \theta'_x \\ \theta'_y \\ \theta'_z \end{bmatrix}_2 = \left[-\begin{bmatrix} C_1 \\ S_1 \\ 0 \end{bmatrix}(0) + \begin{bmatrix} 0 \\ 0 \\ -1 \end{bmatrix}(0) \\ \begin{bmatrix} -S_1 \\ C_1 \\ 0 \end{bmatrix} \right]\theta'_2
\tag{5.44}
$$

And simplification yields:

$$
^0\begin{bmatrix} X' \\ Y' \\ Z' \\ \theta'_x \\ \theta'_y \\ \theta'_z \end{bmatrix}_2 = \begin{bmatrix} 0 \\ 0 \\ 0 \\ -S \\ C_1 \\ 0 \end{bmatrix}\theta'_2
\tag{5.45}
$$

Column 3 is also found in a similar manner, except $^0T_2$ is equal to the product of A_1A_2. Column 3 is then equal to:

$$
{}^{0}\begin{bmatrix} X' \\ Y' \\ Z' \\ \theta'_x \\ \theta'_y \\ \theta'_z \end{bmatrix}_3 = \begin{bmatrix} -\begin{bmatrix} C_1C_2 \\ S_1C_2 \\ -S_2 \end{bmatrix}(0) + \begin{bmatrix} -S_1 \\ C_1 \\ 0 \end{bmatrix}(0) \\ \begin{bmatrix} C_1S_2 \\ S_1S_2 \\ C_2 \end{bmatrix} \end{bmatrix}\theta'_3 \qquad (5.45)
$$

And simplification yields:

$$
{}^{0}\begin{bmatrix} X' \\ Y' \\ Z' \\ \theta'_x \\ \theta'_y \\ \theta'_z \end{bmatrix}_3 = \begin{bmatrix} 0 \\ 0 \\ 0 \\ C_1S_2 \\ S_1S_2 \\ C_2 \end{bmatrix}\theta'_3 \qquad (5.46)
$$

Combining these three columns into the Jacobian for the total system yields:

$$
J_{\text{Euler}} = \begin{bmatrix} 0 & 0 & 0 \\ 0 & 0 & 0 \\ 0 & 0 & 0 \\ 0 & -S_1 & C_1S_2 \\ 0 & C_1 & S_1S_2 \\ 1 & 0 & C_2 \end{bmatrix} \qquad (5.47)
$$

RPY wrist. Columns 1 and 2 of the RPY wrist are found in the same way as those of the Euler wrist. Column 3 is found using Eqn. (5.40), since it is a rotation about the X coordinate axis. The ${}^{0}T_2$ component is equal to the product of the A_1A_2 matrices. Column 3 is then equal to:

$$
{}^{0}\begin{bmatrix} X' \\ Y' \\ Z' \\ \theta'_x \\ \theta'_y \\ \theta'_z \end{bmatrix}_3 = \begin{bmatrix} -\begin{bmatrix} C_1C_2 \\ A_1C_2 \\ -S_2 \end{bmatrix}(0) + \begin{bmatrix} -S_1 \\ C_1 \\ 0 \end{bmatrix}(0) \\ \begin{bmatrix} C_1S_2 \\ S_1S_2 \\ C_2 \end{bmatrix} \end{bmatrix}\alpha'_3 \qquad (5.48)
$$

And simplification yields:

$$
{}^{0}\begin{bmatrix} X' \\ Y' \\ Z' \\ \theta'_x \\ \theta'_y \\ \theta'_z \end{bmatrix}_3 = \begin{bmatrix} 0 \\ 0 \\ 0 \\ C_1C_2 \\ S_1C_2 \\ -S_2 \end{bmatrix}\alpha'_3 \qquad (5.49)
$$

Combining these three columns into the Jacobian for the total system yields:

$$J_{RPY} = \begin{bmatrix} 0 & 0 & 0 \\ 0 & 0 & 0 \\ 0 & 0 & 0 \\ 0 & -S_1 & C_1C_2 \\ 0 & C_1 & S_1C_2 \\ 1 & 0 & -S_2 \end{bmatrix} \tag{5.50}$$

Solution for arm Jacobian

Cartesian arm. The first prismatic joint in the Cartesian arm has its motion along the X coordinate axis. To find its column in the Jacobian matrix, use Eqn. (5.24). Since $i = 1$, the $^0T_0$ orientation matrix is equal to the identity matrix. For column 1,

$$^0\begin{bmatrix} X' \\ Y' \\ Z' \\ \theta'_x \\ \theta'_y \\ \theta'_z \end{bmatrix}_1 = \begin{bmatrix} 1 \\ 0 \\ 0 \\ 0 \\ 0 \end{bmatrix} a'_1 \tag{5.51}$$

The second joint is a prismatic joint along the Z coordinate axis, so Eqn. (5.23) is used to calculate its column. The $^0T_1$ orientation matrix is equal to the A_1 matrix. Column 2 is found to be:

$$^0\begin{bmatrix} X' \\ Y' \\ Z' \\ \theta'_x \\ \theta'_y \\ \theta'_z \end{bmatrix}_2 = \begin{bmatrix} 0 \\ 1 \\ 0 \\ 0 \\ 0 \\ 0 \end{bmatrix} d'_2 \tag{5.52}$$

The third joint is the same as the second joint except the $^0T_2$ orientation matrix is equal to the product of the A_1A_2 matrices. Column 3 is then found to be:

$$^0\begin{bmatrix} X' \\ Y' \\ Z' \\ \theta'_x \\ \theta'_y \\ \theta'_z \end{bmatrix}_3 = \begin{bmatrix} 0 \\ 0 \\ 1 \\ 0 \\ 0 \\ 0 \end{bmatrix} d'_3 \tag{5.53}$$

Combining these three columns forms the total Jacobian matrix of the Cartesian arm:

$$
J_{Cart.} =
\begin{bmatrix}
1 & 0 & 0 \\
0 & 1 & 0 \\
0 & 0 & 1 \\
0 & 0 & 0 \\
0 & 0 & 0 \\
0 & 0 & 0
\end{bmatrix}
\qquad (5.54)
$$

Cylindrical arm. The first joint in the cylindrical arm is a prismatic joint with motion along the Z coordinate axis. To find its column in the Jacobian matrix, use Eqn. (5.23). Since $i = 1$, the $^0T_0$ orientation matrix is equal to the identity matrix. For column 1,

$$
{}^0\begin{bmatrix}
X' \\
Y' \\
Z' \\
\theta'_x \\
\theta'_y \\
\theta'_z
\end{bmatrix}_1 =
\begin{bmatrix}
0 \\
0 \\
1 \\
0 \\
0 \\
0
\end{bmatrix} d'_1
\qquad (5.55)
$$

The second joint is a revolute joint about the Z coordinate axis so Eqn. (5.39) is used to calculate its column. The $^0T_1$ orientation matrix is equal to the A_1 matrix. The $^1(\bar{P})_3$ matrix is the position portion of the product of the A_2A_3 matrices. Column 2 is found to be:

$$
{}^0\begin{bmatrix}
X' \\
Y' \\
Z' \\
\theta'_x \\
\theta'_y \\
\theta'_z
\end{bmatrix}_2 =
\left[-\begin{bmatrix} 1 \\ 0 \\ 0 \end{bmatrix}(C_2d_3 + a_2S_2) + \begin{bmatrix} 0 \\ 1 \\ 0 \end{bmatrix}(-S_2d_3 + a_2C_2) \; \begin{bmatrix} 0 \\ 0 \\ 1 \end{bmatrix} \right] \theta'_2
\qquad (5.56)
$$

And simplification yields:

$$
{}^0\begin{bmatrix}
X' \\
Y' \\
Z' \\
\theta'_x \\
\theta'_y \\
\theta'_z
\end{bmatrix}_2 =
\begin{bmatrix}
-C_2d_3 - a_2S_2 \\
-S_2d_3 + a_2C_2 \\
0 \\
0 \\
0 \\
1
\end{bmatrix} \theta'_2
\qquad (5.57)
$$

The third joint is a prismatic joint with motion along the Z coordinate axis. The $^0T_2$ orientation matrix is equal to the orientation portion of the product of the A_1A_2 matrices. Column 3 is then found to be:

$$
{}^0\begin{bmatrix} X' \\ Y' \\ Z' \\ \theta'_x \\ \theta'_y \\ \theta'_z \end{bmatrix}_3 = \begin{bmatrix} -S_2 \\ C_2 \\ 0 \\ 0 \\ 0 \\ 0 \end{bmatrix} d'_3
\tag{5.58}
$$

Combining these three columns forms the total Jacobian matrix of the cylindrical arm:

$$
J_{Cyl.} = \begin{bmatrix} 0 & -C_2d_3 - a_2S_2 & -S_2 \\ 0 & -S_2d_3 + a_2C_2 & C_2 \\ 1 & 0 & 0 \\ 0 & 0 & 0 \\ 0 & 0 & 0 \\ 0 & 1 & 0 \end{bmatrix}
\tag{5.59}
$$

Spherical arm. The first joint in the spherical arm is a revolute joint about the Z coordinate axis. Eqn. (5.39) is used to calculate its column in the Jacobian matrix. The $^0T_0$ orientation matrix is equal to the identity matrix and the $^0(\bar{P})_3$ matrix is equal to the position portion of the product of the $A_1A_2A_3$ matrices. Column 1 is then equal to:

$$
{}^0\begin{bmatrix} X' \\ Y' \\ Z' \\ \theta'_x \\ \theta'_y \\ \theta'_z \end{bmatrix}_1 = \left[-\begin{bmatrix} 1 \\ 0 \\ 0 \end{bmatrix} S_1S_2d_3 \begin{bmatrix} 0 \\ 0 \\ 1 \end{bmatrix} + \begin{bmatrix} 0 \\ 1 \\ 0 \end{bmatrix} C_1S_2d_3 \right] \theta'_1
\tag{5.60}
$$

And simplification yields:

$$
{}^0\begin{bmatrix} X' \\ Y' \\ Z' \\ \theta'_x \\ \theta'_y \\ \theta'_z \end{bmatrix}_1 = \begin{bmatrix} -S_1S_2d_3 \\ C_1S_2d_3 \\ 0 \\ 0 \\ 0 \\ 1 \end{bmatrix} \theta'_1
\tag{5.61}
$$

The second joint is a revolute joint about the Z coordinate axis so Eqn. (5.39) is used to calculate its column. The $^0T_1$ orientation matrix is equal to the orientation portion of the A_1 matrix. The $^1(\bar{P})_3$ matrix is the position portion of the product of the A_2A_3 matrices. Column 2 is found to be:

$$
{}^0\begin{bmatrix} X' \\ Y' \\ Z' \\ \theta'_x \\ \theta'_y \\ \theta'_z \end{bmatrix}_2 = \left[-\begin{bmatrix} C_1 \\ S_1 \\ 0 \end{bmatrix} (-C_2 d_3) + \begin{bmatrix} 0 \\ 0 \\ -1 \end{bmatrix} S_2 d_3 \quad \begin{bmatrix} -S_1 \\ C_1 \\ 0 \end{bmatrix} \right] \theta'_2 \tag{5.62}
$$

And simplification yields:

$$
{}^0\begin{bmatrix} X' \\ Y' \\ Z' \\ \theta'_x \\ \theta'_y \\ \theta'_z \end{bmatrix}_2 = \begin{bmatrix} C_1 C_2 d_3 \\ S_1 C_2 d_3 \\ -S_2 d_3 \\ -S_1 \\ C_1 \\ 0 \end{bmatrix} \theta'_2 \tag{5.63}
$$

The third joint is a prismatic joint with motion along the Z coordinate axis. The ${}^0T_2$ orientation matrix is equal to the orientation portion of the product of the $A_1 A_2$ matrices. Column 3 is equal to:

$$
{}^0\begin{bmatrix} X' \\ Y' \\ Z' \\ \theta'_x \\ \theta'_y \\ \theta'_z \end{bmatrix}_3 = \begin{bmatrix} C_1 S_2 \\ S_1 S_2 \\ C_2 \\ 0 \\ 0 \\ 0 \end{bmatrix} d'_3 \tag{5.64}
$$

Combining these three columns forms the total Jacobian matrix of the spherical arm:

$$
J_{Sph.} = \begin{bmatrix} -S_1 S_2 d_3 & C_1 C_2 d_3 & C_1 S_2 \\ C_1 S_2 d_3 & S_1 C_2 d_3 & S_1 S_2 \\ 0 & -S_2 d_3 & C_2 \\ 0 & -S_1 & 0 \\ 0 & C_1 & 0 \\ 1 & 0 & 0 \end{bmatrix} \tag{5.65}
$$

Revolute arm. The first joint is a revolute joint about the Z coordinate axis so Eqn. (5.39) is used to calculate its column. The ${}^0T_0$ orientation matrix is equal to the identity matrix. The ${}^0(\bar{P})_3$ matrix is the position portion of the product of the $A_1 A_2 A_3$ matrices. Column 1 is found to be:

$$
{}^0\begin{bmatrix} X' \\ Y' \\ Z' \\ \theta'_x \\ \theta'_y \\ \theta'_z \end{bmatrix}_1 = \begin{bmatrix} -S_1(a_3 C_{23} + a_2 C_2) \\ C_1(a_3 C_{23} + a_2 C_2) \\ 0 \\ 0 \\ 0 \\ 1 \end{bmatrix} \theta'_1 \tag{5.66}
$$

where C_{ab} is equal to the cosine of $(\theta_a + \theta_b)$, and S_{ab} is equal to the sine of $(\theta_a + \theta_b)$.

The second joint is also a revolute joint about the Z coordinate axis so Eqn. (5.39) is used to calculate its column. The $^0T_1$ orientation matrix is equal to the orientation portion of the A_1 matrix. The $^1(\bar{P})_3$ matrix is the position portion of the product of the A_2A_3 matrices. Column 2 is equal to:

$$
^0\begin{bmatrix} X' \\ Y' \\ Z' \\ \theta'_x \\ \theta'_y \\ \theta'_z \end{bmatrix}_2 = \begin{bmatrix} -C_1(a_3S_{23} + a_2S_2) \\ -S_1(a_3S_{23} + a_2S_2) \\ a_3C_{23} + a_2C_2 \\ S_1 \\ -C_1 \\ 0 \end{bmatrix} \theta'_2 \tag{5.67}
$$

The third joint is again a revolute joint about the Z coordinate axis and Eqn. (5.39) is used to calculate its column. The $^0T_2$ orientation matrix is equal to the orientation portion of the product of the A_1A_2 matrices. The $^2(\bar{P})_3$ matrix is equal to the position portion of the A_3 matrix. Column 3 is found to be:

$$
^0\begin{bmatrix} X' \\ Y' \\ Z' \\ \theta'_x \\ \theta'_y \\ \theta'_z \end{bmatrix}_3 = \begin{bmatrix} -a_3C_1C_{23} \\ -a_3S_1C_{23} \\ -a_3S_{23} \\ S_1 \\ -C_1 \\ 0 \end{bmatrix} \theta'_3 \tag{5.68}
$$

Combining these three columns forms the total Jacobian matrix of the revolute arm:

$$
J_{\text{Rev.}} = \begin{bmatrix} -S_1(a_3C_{23} + a_2C_2) & -C_1(a_3S_{23} + a_2S_2) & -a_3C_1C_{23} \\ C_1(a_3C_{23} + a_2C_2) & -S_1(a_3S_{23} + a_2S_2) & -a_3S_1C_{23} \\ 0 & a_3C_{23} + a_2C_2 & -a_3S_{23} \\ 0 & S_1 & S_1 \\ 0 & -C_1 & -C_1 \\ 1 & 0 & 0 \end{bmatrix} \tag{5.69}
$$

The PUMA–560. The Jacobian for the PUMA–560 manipulator is presented here although the detailed development is omitted for the sake of brevity. Each joint is calculated in the same manner as the preceding arms and wrists. The Jacobian element for each joint is as follows.

The first joint is a revolute joint about the Z coordinate axis so Eqn. (5.39) is used to calculate its column. The $^0T_0$ orientation matrix is equal to the identity matrix. The $^0(\bar{P})_6$ matrix is the position portion of the product of the $A_1A_2A_3A_4A_5A_6$ matrices. Column 1 is calculated to be:

$$
{}^{0}\begin{bmatrix} X' \\ Y' \\ Z' \\ \theta'_x \\ \theta'_y \\ \theta'_z \end{bmatrix}_1 = \begin{bmatrix} -S_1(C_2(C_3C_4C_5d_6+d_4))-S_2(S_3C_4C_5d_6-C_3(C_5d_6+d_4)))-C_1(S_4S_5d_6+d_2) \\ C_1(C_2(C_3C_4C_5d_6+d_4))-S_2(S_3C_4C_5d_6-C_3(C_5d_6+d_4)))-S_1(S_4S_5d_6+d_2) \\ 0 \\ 0 \\ 0 \\ 1 \end{bmatrix} \theta'_1
$$

(5.70)

The second joint is also a revolute joint about the Z coordinate axis and again Eqn. (5.39) is used to calculate the column. The ${}^{0}T_1$ orientation matrix is equal to the orientation portion of the A_1 matrix. The ${}^{1}(\bar{P})_6$ matrix is equal to the position portion of the product of the $A_2A_3A_4A_5A_6$ matrices. Column 2 is equal to:

$$
{}^{0}\begin{bmatrix} X' \\ Y' \\ Z' \\ \theta'_x \\ \theta'_y \\ \theta'_z \end{bmatrix}_2 = \begin{bmatrix} -C_1(S_2(C_3C_4C_5d_6+S_3(C_5d_6+d_4))+C_2(S_3C_4C_5d_6-C_3(C_5d_6+d_4))) \\ -S_1(S_2(C_3C_4C_5d_6+S_3(C_5d_6+d_4))+C_2(S_3C_4C_5d_6-C_3(C_5d_6+d_4))) \\ -(C_2(C_3C_4C_5d_6+S_3(C_5d_6+d_4))-S_2(S_3C_4C_5d_6-C_2(C_5d_6+d_4))) \\ -S_1 \\ C_1 \\ 0 \end{bmatrix} \theta'_2
$$

(5.71)

The third joint is calculated the same way as the second joint, except the ${}^{0}T_2$ orientation matrix is equal to the orientation portion of the product of the A_1A_2 matrices and the ${}^{2}(\bar{P})_6$ matrix is the position portion of the product of the $A_3A_4A_5A_6$ matrices. Column 3 is then equal to:

$$
{}^{0}\begin{bmatrix} X' \\ Y' \\ Z' \\ \theta'_x \\ \theta'_y \\ \theta'_z \end{bmatrix}_3 = \begin{bmatrix} -C_1C_2(S_3C_4S_5d_6-C_3(C_5d_6+d_4))-C_1S_2(C_3C_4S_5d_6+S_3(C_5d_6+d_4)) \\ -S_1C_2(S_3C_4S_5d_6-C_3(C_5d_6+d_4))-S_1S_2(C_3C_4S_5d_6+S_3(C_5d_6+d_4)) \\ S_2(S_3C_4S_5d_6-C_3(C_5d_6+d_4))-C_2(C_3C_4S_5d_6+S_3(C_5d_6+d_4)) \\ -S_1 \\ C_1 \\ 0 \end{bmatrix} \theta'_3
$$

(5.72)

The fourth joint is the same as the previous joints except the ${}^{0}T_3$ orientation matrix is equal to the orientation portion of the product of the $A_1A_2A_3$ matrices and the ${}^{3}(\bar{P})_6$ matrix is the position portion of the product of the $A_4A_5A_6$ matrices. Column 4 is found to be:

$$
{}^{0}\begin{bmatrix} X' \\ Y' \\ Z' \\ \theta'_x \\ \theta'_y \\ \theta'_z \end{bmatrix}_4 = \begin{bmatrix} -C_1C_{23}S_4S_5d_6-S_1C_4S_5d_6 \\ -S_1C_{23}S_4S_5d_6+C_1C_4S_5d_6 \\ S_{23}S_4S_5d_6 \\ C_1S_{23} \\ S_1S_{23} \\ C_{23} \end{bmatrix} \theta'_4
\qquad (5.73)
$$

The fifth joint again is the same except the $^0T_4$ orientation matrix is equal to the orientation portion of the product of the $A_1A_2A_3A_4$ matrices and the $^4(\bar{P})_6$ is the position portion of the product of the A_5A_6 matrices. Column 5 is equal to:

$$
^0\begin{bmatrix} X' \\ Y' \\ Z' \\ \theta'_x \\ \theta'_y \\ \theta'_z \end{bmatrix}_5 = \begin{bmatrix} C_5d_6(C_1C_{23}C_4-S_1S_4)-C_1S_{23}S_5d_6 \\ C_5d_6(S_1C_{23}C_4+C_1S_4)-S_1S_{23}S_5d_6 \\ -S_{23}C_4C_5d_6-C_{23}S_5d_6 \\ -C_1C_{23}S_4-S_1C_4 \\ -S_1C_{23}S_4+C_1C_4 \\ S_{23}S_4 \end{bmatrix} \theta'_5 \quad (5.74)
$$

The last joint has the $^0T_5$ orientation matrix equal to the orientation portion of the product of the $A_1A_2A_3A_4A_5$ matrices and the $^5(\bar{P})_6$ is the position portion of the A_6 matrix. Column 6 is equal to:

$$
^0\begin{bmatrix} X' \\ Y' \\ Z' \\ \theta'_x \\ \theta'_y \\ \theta'_z \end{bmatrix}_6 = \begin{bmatrix} 0 \\ 0 \\ 0 \\ S_5(C_1C_{23}C_4-S_1S_4)+C_1S_{23}C_5 \\ S_5(S_1C_{23}C_4+C_1S_4)+S_1S_{23}C_5 \\ -S_{23}C_4S_5+C_{23}C_5 \end{bmatrix} \theta'_6 \quad (5.75)
$$

The complete Jacobian for the PUMA–560 is formed by combining these six columns in matrix form.

5.4 Speed control and the pseudo-inverse Jacobian matrix

For the remainder of this chapter we will be studying the use of the Jacobian matrix in solving the speed control problem. Let the vector $\bar{X}$ denote a set of command variables in an arbitrary coordinate system, and let vector $\bar{q}$ be a set of joint angles. The command variables are chosen so that:

$$\bar{X} = F(\bar{q}) \quad (5.76)$$

Or expressing this relationship in matrix form:

$$
\begin{bmatrix} X_1 \\ X_2 \\ - \\ - \\ X_m \end{bmatrix} = \begin{bmatrix} F_1(q_1, q_2, \ldots\ldots, q_n) \\ F_2(q_1, q_2, \ldots\ldots, q_n) \\ - \quad - \quad - \\ - \quad - \quad - \\ F_m(q_1, q_2, \ldots\ldots, q_n) \end{bmatrix} \quad (5.77)
$$

where m is the number of command variables and n is the number of degrees of freedom.

Differentiating Eqn. (5.76) with respect to time yields:

$$\bar{X}' = (\delta F(\bar{q})/\delta \bar{q})(d\bar{q}/dt) = J(\bar{q})\bar{q}' \quad (5.78)$$

Representing this equation in matrix form yields:

$$\begin{bmatrix} X'_1 \\ X'_2 \\ - \\ - \\ X'_m \end{bmatrix} = \begin{bmatrix} \delta F_1/\delta q_1 & \delta F_1/\delta q_2 & - - - & \delta F_1/\delta q_n \\ \delta F_2/\delta q_1 & \delta F_2/\delta q_2 & - - - & \delta F_2/\delta q_n \\ - & - & - - - & - \\ - & - & - - - & - \\ \delta F_m/\delta q_1 & \delta F_m/\delta q_2 & - - - & \delta F_m/\delta q_n \end{bmatrix} \begin{bmatrix} q'_1 \\ q'_2 \\ - \\ - \\ q'_n \end{bmatrix} \quad (5.79)$$

where $J(\bar{q})$ is the same as the Jacobian matrix that was outlined previously. Substituting this relationship into Eqn. (5.78) yields:

$$\bar{X}' = J\bar{q}' \quad (5.80)$$

In order to obtain the rate of the motion of each of the joint coordinates, multiply the above equation by J^{-1} which yields:

$$\bar{q}' = J^{-1}\bar{X}' \quad (5.81)$$

Unfortunately, this equation is only valid if $m = n$, that is the Jacobian matrix is a square matrix and its inverse exists. When m is not equal to n, then special techniques must be used to solve for the command variables. To solve these special cases, a matrix called the pseudo-inverse matrix[7] is used.

The over-determined case

The first case to be considered is when $m > n$ or the number of command inputs is greater than the number of degrees of freedom. In this case the number of equations is greater than the number of unknowns in the system. We call this case 'over-determined' and there is no guarantee of the existence of a certain $\bar{q}'$ such that $J\bar{q}' = \bar{X}'$ is satisfied. It is possible, however, to satisfy the system with the minimum possible error.

Assume a set of joint variables $\bar{q}'^*$ which will satisfy the least square error condition:

$$E = \text{Error}^2 = (J\bar{q}'^* - \bar{X}')^T (J\bar{q}'^* - \bar{X}') \quad (5.82)$$

Expanding this equation yields:

$$E = (J\bar{q}'^*)^T (J\bar{q}'^*) - 2(\bar{X}')^T J\bar{q}'^* + (\bar{X}')^T(\bar{X}') \quad (5.83)$$

In order to minimise the minimum error, set $\delta E/\delta \bar{q}_i = 0$, where $i = 1, 2, \ldots, n$. This results in:

$$0 = \delta/\delta\bar{q}(\bar{q}'^{*T}(J^T J)\bar{q}'^*) - \delta/\delta\bar{q}(2\bar{X}'^T J\bar{q}'^*) \quad (5.84)$$

By manipulation and simplification it can be shown that:

$$(J^T J)\bar{q}'^* = J^T \bar{X}' \quad (5.85)$$

or,

$$\bar{q}'^* = (J^T J)^{-1} J^T \bar{X}' \quad (5.86)$$

provided that $(J^T J)^{-1}$ exists.

The matrix $(J^TJ)^{-1}J^T$ is defined to be the 'psuedo-inverse Jacobian matrix' J^I. Substituting into Eqn. (5.86) yields:

$$\bar{q}'^* = J^I\bar{X}' \tag{5.87}$$

from which the set of joint command variables can be obtained which have a minimum error associated with them.

The under-determined case

The second case to be considered is where $m < n$ or the number of the input commands are less than the number of degrees of freedom. In this case the number of equations is less than the number of unknowns. Under these conditions, the number of solutions is infinite. Among these infinite solutions, however, there exists one whose Euclidean norm is a minimum. This optimal solution can be found using the Lagrange multiplier method described here.

Let,

$$\bar{\lambda} = \begin{bmatrix} \lambda_1 \\ \lambda_2 \\ - \\ - \\ \lambda_m \end{bmatrix} \tag{5.88}$$

The new objective form becomes:

$$\psi(\bar{q}') = (\bar{q}')^T(\bar{q}') + \bar{\lambda}^T(J\bar{q}' - \bar{X}') \tag{5.89}$$

$\psi(\bar{q}')$ is a minimum when $\delta\psi/\delta\bar{q} = 0$. With this in mind, Eqn. (5.89) becomes:

$$0 = 2\bar{q}' + J^T\bar{\lambda} \tag{5.90}$$

Simplifying yields:

$$\bar{q}' = -J^T\bar{\lambda}/2 \tag{5.91}$$

Substituting Eqn. (5.91) into $J\bar{q}' = \bar{X}'$ yields:

$$J(-J^T\bar{\lambda}/2) = \bar{X}' \tag{5.92}$$

This equation can be simplified to:

$$\bar{\lambda} = -2(JJ^T)^{-1}\bar{X}' \tag{5.93}$$

provided that $(JJ^T)^{-1}$ exists.

Then substituting this value of $\bar{\lambda}$ into Eqn. (5.91) and simplifying yields:

$$\bar{q}' = J^T(JJ^T)^{-1}\bar{X}' \tag{5.94}$$

The term $J^T(JJ^T)^{-1}$ is recognised as the 'psuedo-inverse Jacobian matrix.' Substituting into Eqn. (5.94) yields:

$$\bar{q}' = J^I\bar{X}' \tag{5.95}$$

from which the joint command variables can be found.

Finding the psuedo-inverse Jacobian matrix is a very complex calculation. In order to avoid this calculation the following approach is used. Let,

$$(JJ^T)^{-1}\bar{X}' = \bar{Y} \tag{5.96}$$

Then rearrange to obtain:

$$(JJ^T)\bar{Y} = \bar{X}' \tag{5.97}$$

Equation (5.97) is a system of linear equations. Solving for $\bar{Y}$ can be accomplished by any of several well known methods such as the 'successive over relaxation' method, the QR method, or the Guass-Seidel method. After obtaining $\bar{Y}$, the joint command variables are obtained by multiplying J^T by $\bar{Y}$, which yields the solution.

References

[1] Angeles, J. 1982. *Spatial Kinematic Chains*. Springer-Verlag, Berlin.
[2] Bottema, O. and Roth B. 1979. *Theoretical Kinematics*. North-Holland, Amsterdam.
[3] Duffy, J. 1980. *Analysis of Mechanisms and Robot Manipulators*. Wiley-Interscience, New York.
[4] Whitney, D. E. 1972. The mathematics of coordinated control of prosthetic arm and manipulators. *Trans. ASME, J. Dynamic System, Measurement and Control,* December 1972: 303–309.
[5] Featherstone, R. 1983. Position and velocity transformations between robot end-effector coordinates and joint angles. *Int. J. Robotics Research,* 2(2).
[6] Paul, R. P. 1981. *Robot Manipulators: Mathematics, Programming and Control*. MIT Press, Cambridge, MA, USA.
[7] Rao and Mitra. 1971. *Generalized Inverse of Matrices and Its Applications*. John Wiley & Sons, New York.

Chapter Six

TRAJECTORY CONTROL

IN ROBOT applications it is frequently necessary for the manipulator to follow a planned path between goal points. In applications such as painting and welding, it is necessary for the manipulator to follow the shape of the object on which it is working. In other applications, it may be necessary for the manipulator to avoid obstacles between goal points. In order to define the path for the manipulator to follow, the operator specifies critical points along the desired path. These points, which are known as 'way points', may include safe points for the avoidance of an obstacle.

In general, the operator would like to specify these points in a coordinate system in which he can easily visualise and measure, such as the Cartesian coordinate system. However, the robot understands only joint coordinates. Therefore, we must decide whether to plan the trajectory in Cartesian coordinates or convert the way points to joint coordinates and plan the trajectory in joint coordinates.

The first method has the burden of converting between Cartesian and joint coordinates during each sampling period[1]. This is necessary because the current joint positions must be converted to their equivalent Cartesian coordinates using Jacobian transformations so that the errors from the desired Cartesian path may be calculated. The equivalent errors in joint coordinates must then be determined using inverse Jacobians. Then the necessary torques for the joints may be calculated using the robot's dynamic equations. The complexity of these calculations severely limits the sampling frequency for the robot controller.

Due to the large amount of on-line computation required by the above method, the second method is adopted, i.e. planning the trajectory in joint coordinates. Here, the desired Cartesian path is converted into a functionally equivalent path in joint coordinates. However, there is no known transformation of a function from Cartesian to joint coordinates since the Jacobian transformation works only for individual points. Therefore, a curve-fitting method is used to approximate the desired Cartesian path by functions of joint variables. This method requires that enough points on the Cartesian path be transformed to joint coordinates to ensure a good approximation[1].

The next decision which must be made is whether the trajectory planning should take place on-line or off-line[2]. The on-line method has the advantage of allowing the robot to respond to external stimuli which may cause it to modify its path. However, since the curve-fitting calculations are quite lengthy, this method limits how closely in time the way points may be specified and thus degrades the accuracy of the trajectory.

The off-line method has the advantage of allowing way points to be specified more closely in time, thus increasing the accuracy of the trajectory and allowing a shorter sampling period. Since most robot applications involve very repetitive operations, this method also greatly reduces the amount of computing time needed for an application. All the data for an application is calculated only once, before the robot runs the application for the first time. This data is stored and retrieved by the robot controller at run time. If the robot is executing some preplanned trajectory and an external stimulus is received, it may simply switch to the data for another preplanned trajectory. Thus, we choose to plan our trajectories off-line.

6.1 Trajectory planning

In order to approximate a Cartesian path, m functions of approximation are needed – one for each joint. The function of approximation for a particular joint must pass through the value calculated for that joint at each way point. In addition, the function must be continuous in position, velocity and acceleration in order to remain within the physical limitations of the robot. These conditions could be met by deriving a single polynomial which passed through the way points, but it would be likely to contain extrema between way points which would have to be checked against the robot's limits and might cause large deviations from the desired Cartesian path.

A better solution is to define a separate polynomial for each segment of the path and ensure the continuity constraints at each way point. This requires third-order polynomials for the intermediate segments and fourth-order polynomials for the beginning and end segments since we want the velocity and acceleration at the start- and end-points to be zero. Thus, to develop a trajectory through n points requires $n-1$ polynomials: a fourth-order polynomial, $n-3$ third-order polynomials, and a final fourth-order polynomial[2].

Several methods of polynomial curve fitting are described elsewhere[3–6], but we will use a method widely used in the field of computer graphics for connecting data points with smooth curves. These curves, known as spline functions, have been thoroughly studied and investigated and found to provide the shortest path which satisfies our continuity constraints[7]. Spline functions are also relatively straight-forward to calculate on small to medium-size computers. We will derive our results following the method of Ho and Cook[8].

The equation of the spline segment between two intermediate points F_k and F_{k+1} $(2 < k < n-2)$ of an n-point trajectory consisting of $n-1$ spline segments may be written for a single joint as:

$$F(t) = B_1 + B_2 t + B_3 t^2 + B_4 t^3 \tag{6.1}$$

where $F(t)$ represents the position of the joint as a function of time. Letting t for this segment run from zero to t_{k+1}, we may express the boundary conditions for this segment as:

$$F(0) = F_k \qquad\qquad F(t_{k+1}) = F_{k+1} \tag{6.2}$$
$$F'(0) = F'_k \qquad\qquad F'(t_{k+1}) = F'_{k+1}$$

where F'_k and F'_{k+1} represent the velocities of the joint at the points F_k and F_{k+1}, respectively. Substituting Eqn. (6.2) into Eqn. (6.1) yields:

$$F(0) = B_1 = F_k \tag{6.3}$$

$$F'(0) = B_2 = F'_k \tag{6.4}$$

$$F(t_{k+1}) = B_1 + B_2 t_{k+1} + B_3 t_{k+1}^2 + B_4 t_{k+1}^3 = F_{k+1} \tag{6.5}$$

$$F'(t_{k+1}) = B_2 + 2B_3 t_{k+1} + 3B_4 t_{k+1}^2 = F'_{k+1} \tag{6.6}$$

Solving Eqns. (6.5) and (6.6) for B_3 and B_4, and substituting Eqns. (6.3) and (6.4) for B_1 and B_2 yields:

$$B_3 = \frac{1}{t_{k+1}} \left[\frac{3(F_{k+1}-F_k)}{t_{k+1}} - 2F'_k - F'_{k+1} \right] \tag{6.7}$$

$$B_4 = \frac{1}{t_{k+1}^2} \left[\frac{2(F_k-F_{k+1})}{t_{k+1}} + F'_k + F'_{k+1} \right] \tag{6.8}$$

For $2 < k < n-3$, Eqns. (6.3), (6.4), (6.7) and (6.8) may be expressed in matrix form as:

$$
\begin{bmatrix} B_1 \\ B_2 \\ B_3 \\ B_4 \end{bmatrix}
=
\begin{bmatrix}
1 & 0 & 0 & 0 \\
0 & 0 & 1 & 0 \\
\dfrac{-3}{t_{k+1}^2} & \dfrac{3}{t_{k+1}^2} & \dfrac{-2}{t_{k+1}} & \dfrac{-1}{t_{k+1}} \\
\dfrac{2}{t_{k+1}^3} & \dfrac{-2}{t_{k+1}^3} & \dfrac{1}{t_{k+1}^2} & \dfrac{1}{t_{k+1}^2}
\end{bmatrix}
\begin{bmatrix} F_k \\ F_{k+1} \\ F'_k \\ F'_{k+1} \end{bmatrix}
\tag{6.9}
$$

In order to calculate the spline coefficients, we must first know the values of t_{k+1}, F'_k and F'_{k+1}. We will select the value t_{k+1} to be the length of the chord between the way points F_k and F_{k+1} in joint coordinates.

$$t_{k+1} = \left[\sum_{i=1}^{m} (q_{i,k+1} - q_{i,k})^2 \right]^{1/2} \tag{6.10}$$

where t_{k+1} is measured in some unit of time which will be determined later, m is the number of joints in the manipulator, and $q_{i,k}$ represents the value of the ith joint at the kth way point.

The velocities F'_k and F'_{k+1} may now be determined using the continuity constraint on the acceleration at the way points. Given three way points F_k, F_{k+1} and F_{k+2} ($2 \le k \le n-3$) with two spline segments connecting them with times $0 \le t \le t_{k+1}$ and $0 \le t \le t_{k+2}$, the acceleration at the end of the first segment can be determined as:

$$F''(t_{k+1}) = 2B_3 + 6B_4 t_{k+1}$$

$$= \frac{6}{t_{k+1}} \left[\frac{2(F_k - F_{k+1})}{t_{k+1}} + F'_k + F'_{k+1} \right]$$

$$+ \frac{2}{t_{k+1}} \left[\frac{3(F_{k+1} - F_k)}{t_{k+1}} - 2F'_k - F'_{k+1} \right] \tag{6.11}$$

Similarly, the acceleration at the beginning of the second segment may be calculated as:

$$F''(0) = 2B_3 = \frac{2}{t_{k+2}} \left[\frac{3(F_{k+2} - F_{k+1})}{t_{k+2}} - 2F'_{k+1} - F'_{k+2} \right] \tag{6.12}$$

The continuity constraint on acceleration requires Eqns. (6.11) and (6.12) to be equal. Equating them and rearranging terms gives:

$$t_{k+2}F'_k + 2(t_{k+2} + t_{k+1})F'_{k+1} + t_{k+1}F'_{k+2}$$

$$= \frac{3}{t_{k+1}t_{k+2}} \left[t_{k+1}^2(F_{k+2} - F_{k+1}) + t_{k+2}^2(F_{k+1} - F_k) \right] \tag{6.13}$$

Expressing Eqn. (6.13) in matrix form for $2 \le k \le n-3$ yields:

$$\begin{bmatrix} t_4 & 2(t_3+t_4) & t_3 & 0 & 0 \\ 0 & t_5 & 2(t_4+t_5) & t_4 & 0 \\ \cdot & \cdot & \cdot & \cdot & \cdot \\ \cdot & \cdot & \cdot & \cdot & \cdot \\ \cdot & \cdot & \cdot & \cdot & \cdot \\ 0 & 0 & t_{n-1} & 2(t_{n-2}+t_{n-1}) & t_{n-2} \end{bmatrix} \begin{bmatrix} F'_2 \\ F'_3 \\ \cdot \\ \cdot \\ \cdot \\ F'_{n-1} \end{bmatrix}$$

$$
= \begin{bmatrix}
\dfrac{3}{t_3 t_4} \left[t_3^2 (F_4 - F_3) + t_4^2 (F_3 - F_2) \right] \\[2ex]
\dfrac{3}{t_4 t_5} \left[t_4^2 (F_5 - F_4) + t_5^2 (F_4 - F_3) \right] \\[1ex]
\vdots \\[1ex]
\dfrac{3}{t_{n-2} t_{n-1}} \left[t_{n-2}^2 (F_{n-1} - F_{n-2}) + t_{n-1}^2 (F_{n-2} - F_{n-3}) \right]
\end{bmatrix}
\tag{6.14}
$$

or symbolically, $[m](F') = [a]$, where $[m]$ is an $(n-4) \times (n-2)$ matrix, $[F']$ is an $(n-2) \times 1$, and $[a]$ is an $(n-4) \times 1$. Solving Eqn. (6.14) for $[F']$ allows a unique set of coefficients for each third-order spline segment to be calculated using Eqn. (6.9). In order to obtain the solution from Eqn. (6.14), it is necessary to state the boundary conditions at the end segments of the lift-off and set-down of the trajectory.

For the first and last segments, we have the additional constraints $F'_1 = F''_1 = F'_n = F''_n = 0$. This requires a fourth-order spline segment of the form:

$$
F(t) = B_1 + B_2 t + B_3 t^2 + B_4 t^3 + B_5 t^4
\tag{6.15}
$$

For the first segment, we let $0 \leqslant t \leqslant t_2$ and have the following boundary conditions:

$$
F(0) = B_1 = F_1
\tag{6.16}
$$

$$
F'(0) = B_2 = 0
\tag{6.17}
$$

$$
F''(0) = 2B_3 = 0
\tag{6.18}
$$

$$
F(t_2) = F_1 + B_4 t_2^3 + B_5 t_2^4 = F_2
\tag{6.19}
$$

$$
F'(t_2) = 3B_4 t_2^2 + 4B_5 t_2^3 = F'_2
\tag{6.20}
$$

Solving Eqns. (6.19) and (6.20) for B_4 and B_5, yields:

$$
B_4 = \frac{4}{t_2^3} (F_2 - F_1) - \frac{1}{t_2^2} F'_2
\tag{6.21}
$$

$$
B_5 = \frac{3}{t_2^4} (F_1 - F_2) + \frac{1}{t_2^3} F'_2
\tag{6.22}
$$

For the last segment, we let $0 \leqslant t \leqslant t_n$ and have the following boundary conditions:

$$
F(0) = B_1 = F_{n-1}
\tag{6.23}
$$

$$
F'(0) = B_2 = F'_{n-1}
\tag{6.24}
$$

$$
F(t_n) = F_n = F_{n-1} + F'_{n-1} t_n + B_3 t_n^2 + B_4 t_n^3 + B_5 t_n^4
\tag{6.25}
$$

$$
F'(t_n) = 0 = F'_{n-1} + 2B_3 t_n + 3B_4 t_n^2 + 4B_5 t_n^3
\tag{6.26}
$$

$$
F''(t_n) = 0 = 2B_3 + 6B_4 t_n + 12B_5 t_n^2
\tag{6.27}
$$

Solving Eqns. (6.25), (6.26) and (6.27) for B_3, B_4 and B_5, yields:

$$B_3 = \frac{1}{t_n^2} (6F_n - 6F_{n-1} - 3F'_{n-1}t_n) \tag{6.28}$$

$$B_4 = \frac{1}{t_n^3} (-8F_n + 8F_{n-1} + 3F'_{n-1}t_n) \tag{6.29}$$

$$B_5 = \frac{1}{t_n^4} (3F_n - 3F_{n-1} - F'_{n-1}t_n) \tag{6.30}$$

Thus, Eqns. (6.16), (6.17), (6.18), (6.21) and (6.22) define the coefficients for the first segment, Eqn. (6.9) defines the coefficients for the intermediate segments 2 through $n-2$, and Eqns. (6.23), (6.24), (6.28), (6.29) and (6.30) define the coefficients for the last segment. However, we must redefine Eqn. (6.14) for finding $[F']$ since we are now including fourth-order segments at the beginning and end. Finding the acceleration at the end of the first segment yields:

$$F''(t_2) = 6B_4t_2 + 12B_5t_2^2$$

$$= \frac{6}{t_2^2} [4(F_2-F_1) - t_2F'_2] + \frac{12}{t_2^2} [3(F_1-F_2) + t_2F'_2] \tag{6.31}$$

Equating Eqn. (6.31) with Eqn. (6.12) yields:

$$(\frac{2}{t_3} + \frac{3}{t_2})F'_2 + \frac{1}{t_3} F'_3 = \frac{3}{t_3^2} (F_3-F_2) + \frac{6}{t_2^2} (F_2-F_1) \tag{6.32}$$

Performing the same operations for the beginning of the last segment yields:

$$(\frac{2}{t_{n-1}} + \frac{3}{t_n}) F'_{n-1} + \frac{1}{t_{n-1}} F'_{n-2}$$

$$= \frac{3}{t_{n-1}^2} (F_{n-1}-F_{n-2}) + \frac{6}{t_n^2} (F_n-F_{n-1}) \tag{6.33}$$

Combining the results of Eqns. (6.32) and (6.33) with Eqn. (6.14) allows us to express the matrix equation for solving for the $n-2$ unknown velocities as $[M][F'] = [A]$, where $[M]$ is $(n-2) \times (n-2)$, $[F']$ is $(n-2) \times 1$, and $[A]$ is $(n-2) \times 1$ in the form:

$$
\begin{bmatrix}
M_{22} & M_{23} & 0 & & \cdot & & \cdot & \cdot & \cdot & & \cdot & 0 \\
t_4 & 2(t_3+t_4) & t_3 & 0 & & \cdot & \cdot & \cdot & & & \cdot & \\
0 & & & & & & & & & & & \\
\cdot & & & & & & & & & & 0 & \\
\cdot & \cdot & \cdot & & & t_{n-1} & 2(t_{n-2}+t_{n-1}) & t_{n-2} & & & & \\
0 & \cdot & \cdot & & & 0 & M_{n-1,n-2} & M_{n-1,n-1}
\end{bmatrix}
\begin{bmatrix}
F'_2 \\
F'_3 \\
\cdot \\
\cdot \\
F'_{n-2} \\
F'_{n-1}
\end{bmatrix}
$$

$$a_2$$

$$
= \begin{bmatrix} \dfrac{3}{t_3 t_4} \ [t_3^2 F_4 - F_3) + t_4^2 (F_3 - F_2)] \\ \cdot \\ \cdot \\ \cdot \\ \dfrac{3}{t_{n-2} t_{n-1}} \ [t_{n-2}^2 (F_{n-1} - F_{n-2}) + t_{n-1}^2 (F_{n-2} - F_{n-3})] \\ a_{n-1} \end{bmatrix} \tag{6.34}
$$

where

$$
M_{22} = \frac{2}{t_3} + \frac{3}{t_2} \qquad\qquad M_{23} = \frac{1}{t_3}
$$

$$
M_{n-1,n-2} = \frac{1}{t_{n-1}} \qquad\qquad M_{n-1,n-1} = \frac{2}{t_{n-1}} + \frac{3}{t_n}
$$

$$
a_2 = (F_3 - F_2) + \frac{6}{t_2^2} (F_2 - F_1)
$$

$$
a_{n-1} = \frac{3}{t_{n-1}^2} (F_{n-1} - F_{n-2}) + \frac{6}{t_n^2} (F_n - F_{n-1})
$$

Solving this equation for $[F']$ allows us to find spline functions with the desired properties. To extend these results for a single joint to all the joints of a robot, we simply duplicate the spline function generation procedure for each joint, using the same values of t_k. Thus, a path through n points for a robot with m joints will consist of $(n-1) \times m$ unique spline functions.

Once the spline functions have been defined for all the segments of all the joints of a manipulator, the parametric variable t must be associated with some unit of physical time. To do this, a method similar to the 'feasible solution converter'[1] is used, replacing t in the spline segments by T/S, where T is the elapsed physical time since the beginning of the spline and S is a scale factor. Thus, each function $F(t)$ becomes $F(T/S)$. The value S must be large enough to prevent the maximum velocity or acceleration for any of the robot's joints being exceeded. However, this value should be as small as possible in order to complete the motion in the shortest possible time. Therefore, the maximum value of S necessary to stay within the physical constraints of the robot's joints should be found.

In order to stay within the velocity constraints, the velocity of $F(T/S)$, $(1/S)F'(t)$, for a joint must stay below the velocity constraint, VC, of that joint. Thus S must be greater than or equal to $F'(t)/VC$. Therefore, the maximum value of S necessary to stay within the velocity constraints is:

$$
S_v = \max_j \left[\frac{\max_i \left(|F'_{ij}(t)| : 0 \leqslant t \leqslant t_{i+1} \right)}{VC_j} \right] \tag{6.35}
$$

where $F'_{ij}(t)$ represents the velocity of the ith spline segment for the jth joint, and VC_j represents the velocity constraint for the jth joint. Similarly, for acceleration:

$$S_a^2 = \max_j \left[\frac{\max_i (|F''_{ij}(t)| : 0 \leqslant t \leqslant t_{i+1})}{AC_j} \right] \qquad (6.36)$$

where $F''_{ij}(t)$ represents the acceleration of the ith spline segment for the jth joint, and AC_j represents the acceleration constraint for the jth joint. Thus, the desired value for S is:

$$S = \max (S_v, S_a) \qquad (6.37)$$

Now that continuous functions for each of the joints have been found, the question may arise as to whether the motion of the tool tip is necessarily continuous in position, velocity and acceleration. The answer to this question is 'yes'. The motion of the tool tip is determined by the motion of all the joints. There exists a fundamental mathematical theorem which states that 'any composite continuous function is continuous'[9]. Therefore, the tool tip motion is continuous.

6.2 Example

Given the following six points in joint coordinates for a three degree of freedom robot, find the optimal trajectory through these points.

$F_1 = [0,0,0]$ $F_2 = [1,1,1]$
$F_3 = [2,2.5,2]$ $F_4 = [4,3.5,2.5]$
$F_5 = [5,2,1.75]$ $F_6 = [6,0,0]$

We begin by using Eqn. (6.10) to calculate the values for t_k,

$t_2 = (1^2 + 1^2 + 1^2)^{1/2} = 1.732$
$t_3 = (1^2 + 1.5^2 + 1^2)^{1/2} = 2.062$
$t_4 = (2^2 + 1^2 + 0.5^2)^{1/2} = 2.291$
$t_5 = (1^2 + -1.5^2 + -0.75^2)^{1/2} = 1.953$
$t_6 = (1^2 + -2^2 + -1.75^2)^{1/2} = 2.839$

Next, these values are used to find the matrix [M] for Eqn. (6.34),

$$[M] = \begin{bmatrix} 2.702 & 0.485 & 0 & 0 \\ 2.291 & 8.706 & 2.062 & 0 \\ 0 & 1.953 & 8.488 & 2.291 \\ 0 & 0 & 0.512 & 2.081 \end{bmatrix}$$

Since t_k is the same for all joints, [M] is also the same. Using this and the fact that we go through the same operations for each joint, we can solve for all joints simultaneously. This is done by expanding the matrices [A] and [F'] to three columns, one for each joint. Thus, the values for the matrix [A] are found to be:

$$[A] = \begin{bmatrix} 2.706 & 3.059 & 2.706 \\ 8.732 & 7.700 & 4.683 \\ 8.633 & -2.724 & -1.362 \\ 1.531 & -2.669 & -1.892 \end{bmatrix}$$

From Eqn. (6.34), by Gaussian elimination:

$$\begin{bmatrix} 2.702 & 0.485 & 0 & 0 \\ 0 & 8.295 & 2.062 & 0 \\ 0 & 0 & 8.003 & 2.291 \\ 0 & 0 & 0 & 1.934 \end{bmatrix} \begin{bmatrix} F'_2 \\ F'_3 \\ F'_4 \\ F'_5 \end{bmatrix}$$

$$= \begin{bmatrix} 2.706 & 3.059 & 2.706 \\ 6.483 & 5.106 & 2.389 \\ 7.117 & -3.926 & -1.924 \\ 1.076 & -2.418 & -1.769 \end{bmatrix}$$

Solving for the F's gives:

$$F'_5 = [\, 0.556, -1.250, -0.914 \,]$$
$$F'_4 = [\, 0.730, -0.133, 0.021 \,]$$
$$F'_3 = [\, 0.595, 0.649, 0.283 \,]$$
$$F'_2 = [\, 0.895, 1.016, 0.951 \,]$$

For the first segments we find:

$$
\begin{aligned}
B_1 &= [\, 0, 0, 0 \,] & &\text{from Eqn. (6.16)} \\
B_2 &= [\, 0, 0, 0 \,] & &\text{from Eqn. (6.17)} \\
B_3 &= [\, 0, 0, 0 \,] & &\text{from Eqn. (6.18)} \\
B_4 &= [\, 0.472, 0.431, 0.453 \,] & &\text{from Eqn. (6.21)} \\
B_5 &= [\, -0.161, -0.138, -0.150 \,] & &\text{from Eqn. (6.22)}
\end{aligned}
$$

Thus, the equations for the first segments are

$$F_{11}(t) = 0.472t^3 - 0.161t^4$$
$$F_{12}(t) = 0.431t^3 - 0.138t^4$$
$$F_{13}(t) = 0.453t^3 - 0.150t^4$$

for $0 \leqslant t \leqslant 1.732$.

Using Eqn. (6.9), we find the following functions for the intermediate segments:

$$F_{21}(t) = 1 + 0.895t - 0.451t^2 + 0.122t^3$$
$$F_{22}(t) = 1 + 1.016t - 0.242t^2 + 0.049t^3$$
$$F_{23}(t) = 1 + 0.951t - 0.354t^2 + 0.062t^3$$

for $(\, 0 \leqslant t \leqslant 2.062)$.

$$F_{31}(t) = 2 + 0.595t + 0.305t^2 - 0.080t^3$$
$$F_{32}(t) = 2.5 + 0.649t + 0.063t^2 - 0.068t^3$$
$$F_{33}(t) = 2 + 0.283t + 0.030t^2 - 0.025t^3$$

for $(\, 0 \leqslant t \leqslant 2.291)$.

$$F_{41}(t) = 4 + 0.730t - 0.246t^2 + 0.069t^3$$
$$F_{42}(t) = 3.5 - 0.133t - 0.404t^2 + 0.040t^3$$
$$F_{43}(t) = 2.5 + 0.021t - 0.143t^2 - 0.033t^3$$

for $(0 \leqslant t \leqslant 1.953)$.

Using Eqns. (6.23), (6.24), (6.28), (6.29) and (6.30) we find the equations of the last segments to be:

$$F_{51}(t) = 5 + 0.556t + 0.157t^2 - 0.143t^3 + 0.022t^4$$
$$F_{52}(t) = 2 - 1.250t - 0.168t^2 + 0.234t^3 - 0.038t^4$$
$$F_{53}(t) = 1.75 - 0.914t - 0.337t^2 + 0.272t^3 - 0.041t^4$$

for $(0 \leqslant t \leqslant 2.839)$.

To determine the optimum value of the scale factor S, we must first find the maximum velocity and acceleration for each spline segment for each joint. The maximum velocity for each spline segment is found at the roots of the second derivative of the spline function. For our example, these values are found to be:

$$\max \left| F'_1 \right| = [1.014, 1.052, 1.039]$$
$$\max \left| F'_2 \right| = [0.339, 0.618, 0.277]$$
$$\max \left| F'_3 \right| = 0.983, 0.668, 0.295]$$
$$\max \left| F'_4 \right| = [0.438, \text{———}, \text{———}]$$
$$\max |F'_5| = [0.618, 1.293, 1.068]$$

where the missing entries for the fourth segment indicate that the maxima for those functions fell outside the time range for that segment. We do not need to check the end-points of a segment for maximum velocity since the continuity constraint on acceleration guarantees that the velocity will continue to change in the same direction as the manipulator passes through the way points.

Next, the maximum acceleration of each joint must be found. Since the acceleration of the cubic spline segments is linear, the maximum acceleration for the intermediate segments will occur at their end-points. The maximum acceleration for the beginning and end segments can be found at the roots of the third derivative of the spline functions for these segments. Thus the maximum acceleration for each segment is found to be:

$$\max \left| F''_1 \right| = [1.038, 0.820, 1.026]$$
$$\max \left| F''_2 \right| = [0.902, 0.484, 0.708]$$
$$\max \left| F''_3 \right| = [0.610, 0.126, 0.060]$$
$$\max \left| F''_4 \right| = [0.492, 0.808, 0.286]$$
$$\max |F''_5| = [0.555, 0.745, 0.669]$$

And the maximum velocity and acceleration for each joint is:

$$\max \left| F' \right| = [\,0.983, 1.293, 1.068\,]$$
$$\max | F'' | = [\,0.902, 1.028, 1.026\,]$$

Given the following constraints on velocity and acceleration, the optimum value of the timescale factor S may be found:

$$VC = [\,100, 90, 100\,] \text{ degrees/second}$$
$$AC = [\,45, 50, 40\,] \text{ degrees/second/second}$$
$$S^v = \max\,(0.01014, 0.01437, 0.01068) = 0.01437$$
$$S_a = \max\,(0.02004, 0.02056, 0.02565)^{1/2} = 0.16017$$

Thus the optimum value for S is 0.16017. Therefore the time required to execute each spline segment is:

$T_2 = 0.2774$ seconds $T_3 = 0.3303$ seconds
$T_4 = 0.3669$ seconds $T_5 = 0.3128$ seconds
$T_6 = 0.4547$ seconds

giving a total time to execute the trajectory of 1.7421 seconds.

To find the position of the joints at some time T after starting the trajectory, we simply determine which segment this time falls in, subtract the time used by the earlier segments and divide by S to find the value of the parameter t for that segment. We then substitute this value of t for the parameter t in the spline functions for that segment. For example, one second after beginning the trajectory in our example above, we find we are 0.0254 seconds into the fourth spline segment. This translates into a parameter value of 0.1586. Substituting this value into functions F_{41}, F_{42} and F_{43} gives a position in joint coordinates of $[4.110, 3.469, 2.499]$.

References

[1] Lin, C.-S., Chang, P.-R. and Luh, J. Y. S. 1982. Formulation and optimization of cubic polynomial joint trajectories for mechanical manipulators. In, *Proc. 21st IEEE Conf. on Decision and Control*, Vol. 1, pp. 330–335. IEEE, New York.

[2] Edwall, C. W., Ho, C. Y. and Pottinger, H. J. 1982. Trajectory generation and control of a robot arm using spline functions. In, *Robots 6*. Society of Manufacturing Engineers, Dearborn, MI, USA.

[3] Paul, R. C. 1972. *Modeling, Trajectory Calculation, and Servoing of a Computer Controlled Arm*. Ph.D. Dissertation, Stanford University, USA.

[4] Paul, R. C. 1975. Manipulator path control. In, *Proc. Int. Conf. on Cybernetics and Society*.

[5] Mujtaba, M. S. 1977. *Discussion of Trajectory Calculation Methods*. Exploratory Study of Computer Integrated Assembly Systems, Stanford Artifical Intelligence Laboratory Progress Report.

[6] Lewis, R. A. 1974. *Autonomous Manipulation of a Robot: Summary of Manipulator Software Functions*. Jet Propulsion Laboratory, NASA Technical Memo 33–679.

[7] Ahlberg, J. H., Nelson, E. N. and Walsh, J. L. 1967. *The Theory of Splines and Their Application.* Academic Press, New York.

[8] Ho, C. Y. and Cook, C. C. 1982. The application of spline functions to trajectory generation for computer controlled manipulators. *Digital Systems for Industrial Automation*, 1(4): 325–333.

[9] Churchill, R. V., Brown, J. W. and Verhey, R. F. 1976. *Complex Variables and Applications.* McGraw-Hill, New York.

Chapter Seven

DYNAMIC AND CONTROL MODEL

IN THIS chapter Lagrangian mechanics are used as a method to obtain the dynamic equations for robotic arm control and to derive the motion equations of a six-link manipulator. Deriving the dynamic model of a manipulator using the Lagrange–Euler (L–E) method is simple and systematic. The resultant equations of motion are a set of second order, coupled, nonlinear differential equations. It should be noted that for simplicity these equations exclude the dynamics of the electronic control device and the gear friction. While other approaches are available to formulate robotic arm dynamics, such as the Newton–Euler, the recursive Lagrange–Euler, and the generalised d'Alembert principle formulations, the L–E method is used here to provide a symbolic solution to manipulator dynamics lending insight into the control problem.

Using the 4 × 4 homogeneous transformation matrix representation of the kinematics chain and the L–E formulation, Bejczy[1] has shown that the dynamic motion equations for a six-joint manipulator (Stanford Arm) are highly nonlinear. The equations consist of inertia loading, coupling reaction forces between joints (Coriolis and centrifugal) and gravity loading effects. The dynamic equations of motion as formulated are usually computationally inefficient, and real-time control based on the 'complete' dynamic model has been difficult to achieve, if not impossible.

To improve the speed of computation, Hollerbach[2] has used simplified sets of equations. In general these 'approximate' models simplify the underlying physics by neglecting second-order terms such as the Coriolis and centrifugal reaction terms. These approximate models, when used in control, result in suboptimal dynamic performance, restricting arm movement to low speeds. At high speeds, the neglected terms become significant, making accurate position control of the arm impossible.

7.1 Lagrangian mechanics

Using the methods of derivation of Symon[3], the Lagrangian function L is defined as the difference between the kinetic energy K and the potential energy P of the system:

$$L = K - P \tag{7.1}$$

The dynamics equations in terms of the Lagrangian function are:

$$\frac{d}{dt}\left(\frac{\delta L}{\delta \dot{q}_i}\right) - \frac{\delta L}{\delta q_i} = F_i \ (i=1,2, \ldots ,n) \tag{7.2}$$

where n is the degrees of freedom of the mechanical system, q_i are the coordinates in which the kinetic and potential energy are expressed, $\dot{q}_i$ are the corresponding velocities, and F_i are the corresponding generalised forces. Since the joint variables form a convenient and readily made coordinate system, they are used for the q_i terms. If q_i describes a displacement along a prismatic joint, F_i represents the force the joint should exert to produce the desired dynamics. If q_i describes an angular displacement of a revolute joint, F_i represents the required torque.

The left-hand side of the dynamic equations can be interpreted as a sum of the forces due to the kinetic and potential energy presently within the system. If $F_i = 0$, it means joint i was not driven by the motor and does not move. If $F_i \neq 0$, the movement is modified by the joint actuators. Given the dynamics of a prescribed movement, it is possible to solve for the necessary actuator adjustments.

7.2 The manipulator dynamics equation

The dynamics equation for any manipulator described by a set of A transformation matrices is now derived. This is performed in five steps:

- Compute the velocity of any point in any link
- Compute the kinetic energy: $K = \frac{1}{2}mv^2$
- Compute the potential energy: $P = mgh$
- Form the Lagrangian function: $L = K - P$

- Find the dynamics equation: $F_i = \dfrac{d}{dt}\left(\dfrac{\delta L}{\delta \dot{q}_i}\right) - \dfrac{\delta L}{\delta q_i}$

The velocity of a point on the manipulator

Given a point $^i\bar{r}$ described with respect to link i, its position in base coordinates is:

$$^0\bar{r} = {}^0T_i{}^i\bar{r} = ({}^0A_1{}^1A_2 \ldots {}^{i-1}A_i)^i\bar{r} \tag{7.3}$$

Its velocity is:

$$\dot{r} = \left(\sum_{j=1}^{i} \left(\frac{\delta T_i}{\delta q_j} \dot{q}_j \right) \right) {}^{i}\bar{r}$$

$$\dot{r} \cdot \dot{r} = \text{Trace} \, (\dot{r}\, \dot{r}^{T})$$

$$\left(\frac{dr}{dt} \right)^2 = \text{Trace} \left[\left(\sum_{j=1}^{i} \left(\frac{\delta T_i}{\delta q_j} \right) \dot{q}_j \,{}^{i}\bar{r} \right) \sum_{k=1}^{i} \left[\left(\frac{\delta T_i}{\delta q_k} \right) \dot{q}_k \,{}^{i}\bar{r} \right]^{T} \right]$$

$$= \text{Trace} \left[\sum_{j=1}^{i} \sum_{k=1}^{i} \frac{\delta T_i}{\delta q_j} \,{}^{i}\bar{r}({}^{i}\bar{r})^{T} \left(\frac{\delta T_i}{\delta q_k} \right)^{T} \dot{q}_j \dot{q}_k \right] \qquad (7.4)$$

The kinetic energy

The kinetic energy of a particle of mass dm located on link i at ${}^{i}\bar{r}$, is:

$$dK_i = \tfrac{1}{2} \, \text{Trace} \left[\sum_{j=1}^{i} \sum_{k=1}^{i} \frac{\delta T_i}{\delta q_j} \,{}^{i}\bar{r}({}^{i}\bar{r})^{T} \left(\frac{\delta T_i}{\delta q_k} \right)^{T} \dot{q}_j \dot{q}_k \right] dm$$

$$= \tfrac{1}{2} \, \text{Trace} \left[\sum_{j=1}^{i} \sum_{k=1}^{i} \frac{\delta T_i}{\delta q_j} \,{}^{i}\bar{r}\, dm({}^{i}\bar{r})^{T} \left(\frac{\delta T_i}{\delta q_k} \right)^{T} \dot{q}_j \dot{q}_k \right] \qquad (7.5)$$

The kinetic energy of link i is then:

$$K_i = \int_{\text{link } i} dK_i = \tfrac{1}{2} \, \text{Trace} \left[\sum_{j=1}^{i} \sum_{k=1}^{i} \frac{\delta T_i}{\delta q_j} \, ({}^{i}\bar{r}\bar{r}^{T} dm) \left(\frac{\delta T_i}{\delta q_k} \right)^{T} \dot{q}_j \dot{q}_k \right] \qquad (7.6)$$

The integral $\int {}^{i}\bar{r}{}^{i}\bar{r}^{T} dm$ is known as the inertia matrix J_i and is given by:

$$J_i = \int {}^{i}\bar{r}{}^{i}\bar{r}^{T} dm \qquad (7.7)$$

$$= \begin{bmatrix} \int x^2 dm & \int xy dm & \int xz dm & \int x dm \\ \int xy dm & \int y^2 dm & \int yz dm & \int y dm \\ \int xz dm & \int yz dm & \int z^2 dm & \int z dm \\ \int x dm & \int y dm & \int z dm & \int dm \end{bmatrix}_{\text{link } i} \qquad (7.8)$$

Recalling that the moments of inertia, cross products of inertia, and first moments of a body are defined as:

$$I_{xx} = \int (y^2 + z^2)dm$$

$$I_{yy} = \int (x^2 + z^2)dm$$

$$I_{zz} = \int (x^2 + y^2)dm$$

$$I_{xy} = \int xydm$$

$$I_{xz} = \int xzdm$$

$$I_{yz} = \int yzdm \tag{7.9}$$

$$\int xdm = m\bar{x}, \int ydm = m\bar{y}, \int zdm = m\bar{z} \tag{7.10}$$

and that,

$$\int x^2 dm = -\tfrac{1}{2}\int (y^2+z^2)dm + \tfrac{1}{2}\int (x^2+z^2)dm + \tfrac{1}{2}\int (x^2+y^2)dm \tag{7.11}$$

$$= \tfrac{1}{2}(-I_{xx}+I_{yy}+I_{zz})$$

$$\int y^2 dm = \tfrac{1}{2}\int (y^2+z^2)dm - \tfrac{1}{2}\int (x^2+z^2)dm + \tfrac{1}{2}\int (x^2+y^2)dm$$

$$= \tfrac{1}{2}(I_{xx} - I_{yy} + I_{zz})$$

$$\int z^2 dm = \tfrac{1}{2}\int (y^2+z^2)dm + \tfrac{1}{2}\int (x^2+z^2)dm - \tfrac{1}{2}\int (x^2+y^2)dm$$

$$= \tfrac{1}{2}(I_{xx} + I_{yy} - I_{zz})$$

Thus J_i can be expressed as:

$$J_i = \begin{bmatrix} \dfrac{-I_{xx}+I_{yy}+I_{zz}}{2} & I_{xy} & I_{xz} & m\bar{x} \\[2mm] I_{xy} & \dfrac{I_{xx}-I_{yy}+I_{zz}}{2} & I_{yz} & m\bar{y} \\[2mm] I_{xz} & I_{yz} & \dfrac{I_{xx}+I_{yy}-I_{zz}}{2} & m\bar{z} \\[2mm] m\bar{x} & m\bar{y} & m\bar{z} & m \end{bmatrix}$$

link i

giving a 4×4 symmetric matrix. Rewriting Eqn. (7.2), the total kinetic energy of the manipulator is then:

$$K = \sum_{i=1}^{6} K_i = \tfrac{1}{2} \sum_{i=1}^{6} \text{Trace} \left[\sum_{j=1}^{6} \sum_{k=1}^{6} \frac{\delta T_i}{\delta q_j} J_i \left(\frac{\delta T_i}{\delta q_k} \right)^T \dot{q}_j \dot{q}_k \right]$$

$$K = \tfrac{1}{2} \sum_{i=1}^{6} \sum_{j=1}^{i} \sum_{k=1}^{i} \text{Trace}\left[\frac{\delta T_i}{\delta q_j} \, J_i \left(\frac{\delta T_i}{\delta q_k} \right)^{\text{T}} \right] \dot{q}_j \dot{q}_k \qquad (7.12)$$

The potential energy

The potential energy of an object of mass m at a height h above some zero reference location in a gravity field g is $P = mgh$. If the acceleration due to gravity is expressed by a vector $\bar{g}$ and the position of the centre of mass of the object by a vector $\bar{r}$ (the centre of mass), then $P_i = -m\bar{g}\bar{r}$. The potential energy of a link whose centre of mass is described by a vector $\bar{r}_i$ with respect to link i coordinate frame T_i is then $P_i = -m_i\bar{g}^{\text{T}}T_i^{i}\bar{r}_i$, where $\bar{g}^{\text{T}} = (g_x, g_y, g_z, 0)$. The total potential energy of the manipulator is:

$$P = - \sum_{i=1}^{6} m_i\bar{g}^{\text{T}}(T_i^{i}\bar{r}_i) \qquad (7.13)$$

The Lagrangian

Forming the Lagrangian $L = K - P$ from Eqns. (7.12) and (7.13) we have,

$$L = \tfrac{1}{2} \sum_{i=1}^{6} \sum_{j=1}^{i} \sum_{k=1}^{i} \text{Trace}\left(\frac{\delta T_i}{\delta q_j} \, J_i \left(\frac{\delta T_i}{\delta q_k} \right)^{\text{T}} \right) \dot{q}_j \dot{q}_k$$

$$+ \sum_{i=1}^{6} m_i\bar{g}^{\text{T}} (T_i^{i}\bar{r}i) \qquad (7.14)$$

The dynamics equations

$$\frac{d}{dt}\left(\frac{\delta L}{\delta \dot{q}_i} \right) - \frac{\delta L}{\delta q_i} = F_i$$

$$\frac{\delta L}{\delta \dot{q}_p} = \sum_{i=1}^{6} \sum_{k=1}^{i} \text{Trace}\left(\frac{\delta T_i}{\delta q_p} \, J_i \left(\frac{\delta T_i}{\delta q_k} \right)^{\text{T}} \right) \dot{q}_k$$

$$\frac{d}{dt}\left(\frac{\delta L}{\delta q_p} \right) = \sum_{i=p}^{6} \sum_{k=1}^{i} \text{Trace}\left(\frac{\delta T_i}{\delta q_k} \, J_i \left(\frac{\delta T_i}{\delta q_p} \right)^{\text{T}} \right) \ddot{q}_k$$

$$+ \sum_{i=p}^{6} \sum_{k=1}^{i} \sum_{m=1}^{i} \text{Trace}\left(\frac{\delta^2 T_i}{\delta q_k \, q_m} \, J_i \left(\frac{\delta T_i}{\delta q_p} \right)^{\text{T}} \right) \dot{q}_k \dot{q}_m$$

$$+ \sum_{i=p}^{6} \sum_{k=1}^{i} \sum_{m=1}^{i} \text{Trace}\left(\frac{\delta^2 T_i}{\delta q_p \delta q_m} \, J_i \left(\frac{\delta T_i}{\delta q_k} \right)^{\text{T}} \right) \dot{q}_k \dot{q}_m \qquad (7.15)$$

$$\frac{\delta L}{\delta q_p} = \sum_{i=p}^{6} \sum_{j=1}^{i} \sum_{k=1}^{i} \text{Trace}\left(\frac{\delta^2 T_i}{\delta q_k \delta q_j} J_i \left(\frac{\delta T_i}{\delta q_k}\right)^{\text{T}}\right)\dot{q}_j \dot{q}_k$$

$$+ \sum_{i=p}^{6} m_i g^{\text{T}}\left(\frac{\delta T_i}{\delta q_p}\, {}^i\bar{r}_i\right) \tag{7.16}$$

Using Eqns. (7.15) and (7.16) we have,

$$F_i = \frac{d}{dt}\left(\frac{\delta L}{\delta \dot{q}_p}\right) - \frac{\delta L}{\delta q_p} = \sum_{j=1}^{6} \sum_{k=1}^{j} \text{Trace}\left(\frac{\delta T_j}{\delta q_k} J_j \left(\frac{\delta T_j}{\delta q_i}\right)^{\text{T}}\right)\ddot{q}_k$$

$$+ \sum_{j=1}^{6} \sum_{k=1}^{j} \sum_{m=1}^{j} \text{Trace}\left(\frac{\delta^2 T_j}{\delta q_k \delta q_m} J_j \left(\frac{\delta T_j}{\delta q_i}\right)^{\text{T}}\right)\dot{q}_k \dot{q}_m$$

$$- \sum_{j=1}^{6} m_j g^{\text{T}}\left(\frac{\delta T_j}{\delta q_i}\, {}^j\bar{r}_j\right) \tag{7.17}$$

These equations are independent of order of summation, so Eqn. (7.17) may be written as:

$$F_i = \sum_{j=1}^{6} D_{ij}\ddot{q}_j + \sum_{j=1}^{6} \sum_{k=1}^{6} D_{ijk}\dot{q}_j \dot{q}_k + D_i \tag{7.18}$$

where

$$D_{ij} = \sum_{p=\max i,j}^{6} \text{Trace}\left(\frac{\delta T_p}{\delta q_j} J_p \left(\frac{\delta T_p}{\delta q_i}\right)^{\text{T}}\right)$$

$$D_{ijk} = \sum_{p=\max i,j,k}^{6} \text{Trace}\left(\frac{\delta^2 T_p}{\delta q_j \delta q_k} J_p \left(\frac{\delta T_p}{\delta q_i}\right)\right)$$

$$D_i = -\sum_{p=i}^{6} m_p g^{\text{T}}\left(\frac{\delta T_p}{\delta q_i}\, {}^p\bar{r}_p\right)$$

$$i = 1, 2, 3, \ldots, 6$$

D_{ii} represents the effective inertia at joint i
D_{ij} represents coupling inertia between joint i and joint j
D_{ijj} represents centripetal forces at joint i due to velocity at joint j
D_{ijk} represents Coriolis forces at joint i due to velocities at joints j and k
D_i represents the gravity loading at joint i

The inertial terms and the gravity terms are particularly important in manipulator control as they affect the servo stability and positioning accuracy. The centripetal and Coriolis forces are important only when the manipulator is moving at a high speed, at which the introduced error in ignoring them is small.

7.3 Robotic arm control

Given the motion equations of a manipulator, the purpose of robotic arm control is to maintain a prescribed motion for the arm along a desired trajectory. This can be accomplished by applying corrective compensation torques to the actuators to adjust for any deviations of the arm from the trajectory.

Current industrial practice is to use conventional servomechanisms to control manipulators. However, the motion dynamics of an n degree-of-freedom manipulator are inherently nonlinear. The dynamics can be described only by a set of n highly coupled, nonlinear, second-order ordinary differential equations. The nonlinearities arise from the inertial loading, the coupling between neighbouring joints, and the gravitational loading of the links. Furthermore, the dynamic parameters of a manipulator vary with the position of the joint variables, which are themselves related by complex trigonometric transformations. Hence, the nonlinearity and complexity of the dynamic manipulator model, causes a closed-form solution of the optimal control to be very difficult, if not impossible.

The servomechanism usage models the varying dynamics of a manipulator inadequately as well as neglecting the coupling effects of the joints. As a result, manipulators move at slow speeds with unnecessary vibrations, making them appropriate only for limited-precision tasks.

Near-minimum-time control is based on the linearisation of the equations of motion about the nominal trajectory, linear feedback and/or suboptimal control laws. The solution is obtained analytically. This control method is still too complex to be useful for manipulators with four or more degrees of freedom while neglecting the effects of unknown external loads.

7.4 Nonlinear state equation of dynamic model of mechanical arm

Kahn and Roth [4] attempted to solve the dynamic control equation of the robot manipulator. Here, we establish the control model of robot manipulator to depict the complexity of its nonlinearity. The dynamics of a six-joint manipulator are represented by Eqn. (7.18).

$$\sum_{j=1}^{6} D_{ij}\ddot{q}_j + \sum_{j=1}^{6}\sum_{k=1}^{6} D_{ijk}\dot{q}_j\dot{q}_k + D_i = F_i$$

where

$$D_{ij} = \sum_{p=\max i,j}^{6} \text{Trace}\left(\frac{\delta T_p}{\delta q_j} J_p \left(\frac{\delta T_p}{\delta q_i}\right)^T\right)$$

$$D_{ijk} = \sum_{p=\max i,j,k}^{6} \text{Trace}\left(\frac{\delta^2 T_p}{\delta q_i \delta q_k} J_p \left(\frac{\delta T_p}{\delta q_i}\right)^{\text{T}}\right)$$

$$D_i = - \sum_{p=i}^{6} m_p g^{\text{T}} \left(\frac{\delta T_p}{\delta q_i} {}^{p}\bar{r}_p\right)$$

The above mentioned dynamics equations are nonlinear, second-order, ordinary differential equations.

Now, to the example of the PUMA manipulator,

$$T_3 = \begin{bmatrix} C_1C_{23} & -S_1 & C_1S_{23} & a_2C_1C_2-S_1d_2 \\ S_1C_{23} & C_1 & S_1S_{23} & a_2S_1C_2+C_1d_2 \\ -S_{23} & 0 & C_{23} & -a_2S_2 \\ 0 & 0 & 0 & 1 \end{bmatrix}$$

Consider,

$$D_{ij} = \sum_{p=\max i,j}^{6} \text{Trace}\left(\frac{\delta T_p}{\delta q_j} J_p \left(\frac{\delta T_p}{\delta q_i}\right)^{\text{T}}\right)$$

For $p = 3$,

$$\frac{\delta T_3}{\delta q_1} = \frac{\delta T_3}{\delta \theta_1} = \frac{\delta}{\delta \theta_1} \begin{bmatrix} C_1C_{23} & -S_1 & C_1S_{23} & a_2C_1C_2-S_1d_2 \\ S_1C_{23} & C_1 & S_1S_{23} & a_2S_1C_2+C_1d_2 \\ -S_{23} & 0 & C_{23} & -a_2S_2 \\ 0 & 0 & 0 & 1 \end{bmatrix}$$

$$= \begin{bmatrix} -S_1C_{23} & -C_1 & -S_1S_{23} & -a_2S_1C_2-C_1d_2 \\ C_1C_{23} & -S_1 & C_1S_{23} & a_2C_1D_2-S_1d_2 \\ 0 & 0 & 0 & 0 \\ 0 & 0 & 0 & 0 \end{bmatrix}$$

$\frac{\delta T_3}{\delta \theta_1}$ is nonlinear in terms of $\cos\theta$ and $\sin\theta$.

So,

$$\sum_{j=1}^{6} D_{ij} \ddot{q}_j + \sum_{j=1}^{6} \sum_{k=1}^{6} D_{ijk} \dot{q}_j \dot{q}_k + D_i = F_i$$

is nonlinear in terms of cosine and sine functions, i.e. D_{ij}, D_{ijk} and D_i.

Now, write

$$\sum_{j=1}^{6} D_{ij}\ddot{q}_j + \sum_{j=1}^{6}\sum_{k=1}^{6} D_{ijk}\dot{q}_j\dot{q}_k + D_i = F_i \qquad i = 1, 2, 3, 4, 5, 6$$

in matrix form. We obtain for the PUMA:

$$(q_1 = \theta_1, q_2 = \theta_2, \ldots, q_6 = \theta_6)$$

$$
\begin{bmatrix}
D_{11} & D_{12} & D_{13} & D_{14} & D_{15} & D_{16} \\
D_{21} & D_{22} & D_{23} & D_{24} & D_{25} & D_{26} \\
\cdot \\
\cdot \\
\cdot \\
D_{61} & D_{62} & D_{63} & D_{64} & D_{65} & D_{66}
\end{bmatrix}
\begin{bmatrix}
\ddot{\theta}_1 \\
\ddot{\theta}_2 \\
\cdot \\
\cdot \\
\cdot \\
\ddot{\theta}_6
\end{bmatrix}
$$

$$\qquad\qquad\qquad (a) \qquad\qquad\qquad\qquad (b)$$

$$= [\dot{\theta}_1\ \dot{\theta}_2\ \dot{\theta}_3\ \dot{\theta}_4\ \dot{\theta}_5\ \dot{\theta}_6]\,[-D_{ijk}]
\begin{bmatrix}
\dot{\theta}_1 \\
\dot{\theta}_2 \\
\dot{\theta}_3 \\
\dot{\theta}_4 \\
\dot{\theta}_5 \\
\dot{\theta}_6
\end{bmatrix}
-
\begin{bmatrix}
D_1 \\
D_2 \\
D_3 \\
D_4 \\
D_5 \\
D_6
\end{bmatrix}
+
\begin{bmatrix}
F_1 \\
F_2 \\
F_3 \\
F_4 \\
F_5 \\
F_6
\end{bmatrix}
$$

$$\qquad\qquad 6 \times 6$$
$$\qquad\quad i = 1,2,3,4,5,6$$

$$\qquad (c) \qquad\qquad\qquad\qquad (d) \qquad (e)$$

(a) will be represented by $D(\theta), 6\times6$
(b) will be represented by $\ddot{\theta}, 6\times1$
(c) will be represented by $Q(\theta,\dot{\theta}), 6\times1$
(d) will be represented by $G(\theta), 6\times1$
(e) will be represented by $U(t), 6\times1$

So we have,

$$D(\theta)\ddot{\theta} = Q(\theta,\dot{\theta}) + G(\theta) + U(t)$$

assuming $D^{-1}(\theta)$ exists, we have

$$\ddot{\theta} = D^{-1}(\theta)\{Q(\theta,\dot{\theta}) + G(\theta)\} + D^{-1}(\theta)U(t)$$
$$\theta = (\theta_1\ \theta_2\ \theta_3\ \theta_4\ \theta_5\ \theta_6)$$

Defining the vector $\bar{X}$ by:

$$\bar{X}^T = (\theta_1\ \theta_2\ \theta_3\ \theta_4\ \theta_5\ \theta_6\ \dot{\theta}_1\ \dot{\theta}_2\ \dot{\theta}_3\ \dot{\theta}_4\ \dot{\theta}_5\ \dot{\theta}_6)$$
$$= (x_1\ x_2\ x_3\ x_4\ x_5\ x_6\ x_7\ x_8\ x_9\ x_{10}\ x_{11}\ x_{12})$$

the following state equations are obtained:

$$
\begin{bmatrix} \dot{x}_1 \\ \dot{x}_2 \\ \dot{x}_3 \\ \dot{x}_4 \\ \dot{x}_5 \\ \dot{x}_6 \end{bmatrix} = \begin{bmatrix} 0\ 0\ 0\ 0\ 0\ 0 &\vdots& 1\ 0\ 0\ 0\ 0\ 0 \\ 0\ 0\ 0\ 0\ 0\ 0 &\vdots& 0\ 1\ 0\ 0\ 0\ 0 \\ 0\ 0\ 0\ 0\ 0\ 0 &\vdots& 0\ 0\ 1\ 0\ 0\ 0 \\ 0\ 0\ 0\ 0\ 0\ 0 &\vdots& 0\ 0\ 0\ 1\ 0\ 0 \\ 0\ 0\ 0\ 0\ 0\ 0 &\vdots& 0\ 0\ 0\ 0\ 1\ 0 \\ 0\ 0\ 0\ 0\ 0\ 0 &\vdots& 0\ 0\ 0\ 0\ 0\ 1 \end{bmatrix} \begin{bmatrix} x_1 \\ x_2 \\ x_3 \\ x_4 \\ x_5 \\ x_6 \\ x_7 \\ x_8 \\ x_9 \\ x_{10} \\ x_{11} \\ x_{12} \end{bmatrix} \tag{7.19}
$$

6×1 6×12 12×1

$$
\begin{bmatrix} \dot{x}_7 \\ \dot{x}_8 \\ \dot{x}_9 \\ \dot{x}_{10} \\ \dot{x}_{11} \\ \dot{x}_{12} \end{bmatrix} = D^{-1}(\theta)\{Q(\theta,\dot{\theta} + G(\theta)\} + D^{-1}(\theta)U(t)
$$

            6×6 6×1 6×1 6×6 6×1

6×1

$$
= D^{-1}(x)N(x) + D^{-1}(x)U(t) \tag{7.20}
$$

     6×6 6×1 6×6 6×1

        6×1 6×1

Let $N_1(x) = D^{-1}(\theta)\{Q(\theta,\dot{\theta}) + G(\theta)\} = D^{-1}(x)N(x)$

     $N_1(x) = D^{-1}(x)N(x)$

     $U_1(x,t) = D^{-1}(x)U(t)$

Rewriting Eqns. (7.19) and (7.20), we obtain:

$$
\begin{bmatrix} \dot{x}_1 \\ \dot{x}_2 \\ \dot{x}_3 \\ \dot{x}_4 \\ \dot{x}_5 \\ \dot{x}_6 \\ \dot{x}_7 \\ \dot{x}_8 \\ \dot{x}_9 \\ \dot{x}_{10} \\ \dot{x}_{11} \\ \dot{x}_{12} \end{bmatrix}
=
\begin{bmatrix}
& & \begin{smallmatrix} 1&0&0&0&0&0 \\ 0&1&0&0&0&0 \\ 0&0&1&0&0&0 \\ 0&0&0&1&0&0 \\ 0&0&0&0&1&0 \\ 0&0&0&0&0&1 \end{smallmatrix} \\[2em]
\bigcirc\ 6\times6 & & \\[1em]
& \bigcirc\ 6\times6 & \bigcirc\ 6\times6
\end{bmatrix}_{12\times12}
\begin{bmatrix} x_1 \\ x_2 \\ x_3 \\ x_4 \\ x_5 \\ x_6 \\ x_7 \\ x_8 \\ x_9 \\ x_{10} \\ x_{11} \\ x_{12} \end{bmatrix}_{12\times1}
+
\begin{bmatrix}
\bigcirc\ 6\times6 \\[2em]
\begin{smallmatrix} 1&0&0&0&0&0 \\ 0&1&0&0&0&0 \\ 0&0&1&0&0&0 \\ 0&0&0&1&0&0 \\ 0&0&0&0&1&0 \\ 0&0&0&0&0&1 \end{smallmatrix}
\end{bmatrix}_{12\times6}
$$

$$
\begin{matrix} N_1(x) \\ 6\times1 \end{matrix}
+
\begin{bmatrix}
\bigcirc\ 6\times6 \\[2em]
\begin{smallmatrix} 1&0&0&0&0&0 \\ 0&1&0&0&0&0 \\ 0&0&1&0&0&0 \\ 0&0&0&1&0&0 \\ 0&0&0&0&1&0 \\ 0&0&0&0&0&1 \end{smallmatrix}
\end{bmatrix}_{12\times6}
\begin{matrix} U_1(x,t) \\ 6\times1 \end{matrix}
$$

or

$$\bar{X}(t) = FX(t) + BN_1(x) + BU_1(x,t) \tag{7.21}$$

where,

$$
F = \begin{bmatrix} \bigcirc\ 6\times6 & I\ 6\times6 \\ \bigcirc\ 6\times6 & \bigcirc\ 6\times6 \end{bmatrix}_{12\times12}
\quad \text{and } B = \begin{bmatrix} \bigcirc\ 6\times6 \\ I\ 6\times6 \end{bmatrix}_{12\times6}
$$

usually in robot manipulator case, the output Y is defined to be:

$$Y = C \begin{bmatrix} x_1 \\ x_2 \\ x_3 \\ x_4 \\ x_5 \\ x_6 \end{bmatrix} \tag{7.22}$$

in the most case, $C = I$.

References

[1] Bejczy, A. K. 1974. *Robot Arm Dynamics and Control.* NASA – JPL Technical Memo 33–669.
[2] Hollerbach, J. M. 1980. A recursive Lagrangian formulation of manipulator dynamics and a comparative study of dynamics formulation. *IEEE Trans. Systems, Man. and Cybernetics,* SMC–10 (11): 730–736.
[3] Symon, K. R. 1961. *Mechanics.* Addison-Wesley, New York.
[4] Kahn, M. and Roth, B. 1971. The near-minimum-time control of open loop kinematic chains. *Trans. ASME,* Series G93: 164–172.

PART II

APPLICATION ORIENTATED DISCUSSION OF PROGRAMMING, TOOLING AND ROBOTISED SYSTEM DESIGN

Chapter Eight

ROBOT PROGRAMMING, LANGUAGES AND SAMPLE PROGRAMS

WHEN programing industrial robots, or other computer controlled machines, it is important to always think in terms of a system, consisting of several different subsystems in which the robot itself is only one component.

Generally such a system consists of the robot arm itself, with its controller, the necessary grippers and/or robot tools, and the part feeding, locating, orientating and other mechanical, electronic and sensory based (vision, force and torque sensing, etc.) devices which are interfaced with it.

If the robot cell is integrated into a larger system (e.g. a flexible assembly system, a robotised welding line, etc.) a further important feature is the necessary provision of interfaces and communication facilities between the robot controller and the 'outside world.' It is important to realise that distributed, sensory feedback processing makes robots – and other devices – more intelligent, more reliable and more flexible.

Sensors (devices capable of obtaining information from the physical world) have been the focus of much development in recent years because robots must be able to provide information not only about their arm position, but also about the part being handled, about the status of the gripper, or in more complex cases about the dynamically changing parameters of the environment. Sensors and their signal processors make the robot capable of seeing, hearing, smelling, detecting and analysing force, torque, heat, pressure, colour and other environmental changes or conditions. Simple touch-triggered and complex non-contact sensors can provide information about the presence of a part in a buffer store, in the robot tool, or in the chuck of a lathe, and they can, for example, interact with the robot controller and modify a preprogrammed sequence in real-time.

For these obvious reasons sensory feedback data must be handled from the robot program and is a very important robot control and software problem to be solved in the future. Most, if not all, robots can handle the simplest sensors (e.g. proximity switches, light sensors, inductive sensors, capacitive sensors, etc.) which send contact signals to be analysed by the robot controller and program via the input/output (I/O) port. The hardware and software processing aspects of more sophisticated sensors, such as vision systems, force and torque sensors, etc. are proving to be more complex, but still essential in terms of reliability and easy programming of the robot.

Having discussed the robot arm model and some control problems in Chapters Two to Seven, this chapter concentrates on some of the most important robot programming methods and languages [1]. The main tasks of the robot system software, the computer control system and related interfacing tasks are also explained. Examples are also given of robot programs and related applications.

8.1 Robot programming methods

There are several different ways of programming industrial manipulators and robots. Since most current robot controllers are not intelligent enough, or in other words do not have the ability to use information to modify system behaviour in a preprogrammed and/or real-time sensed way, the selected robot programming method depends in most cases on the limits of its control system.

Open-loop, or non-servo controlled robots move their arms between exact end-positions on each axis, allowing 'simple sequence' control or 'limit switch' control, whereas closed-loop, or servo controlled machines incorporate feedback devices in each joint, which continuously measure the actual position of the arm, allowing 'point-to-point' control and 'continuous path' control.

Each arm of the sequence controlled manipulator or robot moves between two end-positions, A and B, without offering the controller any position feedback data between points A and B. In most cases these robots are operated pneumatically, hydraulically or by means of stepper motors.

Limit-switch control allows more than two points, i.e. A and B to be set along one axis, by using indexable stops inserted or withdrawn automatically by the controller whilst the arm is in motion between points A and B.

Closed-loop control systems can vary greatly. The point-to-point control system moves each joint by an independent position servo. Thus when moving the arm from A to B, all joints move independently, until the robot reaches the desired position (i.e. B). This motion is not limited to one arm. Point-to-point controlled robots can move in 3D, but each joint moves separately.

Continuous path control is used when the path of the end-effector is of primary importance, rather than just its position. In this case each joint control variable is interpolated from its initial value (i.e. from position A where the motion should start) to its desired end-position (i.e. B). During the motion each joint is moved simultaneously the optimum amount required to complete a smooth path.

Robot programming methods may be classified as follows:

- Physical set-up programming.
- Teach-mode programming.
- Teach-by-showing programming.
- Off-line programming.

Physical set-up programming

The physical set-up, in which the operator sets-up programs by fixing limit switches, stops, etc. In most cases these robots (or rather, reprogrammable point-to-point controlled manipulators) are adequate only for simple pick-and-place jobs.

Teach-mode programming

'Teach-mode' is when the operator leads the robot through the desired locations by means of a teach pendant, or teach box, or by means of a voice input device which is interfaced with the robot controller. The taught locations and the approach vectors are recorded in digital format and are used as coordinate values and approach angles to generate the trajectory for the robot during operation.

Teaching a point-to-point controlled robot means moving each joint of the device individually until the combination of these individual arm positioning efforts provide the desired end-effector position. When the desired point has been reached the position values of each joint are stored in the memory of the controller.

Teaching continuous-path controlled robots is much easier, since the taught path represents the actual trajectory the robot will follow when the program is executed. For example, it is practically impossible to teach a robot to follow a straight line in three-dimensional space if it requires the movement of more than one joint and the robot is point-to-point controlled. On the other hand, this task proves to be very simple with the continuous-path controlled robot.

The teach method is widely used when the robot has to follow a path which is hard to define mathematically, but simple to teach by a skilled operator, or when the robot must learn how to insert a delicate component into a subassembly, etc.

Teach-by-showing programming

This is the most commonly known method of programming paint spraying, welding and other robots which involve the operator teaching the robot by using a detachable grip handle. Auxiliary commands, such as spray-gun 'on/off', I/O handling, gripper 'in' and 'out', motion speed, etc. can also be programmed from the teach-pendant.

In this case, the robot arm, the detachable grip handle, and/or optical and tactile sensors, are used to digitise the guided path in three dimensions. The digitised points and approach vectors are recorded and the controller remembers the positions and arm orientation through which the robot arm has been led. Data on trajectories, sequences and positions can be entered by this method and simply played back by the controller during operation.

Off-line programming

Off-line programming involves use of a high-level robot programming language. This allows writing and editing programs in a language which is closer to the operator's language than to the machine's. In some cases such programs can also be prepared remote, or off-line from the actual robot, or generated in a remote computer and downloaded into the robot controller via a communications line.

Off-line programming can be based on an explicit programming language, or on a 'world-modelling' (i.e. implicit or 'model-based') language. Explicit programming deals with motion-orientated commands, whereas world-modelling attempts to provide the programmer with a software environment capable of describing a handling task by giving a problem-orientated description only, and letting the robot discover what kind of actual motion and action statements have to be executed.

The major benefit of the latter method is that the objects to be handled have to be described and stored as models in the database once only, and this world-model does not then have to be altered even if the production process changes. It should be emphasised that such systems are currently under research and only partly operational, but they will have a great impact on the assembly industry, in particular.

As a general comment one must realise that as with the high-level, off-line programming languages of NC machines, or as with the programming of CNC machines interactively, none of these robot programming methods provide a universal or an optimum solution in every case, and so combined programming methods must be used.

8.2 Robot system architecture

Although there is a strong development trend towards the concept that most robots are 'computers with hands', unfortunately they have very restricted computing capabilities. It would also be desirable to be able to choose the most suitable robot programming language on the same robot for different applications, as is done when programming computers for different tasks. Unfortunately this is not yet the case.

It is rather unfortunate but important to mention that low-cost home computers costing around US$ 1000 often provide a higher level user-friendliness than industrial robots costing over US$ 25,000. Is there a case here for giving robot designers home computers to learn about programming, program editing and interfacing techniques?

There is a clear trend towards employing powerful mini and micro-computers as part of the robot control system, but an estimated 95% of the currently employed robots in industry are very 'user-unfriendly', and their programming is time consuming particularly to those who have been using computers and CNC machines before programming robots, if the job requires more than just simple pick-and-place work.

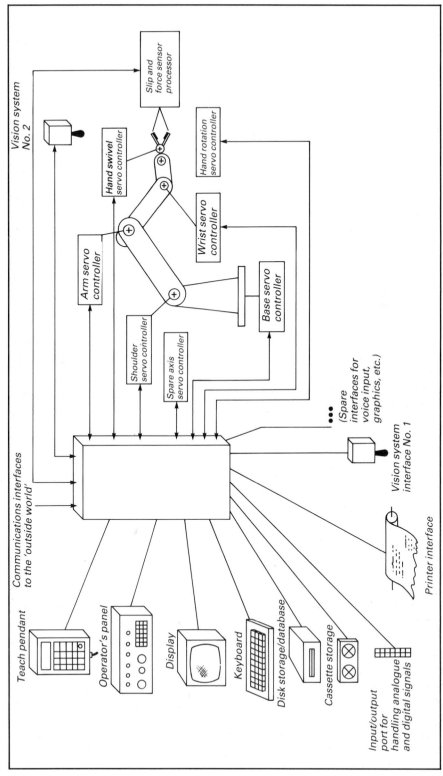

Fig. 8.1 Schematic diagram of a robot controller capable of offering cell supervisory functions as well as controlling the robot arm and the necessary communication tasks

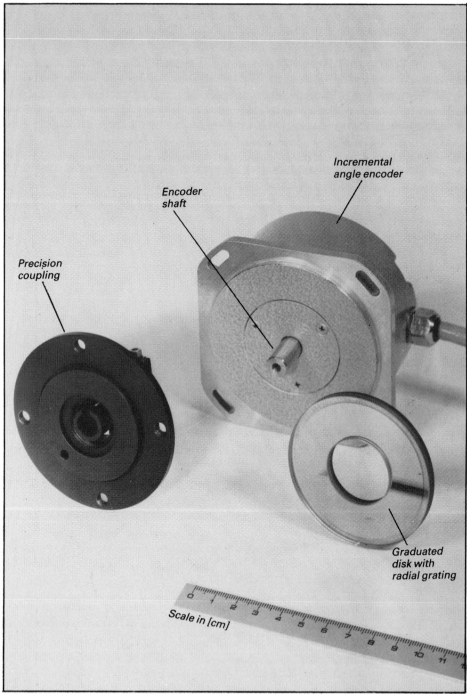

Fig. 8.2 The incremental angular encoder is a widely used measuring system.
(Courtesy of Heidenhain Optics and Electronics Co.)

Robot system control aspects

The robot system controller should in most cases incorporate a powerful mini or microcomputer with sufficient power to execute the system software, and to deal with the control and data communication tasks within the cell as well as with the 'outside world'. This means that the system controller should be able to handle and support:

● Multi-axis control, motion control, acceleration control, dynamic compensation, etc.
● Digital and analogue interfaces.
● Standard communications interfaces (RS-232c, IEEE, etc.) to remote controllers or computers.
● Standard peripherals, such as teach pendant, disk, cassette, printer, display, keyboard, operator's panel, etc.
● Safety devices, interlocks, built-in error detection sensors, leak detection devices in the case of hydraulic robots, power interruption control, extra power sources for sensors requiring power, etc.
● Optionally multiple device and arm control tasks.

The modular system design should also provide different options, relating to the use and interfacing of vision systems, voice input programming systems, graphical input/output peripherals, etc. (This concept is illustrated in Fig. 8.1 explaining the functional relationship between the listed devices.)

Since measuring systems play a very important role in the way the robot is controlled and also reflect on its programming, let us summarise some of their most important aspects.

Displacement in general can be measured in absolute or in incremental modes. The important difference between the two systems is that in the case of the incremental (or relative) measuring system only displacements are measured and each position is determined by the displacement of its predecessor, whereas the absolute system will calculate every motion from a fixed datum.

The measurement principle can be analogue or digital. The simplest method of measuring distances by digital methods is to divide the distance moved into equal small increments.

In robotics the incremental measuring system is of extreme importance. Supposing that the shown scales are scanned by a device called an encoder (Fig. 8.2) and that the signal obtained is amplified and fed into the comparator, the robot will be able to sense its current position as well as being able to measure any movement (and its direction) of the arm or rotation of the joint by adding together the sensed electric pulses.

Feedback devices, such as encoders, resolvers, inductosyns and tachometers are key devices in the robot motion control system. Because of its importance in robotics design and control the incremental angular encoder is discussed in more detail. (Discussions on other types of measuring systems can be found elsewhere [2,3].)

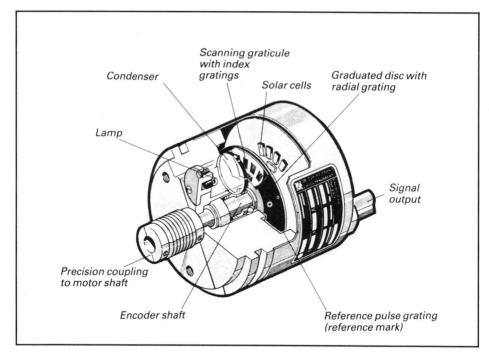

Fig. 8.3 The internal design and the arrangement of components in the incremental angular encoder. (Courtesy of Heidenhain Optics and Electronics Co.)

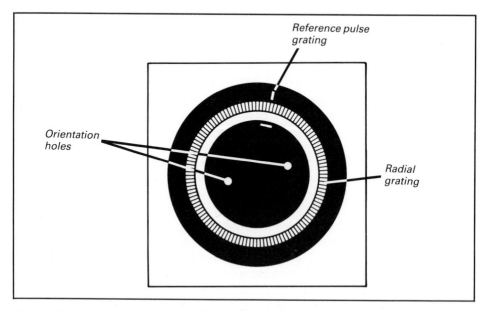

Fig. 8.4 The graduated glass disk of the Heidenhain incremental angular encoder. This extremely thin disk (<0.0001mm) carries the main incremental graduation and the reference pulse or reference mark. The resolution that can be achieved using such a disk is 1000 pulses per degree if quadruple signal evaluation is used (See also Fig. 8.5)

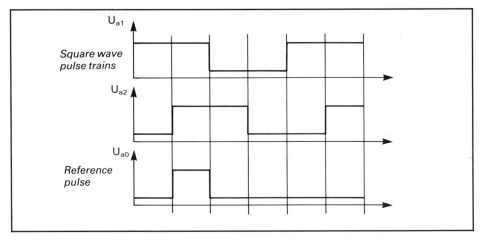

Fig. 8.5 Square wave pulses and the reference pulse in the Heidenhain encoder

The principle of the encoder is that it produces a unique output signal which corresponds to a special pattern put onto a disk or linear scale. In the case of the Heidenhain angular incremental encoder (Fig. 8.3), the resolution of the rotation measurement is determined by the number of lines on the graduated glass disc (Fig. 8.4) mounted on the encoder shaft. (This arrangement is used in robot joints and also in CNC machine tools for measuring angular displacement.) The radial grating is scanned photoelectrically. The light detectors (in this case silicon solar cells) sense the light fluctuations generated by the rotational motion of the graduated glass disc and transform them into electrical signals. The period of the generated signals correspond to the angle of rotation. This encoder provides both square wave signals and their inverted signals as well at a 180° phase shift. This means that when using the reference signal U_{a1} the remaining signal U_{a2} is either behind or preceeds it depending on the direction of the rotation (Fig. 8.5). This status is electronically evaluated in the bidirectional counter of the servo controller.

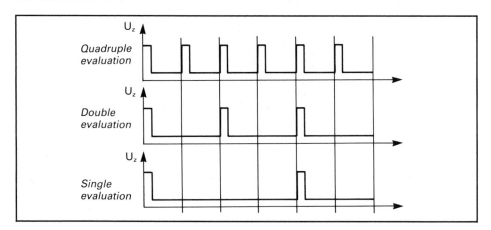

Fig. 8.6 Single, double and quadruple evaluation of encoder signals in the Heidenhain encoder

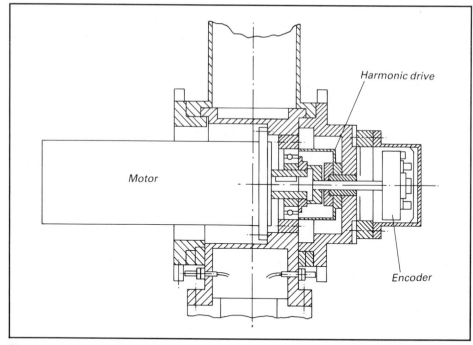

Fig. 8.7 Hitachi robot joint, incorporating the motor, the harmonic drive and the encoder mounted on a common shaft

The resolution of the evaluation can be increased by electronic subdivision, i.e. if not only the rising but also the falling slopes of the counting pulses are utilised in the evaluation. Double evaluation is achieved if the counting step corresponds to half an interval. Further counting signals can be obtained if all slopes of the generated signals are evaluated. This latter case is the quadruple evaluation (Fig. 8.6).

Fig. 8.7 shows a Hitachi robot joint. The way the motor, harmonic drive and angular measurement system are mounted together is indicated, providing a separately controlled module of the arm. Introducing this module also offers the possibility of highlighting the importance of the harmonic drive system in robotics.

The harmonic drive is a speed reducer offering:

● High ratio torque transfer in a single stage (i.e. up to 320:1 with a small size envelope and even higher if units are coupled together).
● High torque transfer (340–6000 Nm).
● Light, small and simple design.
● Operating efficiency as high as 90% since the torque transfer is established by a single low-friction ballrace, eliminating the compound losses arising from multistage gears, shafts, seals and bearings.
● Driving the joint in both directions, so loads may be held in place by braking the input.
● Low backlash, i.e. less than one minute of arc, if required.
● Coaxial input and output shafts, thus simplifying the design and reducing the space requirements.

The harmonic drive system consists of three components: the wave generator, the flexspline, and the circular spline. The wave generator is a thin ballrace fitted onto an elliptical former, and it serves as a high-efficiency torque converter. (Figs. 8.8 and 8.9). The flexspline is a flexible steel can with external teeth fitted over the wave generator and elastically deformed by it into an elliptical shape. The circular spline is an internally toothed solid steel ring engaging the teeth of the flexspline across the major axis of the wave generator. There is a difference of two teeth between the flexspline and the circular spline, so each revolution of the wave generator will move the engagement point by a length of two teeth, generating a relative motion of that amount. At any one time only about 15% of the total number of teeth are engaged at the major axis of the ellipse and the two sets (i.e. the teeth on the wave generator and the flexspline) are of equal pitch.

The harmonic drive operates as follows: as soon as the wave generator starts to rotate, the zone of tooth engagement travels along the same direction as with the major elliptical axis (Fig. 8.9(b)). Due to this travel at 90° of rotation the difference will be exactly 0.5 teeth. When the wave generator has turned through 180° the flexspline has regressed by one tooth relative to the circular spline (Fig. 8.9(c)). After a 360° turn the difference will be two teeth (Fig. 8.9(d)).

System software

The system software requirements cannot be separated from the hardware; most up-to-date system architectures rely equally on both. The system software consists of:

● The operating system.
● The manipulator specific modules.
● The programming language interpreter or compiler.

The operating system module is a set of programs integrated with the control hardware aiming the coordination of the actions taking place inside

Fig. 8.8 Components of the harmonic drive: 1, the wave generator (input); 2, the flexspline (output); and 3, the circular spline (fixed)

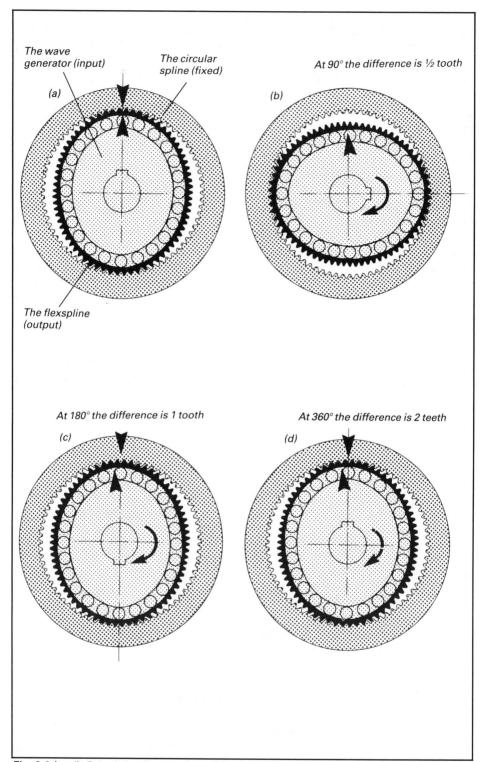

Fig. 8.9 (a–d) Principle of operation of the harmonic drive

the controller, as well as providing a communications link to the outside world. Its tasks are similar to the function of computer operating systems, which include handling the main memory, transferring and receiving data to and from the input/output units, peripherals, sensory input devices and responding to other communication requests. The operating system of the robot controller should be portable, fast enough to react to signals occurring in real-time and be capable of expansion to serve the needs of sophisticated users.

The manipulator specific modules include the software which performs the different coordinate transformations, translates the application specific commands for the manipulator hardware, provides the trajectory generation, takes care of kinetic and kinematic limits, and handles force feedback, speed control, vision and other sensory input and output data and I/O errors at a machine orientated level.

To provide a properly maintainable, self-documented and structured code, the language used to write this module should be Pascal, C, or other suitable commercially available high-level structured languages supporting real-time applications. Machine code or assembler should be used to write only those modules which cannot be implemented in structured real-time languages because of the high speed required.

The robot language is the software interface, by which the robot programmer can direct the manipulator to perform the desired actions. It is essential to provide a language which is user-friendly, provides simple editing facilities and can be extended using macros or subroutines for solving application specific tasks.

In robot applications an interpreter is often preferred to a compiler when writing and debugging the code, because it is easier to edit and partially execute interpreted rather than compiled code. On the other hand it is much more difficult to implement structured programming concepts with an interpreter. The interpreter executes the code as it is typed in, whereas the compiler translates the code through several passes before it generates executable code. This process often takes several minutes but the compiled code is faster when executed.

The programmability of safety signal handling should also be taken into account at this level. It is often overlooked that if a computer program aborts sometimes nothing happens, but if a robot program does the same, machines can be damaged or people can be hurt. Thus when programming robots, among other things, the programming principle, 'if a program must die, it should die gracefully', should be followed. In other words 'take care of safety aspects' – in most cases it is no good simply switching off the power; abort with a clear error message if possible.

It should be possible to use operating system utilities, such as the friendly user interface, soft keys, directory and file handling, and other user utilities at this level. This module of the software should also support application orientated programs written by the users, so that portable subroutine libraries can be written, databases interfaced with the system, and vision systems and other sensory based devices integrated into an existing software frame.

This 'open-ended' design of the robot software system will ensure that the language will be tailored to each individual application by those who best understand the application.

The performance and intelligence level of robots will rely even further on the robot operating system and the power of its controller in the future. Robots which can learn from experience, which can detect and attempt to repair their own faults, which can find, trace and identify missing objects or necessary components for a particular job, and which can solve problems in panic situations, have a potentially great future in industry, particularly in automated assembly.

8.3 Robot programming languages

This discussion on programming languages focuses on those applications where there is a need for high-level or combined robot programming, as well as for a more intelligent and user-friendly robot programming environment.

Robot programming languages are often classified into the following groups:

● *Computer control hardware level*, being the lowest level of control with the possibility of accessing all control features at their source.

● *Point-to-point languages,* enabling the user to guide the robot through a series of points by means of a manual device which activates the robot and storing them in the controller's memory. (Examples include: Cincinnati's T3 and IBM's Funky.)

● *Simple motion level*, offering the 'assembler' compared to the machine code in robotics, allowing simple branching, subroutine definition and calls, often with parameter passing, etc. (Examples include: EMILY, ANORAD, RPL and SIGLA.)

● *Structured programming level*, whose rapid development demonstrates the efforts of combining the most advanced structured programming languages (ALGOL, PL/1, Pascal, etc.) of the computing industry with the requirements in robotics, or the machine control industry in general. (Examples include AL, AML, and to some extent, HELP, MCL, ROBEX and PAL.) Unfortunately people with no experience of structured computer programming find it difficult to understand the structured robot programming concept. Thus they prefer teaching and other non-structured methods.

● *Model-based or task-orientated languages* are recent developments and are still under research. Compared to the structured programming level they are aiming at an even higher level of robot programming, where collision avoidance and real-time decision making can be provided, by letting the robot learn its environment and react to changes in it. (Examples include AUTOPASS and RAPT.)

A brief summary of some of the most interesting robot programming languages is now given.

WAVE

WAVE, the first language allowing the implementation of relatively complex manipulation algorithms, was developed at the Stanford Artificial Intelligence Laboratory in the years 1970-74.

To write an arm control program in WAVE, macros that expand into sequences of arm control instructions to perform simple tasks such as picking up a component, turning a screw driver, etc. must first be defined. Nesting macros and parameter passing between program segments is allowed. Macros can be defined and edited separately or in sequence until they perform as requested. The execution sequence of the predefined macros is controlled by another user-written macro, containing only the sequence of calls to the previously defined macros. (Note the similarity of writing procedures and the main program in compiled ALGOL or Pascal.)

AL

The AL robot programming language is based on the research and development experience of the WAVE system. The core of the AL system consists of the compiler and run-time systems. AL programs are prepared in a file and then compiled and executed.

All data types were chosen for working with three-dimensional objects which have different orientations, locations and distances. AL supports scalar variables which are represented as real numbers, vectors, which represent entities with both direction and magnitude, e.g. acceleration, displacement, etc. The ROT_ation data type represents either an orientation or a rotation about an axis. Rotations also work on vectors without changing their length. The FRAME data type specifies the position and orientation of objects in the real-world, and finally the TRANS_formation data type is used to transform frames and vectors from one coordinate system to another.

The language follows a block structure and also incorporates force sensing and stiffness control, object modelling, and other sophisticated tasks.

AUTOPASS

IBM's AUTOPASS (1972) is a high-level object orientated compiled language for assembly, which allows commands such as 'place object 1 on object 2'. It provides statements such as 'place', tool statements such as 'operate', fastener statements such as 'rivet', and miscellaneous statements such as 'verify'.

Typically the AUTOPASS programmer plans the positions of the objects to be manipulated, the sequence of operations and the usage of end-effectors. The world model used in this language represents assemblies as a graph structure of the object part, subcomponent, attachment and constraint relationship. AUTOPASS statements deal with real-world objects, tools, fasteners and instructions for positioning objects.

POINTY

POINTY, developed at Milan Polytechnic (1975-77) following the basic AL language concept, implements the advanced teaching-by-guiding philosophy with interactive software support. POINTY was designed with emphasis on providing a supportive tool for programmers who wish to write manipulator programs dealing with object modelling.

In this system the programmer defines the object models, but the system offers help in interacting with the external world, in file handling and program editing, allowing the user to concentrate on the task to be solved by the robot. POINTY is an interpreter which is processing AL language instructions, although some special instructions have been added to handle a more flexible input/output data structure as well as to support an interactive operator interface.

From the user's point of view the system considers:

- *Operations on variables and object models,* including variable and type declarations, assignments, arithmetic operations on different data types, composition and decomposition of trees, etc.

- *Interactions with the manipulator,* including operations that are used to move the arm and to read their position data, that describe objects by means of pointers calibrated before program execution, frame rotation (i.e. relative coordinate system transformation) and other manual arm positioning and orientation support, etc.

- *A user interface,* allowing the interpretation of AL files, renaming variables, editing files, accessing bulk stores and other input/output devices, error recovery support, run-time program status report, etc.

The implementation and application of POINTY has proved in several cases that object models could be created and that the combined teach-by-showing and advanced software based programming was possible, and in several cases, was much simpler and faster than other robot programming methods.

SIGLA

SIGLA was developed by Olivetti (1978) for the SIGMA assembly robot (Fig. 8.10). It is a language consisting of the following main elements: operating system, instruction execution control, and user memory handling for actual robot program storage. It allows parallel task processing, has an interpreter structure allowing 'teach-by-doing', and a set of variables and custom instructions.

MAL

MAL was first written in Assembler for the SIGMA robots and implemented on LSI11 and PDP11/34 computers at the Milan Polytechnic (1979). It is an interactive language using BASIC-like programming where the need is for portability and easy programming. It is an algorithmic language and

Fig. 8.10 The dual arm Olivetti SIGMA robot working on a pcb assembly job

any instruction must be explicitly programmed by the user. Assignment, control, I/O statements, robot operations (such as MOVE, INCR, ACT, etc.) and task synchronisation statements can be programmed. The extended version of MAL proposes user-defined control structures tailored to particular applications.

HELP

The core of the HELP language was developed for use in conjunction with DEA's PRAGMA A3000 and BRAVO robots and their inspection machines (Fig. 8.11).

HELP supports macro-definition and calling, allowing the construction of primitives. The operators of this system are purposely left incomplete to allow expansion and modification of particular instructions, e.g. use of special tools by direct use of virtual machine instructions. The virtual machine can support concurrent activities using a priority scheduling method. Activities may be synchronised by using the WAIT and SIGNAL instructions, which are stored as subprograms.

To understand the structure of the language, some important features and statements of HELP as used for robot programming are summarised:

● Assignment statement
    Variable name: = arithmetic expression

Fig. 8.11 The DEA PRAGMA A3000 assembly robot was one of the first robots using a high-level programming language (HELP) and a built-in force sensor monitoring system

● Control statements

        IF–THEN–END
        IF–THEN–ELSE–END
        WHILE–DO–REP
        FOR <var.name> : = TO   DO
                        REP BY
        GOTO <label>
        GOTO <arithmetic expression> <label> . . . <label>
        GOSUB <label>
        GOSUB <arithmetic expression> <label> . . . <label>
        etc.

● Built in subprograms

        MOVE (X,Y,Z), GRASP, UNGRASP, WAIT, SIGNAL, etc.

The design goal of the language has been simple programming and operation, high flexibility, interactive mnemonic programming and editing, parallel task execution for multi-arm control, and variable instruction set for software tailoring and high operating speed. Some of these goals have system-defined primitives.

RAPT

RAPT (Robot APT) programming language for assembly purposes was developed at the University of Edinburgh in 1978. Here, tasks have to be described as a mixed sequence of situations and actions. A situation is the state of the bodies comprising the robot world as described by the programmer specifying spatial relationships between body features. Body descriptions are in terms of their features, currently plane faces, cylindrical shafts, etc.

The body definitions are related to its local embedded coordinate system, as are the body features. The RAPT interpreter therefore does not have to update the feature positions, only that of the body. RAPT has been written in FORTRAN and is currently still under development at the University of Edinburgh, Department of Artificial Intelligence.

ROBEX

The ROBEX (ROBoter EXapt) programming system was designed at the Laboratory of Machine Tools and Production Engineering in Aachen in 1978, specially for loading and unloading machine tools by means of industrial robots. ROBEX's input language and CLDATA file is compatible with the APT based EXAPT system, a widely used high-level NC off-line programming system for machine tools in Europe.

This programming system allows the use of standard EXAPT processor modules for creating the world model and program handling. The world model data is used for both the path generation and the robot dependent NC program data. The user-written program can be edited interactively and is finally stored in the so-called INPUT file. The processor then generates a robot independent pseudo-code which is stored and post-processed for different robot and machine tool hardware.

The end result of the part programming effort is not only the NC tape but also the robot control 'tape' allowing the coordinated control of the machine tool and the robot when loading and unloading components.

MCL

MCL (Machine Control Language) is a structured programming language and an extension of the APT NC language developed by McDonnell Douglas Corporation in 1978 under the ICAM program. Written in FORTRAN it is post-processed to different robots in a similar way in which an APT post-processor interprets a CLDATA file. It is a high-level language designed for off-line programming and can find and identify (locate) objects.

MCL aims to be a standard machine control language with the following capabilities:

- Ability to control any robot by means of a high-level language (i.e. MCL) and different post-processors to different robots. The system also provides a 'verifier post-processor', allowing the user to determine if a given machine control program can run on any of the manufacturing cells defined in the database by simulation.
- Ability to control manufacturing cells where there are robots integrated with other machines.
- MCL overcomes several drawbacks of APT-based languages, since it can process real-time sensory data.

- MCL is capable of modelling two-dimensional binary images by means of a vision system.
- Different frames can be selected at the higher level to define certain points and curves. These definitions are then transformed into the lower level, i.e. post-processed code taking account of the actual hardware it is executed on.

By allowing the support of typical manufacturing cell control tasks, MCL is clearly more than just a robot programming language.

AML

AML (A Manufacturing Language) has been designed to be a well-structured, semantically powerful, interactive language that can be applied to programming IBM manufacturing systems, and of which a subset (AML/ENTRY) is also implemented on the IBM personal computer for programming the IBM 7535 and IBM 7540 manufacturing system robots (1980-83), (Fig. 8.12).

It combines many of the desirable qualities of APL, Pascal and LISP. Featured amongst the design objectives was the idea of providing a base

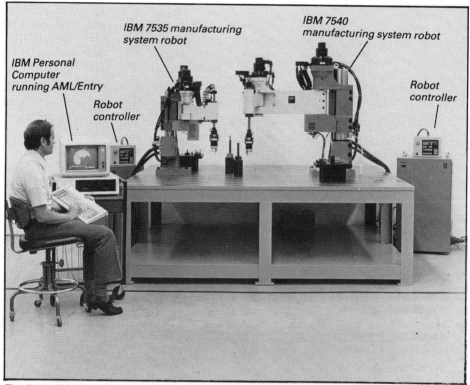

Fig. 8.12 IBM has developed the AML language to be used on a range of robotics equipment. The language can be applied also in robotised assembly where sensory feedback processing is crucial. The photograph shows the IBM 7540 and the IBM 7535 Manufacturing System robots. Both devices are controlled from the IBM Personal Computer using the AML/Entry language (Courtesy of IBM Corp., Boca Raton, Florida)

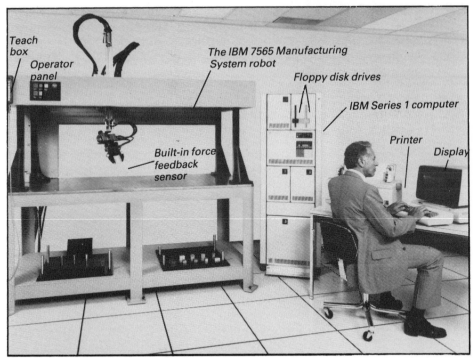

Fig. 8.13 The IBM 7565 Manufacturing System controlled by the Series/1 computer and programmed using the AML language (Courtesy of IBM Corp., Boca Raton, Florida)

language with simple subsets and functional transparency, in the sense that AML routines can be written and executed from user programs as if they were built-in routines. It was also desinged to be an interactive language in order to simplify debugging and to allow on-line programming of the robot.

The core of the system architecture of the manufacturing system is the Series/1 computer, with a variety of data processing terminals (Fig. 8.13). One module of the system software provides an interactive programming environment for AML. Another module performs motion coordination, trajectory planning and sensor monitoring, and the third major module handles the real-time tasks in response to commands from the AML interpreter.

The system can also be interfaced with AML/V, an industrial machine vision programming system designed as a hardware and software extension for the RS1 series robots. The extension of AML is provided in the form of subroutines, divided into camera control and image acquisition, binary region analysis and image arithmetic. These routines support pixel-by-pixel operators which can be composed into more complex functions. AML/V can handle images in the form of bit-per-pixel where each 8-bit character of the STRING contains eight binary pixels. Other representations are also available.

Command structure. AML commands are predefined subroutines that define semantic functions for robotics. An interesting feature of the system is that no syntax distinction is made between the AML command sub-routines and the AML language subroutines; they are the same kind of system-defined primitives.

AML commands can be classified into the following categories:

- *Fundamental subroutines,* which deal with the flow of control in the language interpreter (e.g. RETURN from subroutine, BRANCH is used for unconditional transfer of control, QUIT returns control to a previous level, APPLY calls a subroutine with a program-generated set of argu-ments, etc.).

- *Calculational subroutines,* which perform data transformations on their arguments (e.g. mathematical functions that operate on numbers, such as ATAN, SQRT, etc.; calculational commands which extract attribute information of data, such as LENGTH returns the number of characters to a specified position. For example, MOVE (<1,4>, 12.5) moves routines; and a set of subroutines which implement various matrix and vector operations, borrowed mainly from APL).

- *Robot and sensor I/O commands* are special purpose robotics application routines which provide low-level primitives to move one or more joints to a specified position. For example, MOVE (<1,4>, 12.5) moves joints 1 and 4 to 12.5 position. This is the module from which the SPEED, the ACCEL_eration, and the DECEL_eration is controlled, and from which the robot can be stopped, STOPMOVE, WAITMOVE, FREEZE. STARTUP turns the arm power on and SHUTDOWN turns it off. The MONITOR commands are also part of this set of routines and they tell the real-time system to begin reading a specific sensor or set of sensors at regular time intervals.

- *System interface commands* provide control over the interactive software interface of the AML execution environment (e.g. BREAK invokes the terminal command processor, KEY allows the user to associate interrupt subroutines with console key events, LOAD and UNLOAD determine what objects are in the memory, etc.).

- *Data processing commands* handle the peripherals of the entire comput-ing system.

Language description. The language part of AML is very impressive and easy to use for those who are familiar with Pascal-like, structured langu-ages, but somewhat complicated for operators without any knowledge of data processing. The major components of the AML language are dis-cussed below:

- Like other high-level languages, AML has data objects which can be scalar, such as INTEGER, REAL, STRING, REFERENCE, IDENTI-FIER and LABEL, and aggregate objects, which represent an ordered set of scalar objects.

- AML is expression orientated and evaluates expressions from left-to-right in a manner very similar to other high-level languages. It allows arithmetical, relational, scalar, logical operations, and has different assignment statements. It allows IF-THEN-ELSE, WHILE-DO, RE-PEAT-UNTIL and BEGIN-END type of construction statements, very like Pascal.

- Subroutines are called by name. Subroutines can have parameters, which may be expressions themselves and passed as actual parameters to the subroutine. Variable names have to be declared by a so-called declaration expression, to define, allocate and initialise data variables to be used by a subroutine.

To summarise, AML is a very powerful system programming language for complex robot manufacturing systems, having an open-ended structure which allows further development not only by the vendors but also by the users. If users find it complicated to learn, they can write a simpler interface to it in AML, by constructing a set of even 'higher-level' subroutines which handle many of previously programmable functions as default values.

8.4 Unimation's VAL language

The programming language VAL is a variation of research languages developed at the Stanford Artificial Intelligence Laboratory. It was initially written by V. Scheinman and B. Shimano in 1975, and implemented in a PDP11/45 system. It was developed further and became a commercially available language in 1978. (Within this discussion and examples, Version 12 is used.)

The VAL programming language consists of a full set of English language instructions for writing and editing robot programs. A robot program can be generated by the teach-by-showing method, using the manual controls and the teach box, or via the keyboard. Taught points are stored as transformations, or if increased positioning accuracy is required, then also in the form of joint angles, in the memory of the LSI 11-based controller (Fig. 8.14).

A location in VAL can be described by one of the following methods:

- *Transformation* – These define a location in terms of X, Y, Z Cartesian coordinates and tool orientation (Euler) angles (O,A,T) relative to a reference frame fixed in the base of the robot.

- *Precision point* – Aiming maximum position and orientation accuracy when repeating a location, with the disadvantage of limited variable manipulation possibility during program execution (i.e. most instructions do not work in VAL with precision points).

- *Compound transformation* – Used for applications requiring the definition of locations relative to other locations previously defined.

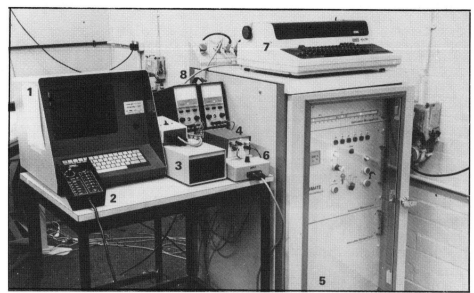

Fig. 8.14 The LSI/11 based VAL controller of the Unimation PUMA 560 robot at Trent Polytechnic, Nottingham. (Note the custom built communications interface and the power supplies for powered tools and the attached printer), 1, display and keyboard; 2, teach pendant; 3, disk drive; 4, power supply; 5, the LSI/11 based Unimation VAL controller; 6, communications link to a microcomputer; 7, DEC LA34 printer; 8, volt/amp meters

VAL is interactive and operates in interpretative mode, i.e. monitor commands are immediately interpreted and executed whilst user program steps are translated and stored in machine code for real-time computations, performed when executing the robot program to convert the stored data to joint angle information. Thus separate steps are not necessary to compile, link and execute VAL programs. Location points can be stored either as a precision point or a transformation and can be taught or given as coordinate values via the keyboard.

Each module of the arm is driven by a permanent-magnet servo motor and the necessary gears. Position sensing is performed by incremental encoders. (New versions of the PUMA robot employ digital encoders.) The correct functioning of the continuous-path controlled arm requires control of the position and the velocity vector of each joint. Because the actual arm position must be measured relative to a known absolute position, the robot has to be initialised when powered up. The encoders on the motor shafts in each joint are used to provide the absolute positions for the system control. The incremental encoders provide both position and velocity feedback signals for the servo control system, which are read approximately 30 times during the 28ms cycle time of the robot controller.

Because of limited space, no attempt is made within this text to provide a full description of the VAL language. Rather, a list is provided of the most important monitor commands and program instructions for Version 12, and self-documented sample programs are given to demonstrate the way in which PUMA robots may be programmed.

Monitor commands

The user interface of the VAL system consists of the monitor commands and the robot program instructions. Leaving aside all necessary details, monitor commands deal with the following tasks.

● *Definition of a location in the 3D space*

PO_int, sets the value of the location variable. Defines a precision point or a transformation and its value is displayed as *X,Y,Z*, world coordinates, and O,A,T Euler angles in degrees. (Note that PO_int means that one has to type only PO and the rest of the command will be understood by the system automatically.) For example, POINT #POS1 allows the definition or modification of the precision point POS1, POINT P1 = P2 sets the value of transformation P1 equal to P2.

H_ere, defines or assigns the value of a transformation or a precision point to be equal to the current robot location. For example, HERE P1 where P1 becomes the current robot location, HERE #POS1 the current location becomes the value of the precision point POS1.

W_here, displays the current robot location and joint variables in the Cartesian world coordinate system.

TE_ach, can be used for recording a series of location values by driving the robot from the teach pendant and pressing the [RECORD] button every time a location should be stored. For example, TEACH LOC1 effects monitor teach mode, so that every time the [RECORD] button is pressed a new location (LOC2, LOC3, etc.) is stored automatically.

B_ase, offsets and rotates the reference frame of the robot as specified. This command can be used subsequently for defining new locations in different reference frames. (Note that each point can be accessed in the reference frame they were defined in, but the command has no effect on locations defined as precision points.) For example, BASE 250,150,-50,5 redefines the world reference frame by shifting *X* by 300mm, *Y* by 150mm, *Z* by 50mm in the negative direction and rotates the robot around *Z* by 30°.

DP_oint, deletes locations, thus helps to recover storage space limitations. For example, DPOINT P1, #POS1 deletes the values of transformation P1 and precision location POS1.

TO_ol, defines robot tool (i.e. gripper, etc.) relative transformation. The tool transformation is set to default (i.e. zero) when the system is powered-up. The 'null tool' has its centre at the surface of the tool mounting flange and its coordinate axis parallel to that of the last joint of the robot ([0,0,0,90,-90,0]). For example, TOOL NEWHAND replaces the current tool transformation to the values given in NEWHAND.

● *Editing part program commands*

ED_it, creates or modifies a program, named in its argument. For example, EDIT TASK1 edits or creates a program called TASK1. ED_it has several options, including:
  C change program name
  D delete a specified number of instructions
  I insert new lines into existing programs
  L display the previous step of the program for editing
  P display the next program line and set it to become the current program line
  R replace characters in the currently displayed line
  S step a program line ahead
  T initiate the joint interpolated teach mode from the program (see also the TEACH command)
  TS initiate the straight-line teach mode from the program (see also the TEACH command)
  E exit from the editor

● *List program and location-data commands*

DIR_ectory, displays user program names in all directories.

LISTP_rogram, displays all lines of the specified user program(s).

LISTL_ocation, displays the values of all stored location data.

● *Program and location-data storage on floppy disk*

FORMAT, creates file directory and erases all previously stored information on the diskette. (Note that VAL can store the same programs and location variables in different files on the same disk. It is advised to take care with this sometimes confusing feature. See also the STORE command.)

LISTF_ile, displays the file directory and the amount of available space on the diskette.

LOAD, loads the specified program (name1.PG) and the related location file (name2.LOC) if specified into the memory.

LOADP_rogram, loads program from disk into memory, or replaces the program in memory by the program stored under the same name on disk.

LOADL_ocation, loads location variables from disk into memory, or replaces the location variables in memory by the program stored under the same name on disk.

STORE, stores the specified program(s) and their location variables in the specified file. For example, STORE PAUL=PALLET stores the program named PALLET and its location variables, if any, into a disk file named PAUL; STOREP_rogram, same as STORE, but stores only the program.

STOREL_ocation, same as STORE, but stores only the location variables, if any.

DE_lete file, deletes specified file, including all programs and location variables stored in the specified file, from the diskette.

ER_ase, erases the contents of the floppy disk and initialises the disk file-directory information.

CO_mpress, crunches specified disk file (i.e. recovers unusable gaps created during earlier file delete operations).

● *Program control commands*

A_bort, terminates program execution after the completion of the step currently being executed.

EX_ecute, runs the specified program n times. If $n < 0$ then the program will be executed infinitely.

SP_eed, sets the monitor speed value between the range of 0.01 (jogging speed) and 327.67 (fast). Monitor speed 100 is the speed value considered to be 100% and 'normal'. (Note that the monitor speed modifies the speed value issued in the program. For example, monitor speed 50 will reduce all speed values issued in the program to their 50% equivalent when executed.)

PR_oceed, proceeds execution of the program interrupted by the PAUSE instruction, the ABORT command, or by a run-time error.

DO, executes the specified program instruction as if it was the next step in a program. For example DO MOVES P5 moves the robot tool along a straight line from the current position to the transformation defined by P5, DO GOTO 100 transfers program execution control to step 100.

R_entry, allows the execution of a previously aborted step.

N_ext, offers single step program execution.

● *System status and control commands*

CA_librate, calibrates joint position encoders. If the power goes down the PUMA becomes uncalibrated. It is advised to calibrate the robot before program execution, although it can be run without.

STA_tus, displays run-time robot program status information, including the number of completed and remaining program loops and the current monitor speed value.

DON_e, aborts the VAL monitor program (i.e. robot operating system). (Type 173000G after the @ prompt to re-enter VAL.)

FR_ee, displays the available memory percentage.

Z_ero, deletes memory contents (i.e. all programs and location variables) and reinitialises VAL.

● *System switches*

System switches are used to control various display message, keyboard and run-time hardware error monitoring features in the VAL system. For example, the CP switch turns off continuous path processing, EHAND effects servo-controlled tools, VISION effects the vision system, DIS_able / EN_able, disables/enables system switches, SW_itch, displays current system switch settings, etc.

● *System diagnostics and modification*

System diagnostics and modification commands allow testing various components of the robot system and can be fed back and used in the VAL program.

Program instructions

VAL program instructions are many and varied and so only some are listed, with the necessary explanation so that robot program examples listed and explained in Section 8.5 can be followed.

It should be noted that one of the major drawbacks of the discussed version (i.e. VAL Version 12) is that real numbers and real variables cannot be defined, nor calculated, that the calculation with integers is also very limited and that data input and parameter passing by value or variable name in procedures is impossible. (Unfortunately the more advanced version of VAL, i.e. VAL II, was still in the debugging phase when this text was written.)

● *Robot configuration control statements*

LE_fty and RI_ghty, establish the human left or right robot arm (i.e. joints 1,2,3) configuration.

AB_ove and BE_low, change the configuration such that the elbow is pointing up or down.

FL_ip and NOF_lip, change the range of operation of joint 5 in the case of the 6-axis PUMA robot to negative or positive angles.

● *Motion control statements*

READ_y moves the robot into a standard configuration. For example, DO READ_y can be used to check robot calibration.

APPRO_ach location (joint interpolated motion).

APPROS, approach location from the current position along a straight line.

DEPART from a location (joint interpolated motion).

DEPARTS, from the current position along a straight line.

AL_ign the tool Z coordinate to the nearest robot world coordinate system axis. (This instruction is very useful before a series of locations are taught, in the form of DO ALIGN.)

MOVE to a location (joint-interpolated motion).

MOVES the robot to the location and orientation specified by the variable name of the location (straight-line motion).

MOVET, same as MOVE but also operates tool open/close. For example, MOVET #P3,1, moves from the current location to the precision transformation defined by P3 in a joint-interpolated mode, closing the tool during the motion. (Note, 1 opens, 0 closes the tool in the case of the pneumatic tool control system.)

MOVEST, same as above, but the tool following a straight-line path.

DRA_w, maintaining the robot configuration and tool orientation the robot tool is moved the specified distances along X, Y and Z. For example, DRAW 60,50,-30, motion in 3D; DRAW 25,55, motion in the X, Z plane, etc.

DRI_ve, operates single joints at the specified speed. For example, DRIVE 3,-51.5,60, drives joint 3 by 51.5° into the negative direction at a speed of 60.

● *Hand (i.e. robot tool) control statements*

CLOSE and OPEN, close or open pneumatic gripper during the next motion.

CLOSEI and OPENI, close or open the pneumatic gripper immediately.

RE_lax, immediately turns off the pneumatic control solenoid valves, and GR_asp closes the servo controlled hand.

● *Integer variable assignment statements*

SETI, assigns an integer variable name similar to the LET statement in BASIC.

TYPEI, displays the current value of the integer on the display terminal.

● *Location assignment and modification statements*

HE_re, sets the value of a transformation or precision point equal to the current robot location. For example, HERE P1 sets transformation P1, or HERE #POS1 precision point POS1 equal to the current position.

SET, is a transformation or precision point assignment statement. For example, SET P1=P2, or SET #POS1 = #POS2.

SH_ift transformation, by the defined amounts in X, Y and Z. For example, SHIFT P1 BY 20,,55 shifts tool by 20mm along X and by 55mm along Z.

TO_ol assigns new tool transformation (see also the TOOL monitor command).

FR_ame transformation, allows the definition of different frame coordinate systems the robot should work in. It is often used to change the BASE frame. For example, FRAME BASE=P1,P2,P3 defines a new frame as shown in Fig. 8.15.

INV_erse transformation, allows the robot programmer to think and program as if he was 'sitting' in a relative coordinate system (Fig. 8.16). More precisely, if for example the transformation on the right side of the instruction (POSA) defines a location relative to another location, ORIG, then the inverse transformation represents the location ORIG relative to POSA.

● *Program control statements*

GOS_ub routine, calls and executes the named subprogram. For example, GOSUB PALLET, executes the PALLET subprogram. When a RETURN instruction is executed in the subprogram, program control is returned to the next executable instruction.

RET_urn, in its simplest form returns control when the execution of a subprogram has finished, or if issued in the main program acts as a STOP instruction.

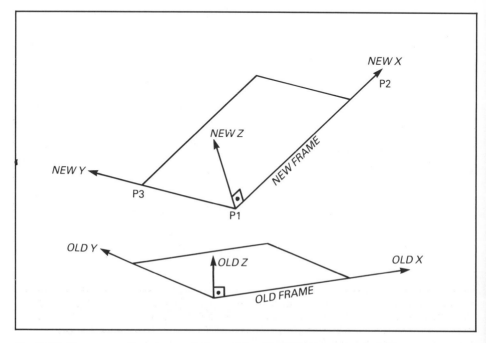

Fig. 8.15 Diagrammatic representation of the FRAME instruction in VAL

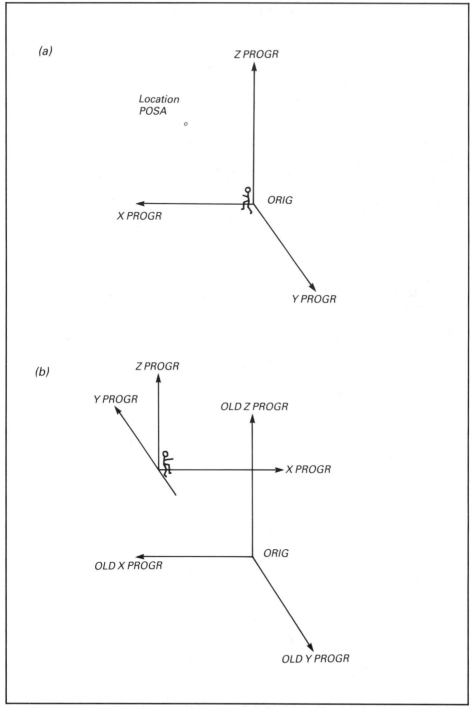

Fig. 8.16 Diagrammatic representation of the INVERSE instruction in VAL: (a) before the execution of the INVERSE instruction, INVERSE POSA.INV = ORIG; and (b) after execution of the INVERSE instruction. (Note, XPROGR, YPROGR and ZPROGR represent the current coordinate system)

ST_op, terminates the execution of the program unless there are more program loops remaining to be completed. If the optional string argument is also used, then sends a message to the output device (e.g. display). For example, STOP END OF LOOP, will display 'END OF LOOP'.

HA_lt, is similar to STOP, but will in any case stop program execution.

GOT_o a label, transfers program control to the specified label. For example, GOTO 100.

IF THEN, similar to the structure used in most computer languages. This instruction compares integer variables and if the result is true then program execution branches to the label specified in the instruction. For example, IF I GT 5 THEN 50, if I integer is greater than 5, transfer program control to label 50, or else the next executable instruction will determine what should happen.

Relationships allowed are as follows: EQ <equal>, NE <not equal>, LT <less than>, TG <greater than>, LE <less than or equal> and GE <greater than or equal>.

SI_gnal at channel . . ., sets the specified output signal high or low. For example, SIGNAL 5,-2 sets output 5 high and output 2 low.

IFSIG_nal is matched (i.e. low or high) at channel . . . THEN . . ., take action . . ., allows conditional program control, depending on the current input signal status at the tested channels. (Note, there must be four channel specifications in the instruction.) For example, IFSIG 5,-2,, THEN 20 will transfer program control to label 20 if the signal at input channel 5 is high and at 2 is low, or else the next executable instruction will determine what should happen.

REACT means continuous monitoring of the specified input channel and the execution of the equivalent of a 'GOSUB program name' instruction as the next instruction whenever the signal is high. For example, REACT 2, HOME will execute the HOME subprogram if the signal at input channel 2 is high, or else the next executable instruction will determine what should happen.

REACTI, is very similar to REACT, but rather than completing the execution of the current instruction and then reacting to it, it will immediately react to the monitored signal.

PAUSE, stops program execution, but allows its continuation if PROCEED is typed in. For example, PAUSE MOUNT NEW GRIPPER, will send the message MOUNT NEW GRIPPER to the output device and wait.

WA_ait, until the specified channel matches the external input signal. For example, WAIT 2 stops program execution until channel 2 becomes high.

IG_nore, disables the REACT or REACTI instructions associated with the specified channel. For example, IGNORE 3 stops monitoring channel 3 high signals.

● *Trajectory control statements*

CO_arse enables a low-tolerance feature in the servo, so that larger errors are permitted when reaching final position.

FI_ne is the default and the opposite of the COARSE statement.

SP_eed, sets programmed speed. (Permitted range equals 0.01 for slow, to 327.4 for very fast).

● *Miscellaneous statements*

DEL_ay time, is interpreted in seconds, whilst the robot does not perform any other activity. The value range is 0.01 – 327.4s.

REM_ark, is the notation of the comment line in the VAL program.

TYPE displays the message specified in its argument.

BASE redefines the reference frame of the robot in the world coordinate system the same way as the monitor command BASE and allows all locations to be shifted by a user-specified value.

8.5 Sample programs written in VAL

The purpose of this section is to provide self-documented sample programs written in VAL and tested on a PUMA 560 robot, enabling the reader to familiarise himself with some problems that can be solved with PUMA robots.

It should be noted that unless the robot is mounted in an inverted position, when executing the sample programs on the six-axis PUMA robot, in most cases it is advised to use the same or similar locations as described in Table 8.1. These locations can be established for example by teaching, using the .HERE monitor command. (For example, move the robot to the desired location, then type .HERE START, . . . CHANGE ?, answer Y and change as required.)

Table 8.1 Location file

Variable	X/JT1	Y/JT2	Z/JT3	O/JT4	A/JT5	T/JT6
START	57.41	633.25	122.25	134.72	81.94	130.53
PICKUP	579.01	528.41	218.25	120.76	32.78	168.02
PLACE	−232.72	794.91	312.34	−174.01	2.32	−178.09

```
.PROGRAM DEMO1
    1.          TYPE PROGRAM DEMO1 RUNNING...
    2.          TYPE
    3.          TYPE ****************************************************************
    4.          TYPE * THIS PROGRAM DEMONSTRATES THE DIFFERENCE BETWEEN          *
    5.          TYPE * THE 'ABOVE' AND THE 'BELOW' CONFIGURATIONS OF THE SIX *
    6.          TYPE * AXIS PUMA ROBOT   (SET SLOW MONITOR SPEED)              *
    7.          TYPE ****************************************************************
    8.          TYPE
    9.          PAUSE TYPE 'PROCEED' AND PRESS RETURN TO CONTINUE
   10.          REMARK *** PROGRAM WILL WAIT AS LONG AS REQUIRED
   11.          MOVE START
   12.          REMARK *** MOVE TO THE TAUGHT LOCATION 'START' (REFER TO THE
   13.          REMARK *** LOCATION FILE)
   14.          ABOVE
   15.          REMARK *** ORIENTATE ARM SO THAT THE ELBOW IS POINTING UPWARDS
   16.          MOVE PICKUP
   17.          REMARK *** MOVE TO THE TAUGHT LOCATION 'PICKUP' (REFER TO THE
   18.          REMARK *** LOCATION FILE)
   19.          TYPE
   20.          TYPE ===== THE ARM IS NOW IN ABOVE CONFIGURATION =====
   21.          PAUSE TYPE 'PROCEED' AND PRESS RETURN TO CONTINUE
   22.          BELOW
   23.          REMARK *** ORIENTATE ARM SO THAT THE ELBOW IS POINTING DOWNWARDS
   24.          MOVE PLACE
   25.          REMARK *** MOVE TO LOCATION 'PLACE'
   26.          TYPE
   27.          TYPE ===== THE ARM IS NOW IN BELOW CONFIGURATION =====
   28.          PAUSE TYPE 'PROCEED' AND PRESS RETURN TO CONTINUE
   29.          ABOVE
   30.          MOVE START
   31.          TYPE
   32.          TYPE THE ARM HAS RETURNED TO ITS NORMAL 'ABOVE' CONFIGURATION
   33.          TYPE AND IS AT THE LOCATION NAMED 'START'
   34.          TYPE
   35.          TYPE ************** END OF PROGRAM ****************
   36.          RETURN 0
   37.          REMARK *** RETURN INDICATES THE END OF A LOOP. THE NUMBER IS
   38.          REMARK *** THE SKIP COUNT, WHICH INDICATES WHICH LINE TO
   39.          REMARK *** RESTART AT. IN THIS CASE EQUALS 0+1 THUS EXECUTION
   40.          REMARK *** CONTINUES AT THE NEXT LINE AFTER THE ONE IN
   41.          REMARK *** WHICH THE ROUTINE WAS CALLED.
.END
```

Fig. 8.17 Sample VAL program: the use of the 'ABOVE' and 'BELOW' configuration instructions

'ABOVE' and 'BELOW' arm configurations

The PUMA robot can operate in the ABOVE or in the BELOW configuration. This can occur when joint 3 (the elbow) is pointing upwards (ABOVE) or downwards (BELOW). It should be noted that neither instruction will do anything until the MOVE command is executed, as only then can the arm configuration change. (The MOVE instruction represents a joint-interpolated motion and is further explained in the DEMO3 program.) Fig. 8.17 provides the program listing to demonstrate these two configurations.

'RIGHTY' and 'LEFTY' arm configurations

This routine demonstrates the 'RIGHTY' (i.e. human right arm) and 'LEFTY' arm configurations. Neither instruction will do anything until the

MOVE command is executed, as only then can the arm configuration change. Fig. 8.18 provides the program listing to demonstrate these two configurations.

'MOVE' and 'MOVES' instructions

This program demonstrates the difference between the joint-interpolated motion (MOVE) and the straight-line motion (MOVES) established by the continuous-path control system of the robot (Fig. 8.19).

Approach/depart locations

This program enables the operator to identify the differences between approaching a location or departing from a location along a straight line, or in joint-interpolated mode (i.e. along an arc). If the APPRO or DEPART instruction has an 'S' (i.e. APPROS or DEPARTS), then the arm moves along a straight-line path and the tool is rotated smoothly to its final orientation (Fig. 8.20). (Configuration changes are not permitted during this motion.)

```
.PROGRAM DEMO2
  1.      TYPE PROGRAM   DEMO2 RUNNING...
  2.      TYPE
  3.      TYPE ***********************************************************
  4.      TYPE * THIS PROGRAM DEMONSTRATES THE DIFFERENCE BETWEEN      *
  5.      TYPE * THE 'RIGHTY' AND 'LEFTY' ARM ORIENTATIONS USING THE   *
  6.      TYPE * SIX AXIS PUMA ROBOT   (SET SLOW MONITOR SPEED)        *
  7.      TYPE ***********************************************************
  8.      TYPE
  9.      PAUSE TYPE 'PROCEED' AND PRESS RETURN TO CONTINUE
 10.      REMARK *** PROGRAM WILL WAIT AS LONG AS REQUIRED
 11.      RIGHTY
 12.      REMARK *** ORIENTATE THE ARM TO RESEMBLE A HUMAN RIGHT ARM
 13.      MOVE PICKUP
 14.      REMARK *** MOVE TO THE LOCATION 'PICKUP'. NOTE THAT THE
 15.      REMARK *** ARM WILL BE MADE 'RIGHTY' DURING THE MOTION
 16.      TYPE
 17.      TYPE ===== THE ARM IS NOW IN 'RIGHTY' ORIENTATION
 18.      PAUSE TYPE 'PROCEED' AND PRESS RETURN TO CONTINUE
 19.      MOVE PLACE
 20.      REMARK *** MOVE TO THE LOCATION 'PLACE'
 21.      PAUSE TYPE 'PROCEED' AND PRESS RETURN TO CONTINUE
 22.      LEFTY
 23.      MOVE PICKUP
 24.      REMARK *** MOVE TO THE LOCATION 'PICKUP'. NOTE THAT THE
 25.      REMARK *** ARM WILL BE MADE 'LEFTY' DURING THE MOTION
 26.      TYPE
 27.      TYPE ===== THE ARM IS NOW IN 'LEFTY' ORIENTATION
 28.      PAUSE TYPE 'PROCEED' AND PRESS RETURN TO CONTINUE
 29.      MOVE PLACE
 30.      REMARK *** MOVE TO THE LOCATION 'PLACE'
 31.      TYPE
 32.      TYPE **************  END OF PROGRAM  ****************
 33.      RETURN 0
.END
```

Fig. 8.18 Sample VAL program: the use of the 'RIGHTY' and 'LEFTY' arm orientation instructions

```
.PROGRAM DEMO3
    1.       TYPE PROGRAM   DEMO3 RUNNING...
    2.       TYPE
    3.       TYPE ********************************************************
    4.       TYPE * THIS PROGRAM DEMONSTRATES THE DIFFERENCE BETWEEN      *
    5.       TYPE * THE JOINT INTERPOLATED 'MOVE' AND THE STRAIGHT LINE   *
    6.       TYPE * INTERPOLATED 'MOVES' INSTRUCTIONS USING THE SIX AXIS  *
    6.       TYPE * PUMA ROBOT      (SET SLOW MONITOR SPEED)              *
    7.       TYPE ********************************************************
    8.       TYPE
    9.       PAUSE TYPE 'PROCEED' AND PRESS RETURN TO CONTINUE
   10.       REMARK *** PROGRAM WILL WAIT AS LONG AS REQUIRED
   11.       MOVE PICKUP
   12.       REMARK *** ROBOT ARM MOVES TO LOCATION 'PICKUP' IN A JOINT
   13.       REMARK *** INTERPOLATED MODE
   14.       MOVE PLACE
   15.       REMARK *** ROBOT ARM MOVES TO LOCATION 'PLACE' IN A JOINT
   16.       REMARK *** INTERPOLATED MODE
   17.       MOVES PICKUP
   18.       REMARK *** ROBOT ARM MOVES TO LOCATION 'PICKUP' IN A
   19.       REMARK *** STRAIGHT LINE INTERPOLATED MODE
   20.       MOVES PLACE
   21.       REMARK *** ROBOT ARM MOVES TO LOCATION 'PLACE' IN A
   22.       REMARK *** STRAIGHT LINE INTERPOLATED MODE
   31.       TYPE
   32.       TYPE ************** END OF PROGRAM **************
   33.       RETURN 0
.END
```

Fig. 8.19 Sample VAL program: the use of the 'MOVE' and 'MOVES' configuration instructions

```
.PROGRAM DEMO4
    1.       TYPE   PROGRAM DEMO4 RUNNING...
    2.       TYPE
    3.       TYPE ********************************************************
    4.       TYPE * THIS PROGRAM DEMONSTRATES THE 'APPRO/APPROS' AND *
    5.       TYPE * 'DEPART/DEPARTS' INSTRUCTIONS USING THE SIX AXIS *
    6.       TYPE * PUMA ROBOT     (SET SLOW MONITOR SPEED)          *
    7.       TYPE ********************************************************
    8.       TYPE
    9.       PAUSE TYPE 'PROCEED' AND PRESS RETURN TO CONTINUE
   10.       REMARK *** PROGRAM WILL WAIT AS LONG AS REQUIRED
   11.       APPRO PICKUP,120
   12.       REMARK *** APPROACH LOCATION 'PICKUP' 120 MM ABOVE, ALONG
   13.       REMARK *** THE ACTUAL Z AXIS IN THE NEGATIVE DIRECTION IN
   14.       REMARK *** JOINT INTERPOLATED MODE
   15.       MOVE PICKUP
   16.       REMARK *** ROBOT ARM MOVES TO LOCATION 'PICKUP' IN A JOINT
   17.       REMARK *** INTERPOLATED MODE
   18.       DEPART 100
   19.       REMARK *** ROBOT ARM DEPARTS FROM THE 'PICKUP' LOCATION
   20.       REMARK *** IN JOINT INTERPOLATED MODE
   21.       APPROS PLACE,155
   22.       REMARK *** APPROACH LOCATION 'PLACE' 155 MM ABOVE, FOLLOWING
   23.       REMARK *** THE SAME Z AXIS LOCATION 'PLACE' WAS TAUGHT IN
   24.       REMARK *** IN THE NEGATIVE DIRECTION ALONG A STRAIGHT LINE
   25.       MOVES PLACE
   26.       REMARK *** ROBOT ARM MOVES TO LOCATION 'PLACE' ALONG A
   27.       REMARK *** STRAIGHT LINE
   28.       DEPARTS 100
   29.       REMARK *** ROBOT ARM DEPARTS FROM THE 'PICKUP' LOCATION
   30.       REMARK *** ALONG A STRAIGHT LINE
   31.       TYPE
   32.       TYPE ************** END OF PROGRAM **************
   33.       RETURN 0
.END
```

Fig. 8.20 Sample VAL program: the difference between the 'APPRO/APPROS' and the 'DEPART/DEPARTS' instructions

```
.PROGRAM DEMO5
    1.        TYPE   PROGRAM DEMO5 RUNNING...
    2.        TYPE
    3.        TYPE ********************************************************
    4.        TYPE * THIS PROGRAM DEMONSTRATES THE 'DRAW' INSTRUCTION      *
    5.        TYPE * USING THE SIX AXIS PUMA ROBOT (SET SLOW MONITOR SPEED) *
    7.        TYPE ********************************************************
    8.        TYPE
    9.        PAUSE TYPE 'PROCEED' AND PRESS RETURN TO CONTINUE
   10.        REMARK *** PROGRAM WILL WAIT AS LONG AS REQUIRED
   11.        MOVE START
   12.        REMARK *** MOVE THE ARM TO LOCATION 'START'
   13.        TYPE
   14.        TYPE ***************************************************
   15.        TYPE THE ROBOT WILL MOVE THE FOLLOWING DISTANCES FIRST
   16.        TYPE IN THE 'X', THEN IN THE 'Y' AND AFTER THAT IN THE
   17.        TYPE 'Z' DIRECTIONS FIRST SEPARATELY:
   18.        TYPE
   19.        TYPE 1)   150 MM IN THE 'X' DIRECTION
   20.        TYPE 2)   200 MM IN THE 'Y' DIRECTION
   21.        TYPE 2) -250 MM IN THE 'Z' (NOTE: NEGATIVE) DIRECTION
   22.        TYPE
   23.        TYPE ***************************************************
   24.        PAUSE TYPE 'PROCEED' AND PRESS RETURN TO CONTINUE
   25.        TYPE
   26.        TYPE ===== MOVE 'X' =====
   27.        DRAW 150.00, 0.00, 0.00
   28.        TYPE
   29.        TYPE ===== MOVE 'Y' =====
   30.        DRAW 0.00, 200.00, 0.00
   31.        TYPE
   32.        TYPE ===== MOVE 'Z' =====
   33.        DRAW 0.00, 0.00, -250.00
   34.        TYPE
   35.        TYPE ***************************************************
   36.        TYPE THE ROBOT WILL MOVE THE FOLLOWING DISTANCES
   37.        TYPE IN ALL THREE AXIS:
   38.        TYPE
   39.        TYPE 1) -150 MM IN THE 'X' (NOTE: NEGATIVE) DIRECTION
   40.        TYPE 2) -200 MM IN THE 'Y' (NOTE: NEGATIVE) DIRECTION
   41.        TYPE 2)  250 MM IN THE 'Z' DIRECTION
   42.        TYPE
   43.        TYPE ***************************************************
   44.        PAUSE TYPE 'PROCEED' AND PRESS RETURN TO CONTINUE
   45.        TYPE
   46.        TYPE ===== MOVE 'X','Y' AND 'Z' TOGETHER =====
   47.        DRAW -150.00, -200.00, 250.00
   48.        TYPE
   49.        TYPE **************** END OF PROGRAM ****************
   50.        RETURN 0
.END
```

Fig. 8.21 Sample VAL program: the use of the 'DRAW' instruction

'DRAW' instruction

The DRAW instruction can be used to move the robot arm through specified distances in the *X, Y* and *Z* directions without changing the orientation of the gripper. The motion generated is always in a straight line and may be specified in terms of distances in the *X, Y* and *Z* directions from the present locations (Fig. 8.21).

```
.PROGRAM DEMO6
    1.          TYPE   PROGRAM DEMO6 RUNNING...
    2.          TYPE
    3.          TYPE ***************************************************************
    4.          TYPE * THIS PROGRAM DEMONSTRATES THE 'DRIVE' INSTRUCTION.          *
    5.          TYPE * IT WILL ROTATE EACH SELECTED JOINT THROUGH A SPECIFIED *
    6.          TYPE * ANGLE IN A SIMILAR WAY THE ROBOT MOVES UNDER MANUAL      *
    7.          TYPE * CONTROL    (DO NOT ALTER DEFAULT MONITOR SPEED)          *
    8.          TYPE ***************************************************************
    9.          TYPE
   10.          PAUSE TYPE 'PROCEED' AND PRESS RETURN TO CONTINUE
   11.          REMARK *** PROGRAM WILL WAIT AS LONG AS REQUIRED
   12.          MOVE START
   13.          REMARK *** MOVE TO LOCATION 'START'
   14.          TYPE
   15.          TYPE ===== ROTATE JOINT 1,   45 DEGREES AT SPEED 30 =====
   16.          DRIVE 1, 45.000, 30.00
   17.          PAUSE TYPE 'PROCEED' AND PRESS RETURN TO CONTINUE
   18.          REMARK *** PROGRAM WILL WAIT AS LONG AS REQUIRED
   19.          TYPE
   20.          TYPE ===== ROTATE JOINT 1, -45 DEGREES AT SPEED 50 =====
   21.          DRIVE 1, -45.000, 50.00
   22.          PAUSE TYPE 'PROCEED' AND PRESS RETURN TO CONTINUE
   23.          REMARK *** PROGRAM WILL WAIT AS LONG AS REQUIRED
   24.          TYPE
   25.          TYPE ===== ROTATE JOINT 2,   15 DEGREES AT SPEED 30 =====
   26.          DRIVE 2, 15.000, 30.00
   27.          PAUSE TYPE 'PROCEED' AND PRESS RETURN TO CONTINUE
   28.          REMARK *** PROGRAM WILL WAIT AS LONG AS REQUIRED
   29.          TYPE
   30.          TYPE ===== ROTATE JOINT 3,   35.82 DEGREES AT SPEED 30 =====
   31.          DRIVE 3, 35.820, 30.00
   32.          PAUSE TYPE 'PROCEED' AND PRESS RETURN TO CONTINUE
   33.          REMARK *** PROGRAM WILL WAIT AS LONG AS REQUIRED
   34.          TYPE
   35.          TYPE ===== ROTATE JOINT 4,   50 DEGREES AT SPEED 20 =====
   36.          DRIVE 4, 50.000, 20.00
   37.          PAUSE TYPE 'PROCEED' AND PRESS RETURN TO CONTINUE
   38.          REMARK *** PROGRAM WILL WAIT AS LONG AS REQUIRED
   39.          TYPE
   40.          TYPE ===== ROTATE JOINT 5,   55.5 DEGREES AT SPEED 30 =====
   41.          DRIVE 5, 55.500, 30.00
   42.          PAUSE TYPE 'PROCEED' AND PRESS RETURN TO CONTINUE
   43.          REMARK *** PROGRAM WILL WAIT AS LONG AS REQUIRED
   44.          TYPE
   45.          TYPE ===== ROTATE JOINT 6, -30 DEGREES AT SPEED 15 =====
   46.          DRIVE 6, -30.000, 15.00
   47.          PAUSE TYPE 'PROCEED' AND PRESS RETURN TO CONTINUE
   48.          REMARK *** PROGRAM WILL WAIT AS LONG AS REQUIRED
   49.          TYPE
   50.          TYPE ************** END OF PROGRAM ****************
   51.          RETURN 0
.END
```

Fig. 8.22 Sample VAL program: the use of the 'DRIVE' instruction

'DRIVE' instruction

This instruction is similar to DRAW but can be used to program changes in specific joint angles, which under manual control would be generated by joint-interpolated mode. (Note that depending on the sign of the indicated angle, joints can be moved in both ways.) Fig. 8.22 demonstrates some possible joint movements.

Shift locations

The program shown in Fig. 8.23 (a) demonstrates the way locations can be shifted relative to a known location, or transformation. The program also incorporates a pick-and-place loop. After the third execution of the loop the position of the location 'PLACE' is moved by 150mm in the X direction and 100mm in the Z direction, using the SHIFT instruction. After the shift, the loop is executed another three times using the new 'PLACE' location before the subroutine stops. To follow the algorithm and the motion diagram see Fig. 8.24 and Fig. 8.25.

Change the robot's reference frame (BASE)

This instruction is very useful for example if for any reason the layout has been changed in a way that the robot has been moved but the locations that it has been taught have not. To save time and effort required to re-teach all locations the 'BASE' instruction can be used. (To return to the default reference frame in our case 'BASE 0,0,0,0' must be issued, rather than 'BASE 0,0,-100,30' as seems to be implied by the Users Guide to VAL.) Fig. 8.26 lists the program, Fig. 8.27 shows the flowchart, and Fig. 8.28 the motion diagram.

```
.PROGRAM DEMO7
   1.        TYPE   PROGRAM DEMO7 RUNNING...
   2.        TYPE
   3.        TYPE ****************************************************************
   4.        TYPE * THIS PROGRAM DEMONSTRATES THE 'SHIFT' INSTRUCTION        *
   5.        TYPE * BY INTRODUCING A PICK-AND-PLACE PROGRAM IN WHICH THE     *
   6.        TYPE * POSITION OF A SECOND 'PLACE' LOCATION IS DEFINED         *
   7.        TYPE * RELATIVE TO THE FIRST USING THE SHIFT INSTRUCTION.       *
   8.        TYPE * (PLEASE DO NOT ALTER THE DEFAULT MONITOR SPEED)          *
   9.        TYPE ****************************************************************
  10.        TYPE
  11.        PAUSE TYPE 'PROCEED' AND PRESS RETURN TO CONTINUE
  12.        REMARK *** PROGRAM WILL WAIT AS LONG AS REQUIRED
  13.        SETI CODE=1
  14.        REMARK *** ASSIGN THE VALUE OF 1 TO AN INTEGER VARIABLE, CALLED
  15.        REMARK *** CODE. THIS WILL INDICATE IN OUR PROGRAM WHICH
  16.        REMARK *** 'PLACE' LOCATION IS BEING USED
  17.        TYPE
  18.        TYPE == PLEASE STUDY THE FLOWCHART BEFORE PROGRAM EXECUTION ==
  19.        PAUSE TYPE 'PROCEED' AND PRESS RETURN TO CONTINUE
  20.        REMARK *** PROGRAM WILL WAIT AS LONG AS REQUIRED
  21.    10 SETI COUNT = 0
  22.        REMARK *** SET THE 'COUNT' VARIABLE TO ZERO
  23.        REMARK *** 'COUNT' STORES THE NUMBER OF COMPONENTS MOVED SINCE
  24.        REMARK *** A 'PLACE' SHIFT
  25.        OPENI
  26.        REMARK *** OPEN GRIPPER IMMEDIATELY
  27.    20 APPROS PICKUP, 200
  28.        REMARK *** APPROACH 'PICKUP' LOCATION, 200 MM ABOVE, IN THE
  29.        REMARK *** NEGATIVE 'Z' AXIS ALONG A STRAIGHT LINE
  30.        MOVES PICKUP
  31.        REMARK *** IT IS ALWAYS A GOOD IDEA TO PICK UP A COMPONENT
  32.        REMARK *** BY FOLLOWING A STRAIGHT LINE MOTION
  33.        CLOSEI 0.00
  34.        REMARK *** CLOSE GRIPPER IMMEDIATELY
  35.        DELAY 1.00
  36.        REMARK *** TO BE ON THE SAFE SIDE WAIT 1 SEC FOR PNEUMATIC
  37.        REMARK *** GRIPPER TO CLOSE
  38.        DEPARTS 200.00
  39.        REMARK *** MOVE BACK ALONG A STRAIGHT LINE
  40.        APPROS PLACE, 150
  41.        REMARK *** APPROACH LOCATION 'PLACE' ALONG A STRAIGHT LINE
  42.        MOVES PLACE
  43.        OPENI
  44.        REMARK *** MOVE TO LOCATION 'PLACE' ALONG A STRAIGHT LINE
```

Fig. 8.23 Sample VAL program: the use of the 'SHIFT' instruction

```
45.       REMARK *** AND OPEN GRIPPER
46.       DELAY 1.00
47.       REMARK *** TO BE ON THE SAFE SIDE WAIT 1 SEC FOR PNEUMATIC
48.       REMARK *** GRIPPER TO OPEN
49.       DEPARTS 150.00
50.       REMARK *** DEPART FROM LOCATION 'PLACE' ALONG A STRAIGHT
51.       REMARK *** LINE, 150 MM ABOVE
52.       SETI COUNT = COUNT + 1
53.       REMARK *** ONE COMPONENT HAS BEEN DELIVERED, INCREASE
54.       REMARK *** COUNT BY 1 (IE. COUNT = 1 AFTER THIS INSTRUCTION)
55.       TYPE
56.       TYPEI COUNT
57.       REMARK *** THE VALUE OF THE 'COUNT' VARIABLE WILL BE DISPLAYED
58.       IF COUNT LT 5 THEN 20
59.       REMARK *** IF LESS THEN 5 COMPONENTS HAVE BEEN PUT AT THE
60.       REMARK *** CURRENT 'PLACE' LOCATION THEN GO TO LABEL 20
61.       REMARK ***     ELSE
62.       SETI CODE = CODE + 1
63.       REMARK *** SINCE 5 COMPONENTS HAVE NOW BEEN MOVED, CHANGE THE
64.       REMARK *** CODE TO INDICATE THAT THE PLACE POSITION SHOULD BE
65.       REMARK *** SHIFTED
66.       IF CODE EQ 3 THEN 30
67.       REMARK *** IF THE CODE VARIABLE HAS A VALUE OF 3, THIS MEANS
68.       REMARK *** THAT BOTH PLACE LOCATIONS HAVE BEEN FILLED, THUS
69.       REMARK *** GO TO LABEL 30 AND SHIFT 'PLACE'
70.       TYPE
71.       TYPE NOTE THAT THE 'PLACE' LOCATION WILL NOW BE SHIFTED
72.       TYPE ****************************************************
73.       TYPE
74.       DELAY 2.00
75.       REMARK *** THE ROBOT WILL WAIT FOR APPROXIMATELY 2 SECONDS
76.       SHIFT PLACE BY 150.00, 0.00, 100.00
77.       REMARK *** MOVE (IE. SHIFT) LOCATION 'PLACE' BY 150 MM IN 'X'
78.       REMARK *** AND 100.00 MM IN 'Z'
79.       GOTO 10
80.       REMARK *** TRANSFER PROGRAM CONTROL TO LABEL 10, WHERE THE
81.       REMARK *** ROUTINE RESTARTS USING THE SECOND 'PLACE' POSITION
82.    30 SHIFT PLACE BY -150.00, 0.00, -100.00
83.       REMARK *** MOVE LOCATION 'PLACE' BACK TO WHERE IT WAS BEFORE
84.       TYPE
85.       TYPE **************  END OF PROGRAM  *****************
86.       RETURN 0
.END
```

Fig. 8.23 continued

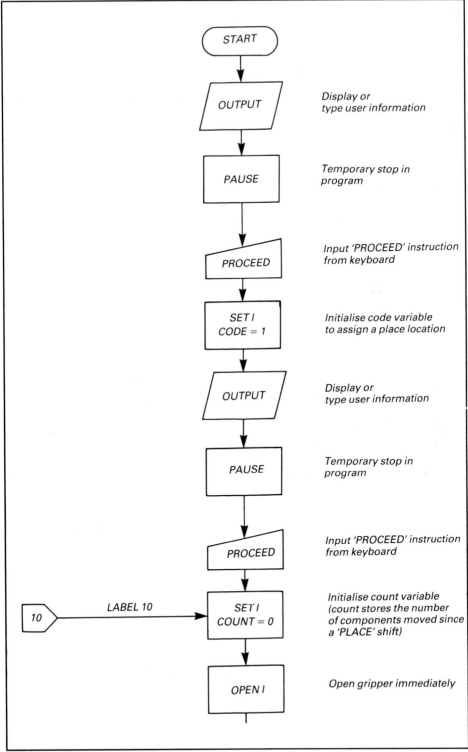

Fig. 8.24 Flowchart of the program shown in Fig. 8.23

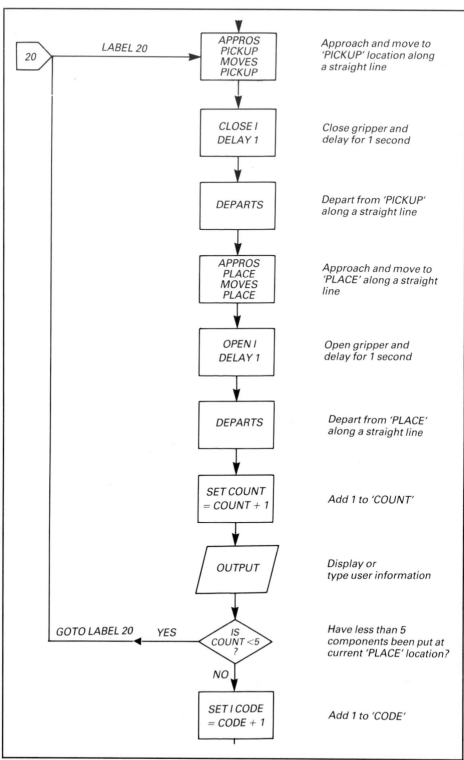

Fig. 8.24 continued

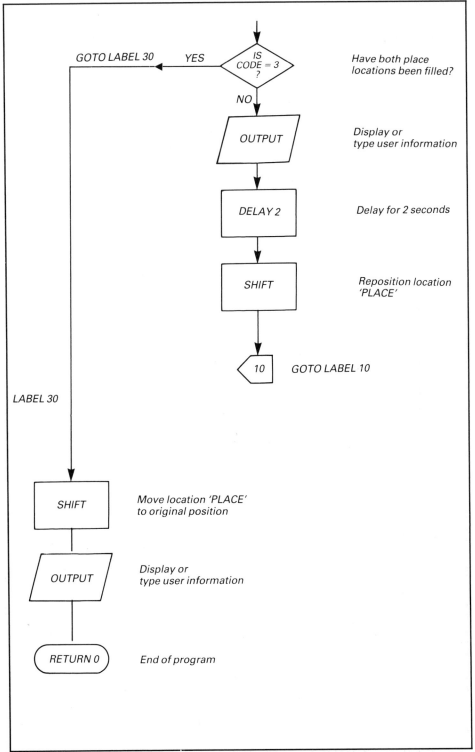

Fig. 8.24 continued

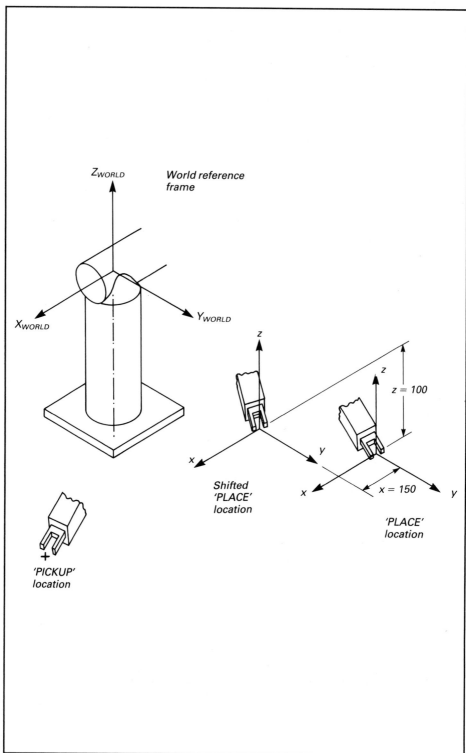

Fig. 8.25 Diagrammatic representation of the program shown in Fig. 8.23

```
.PROGRAM DEMO8
    1.          TYPE    PROGRAM DEMO8 RUNNING...
    2.          TYPE
    3.          TYPE ***************************************************************
    4.          TYPE * THIS PROGRAM DEMONSTRATES THE 'BASE' INSTRUCTION.          *
    5.          TYPE * THE 'BASE' INSTRUCTION IS USED WHEN THE POSITION OF        *
    6.          TYPE * THE ROBOT IS CHANGED IN RELATION TO THE WORKPLACE, SO      *
    7.          TYPE * THAT THE POSITION OF THE ROBOT DO NOT NEED TO BE RE-       *
    8.          TYPE * TAUGHT. NOTE THAT THE ROBOT DISPLACEMENT IN X,Y AND Z      *
    9.          TYPE * MUST BE KNOWN.       (SET SLOW MONITOR SPEED)              *
   10.          TYPE ***************************************************************
   11.          TYPE
   12.          PAUSE TYPE 'PROCEED' AND PRESS RETURN TO CONTINUE
   13.          REMARK *** PROGRAM WILL WAIT AS LONG AS REQUIRED
   14.          MOVE START
   15.          REMARK *** MOVE TO LOCATION 'START'
   16.          SETI CODE = 0
   17.          REMARK *** THE VARIABLE 'CODE' IS USED TO INDICATE WHETHER THE
   18.          REMARK *** REFERENCE BASE IS NORMAL OR NOT.(CODE = 0 IS NORMAL)
   19.     10 SETI COUNT = 0
   20.          REMARK *** A COUNTER IS ALSO SET TO 0. IT IS USED TO MONITOR
   21.          REMARK *** THE NUMBER OF TIMES THE ROBOT COMPLETES THE PICK-
   22.          REMARK *** AND-PLACE JOB BETWEEN LOCATIONS 'PICKUP' AND 'PLACE'
   23.          OPENI
   24.     20 APPRO PICKUP, 150.00
   25.          MOVES PICKUP
   26.          REMARK *** APPROACH AND MOVE TO LOCATION 'PICKUP' WITH AN OPEN
   27.          REMARK *** GRIPPER
   28.          CLOSEI
   29.          DELAY 1.00
   30.          REMARK *** CLOSE GRIPPER AND WAIT 1 SECOND
   31.          DEPARTS 150.00
   32.          REMARK *** DEPART ALONG A STRAIGHT LINE
   33.          APPROS PLACE, 150.00
   34.          MOVES PLACE
   35.          REMARK *** APPROACH AND MOVE TO LOCATION 'PLACE'
   36.          OPENI
   37.          DELAY 1.00
   38.          REMARK *** OPEN GRIPPER AND WAIT 1 SECOND
   39.          DEPARTS 150.00
   40.          SETI COUNT = COUNT + 1
   41.          REMARK *** INCREMENT THE COUNTER BY 1
   42.          TYPE *** CURRENT VALUE OF THE COUNTER ***
   43.          TYPEI COUNT
   44.          REMARK *** DISPLAY THE CURRENT VALUE OF 'COUNT'
   45.          IF COUNT LT 3 THEN 20
   46.          REMARK *** IF THE MOVEMENT SEQUENCE HAS NO BEEN COMPLETED
   47.          REMARK *** THREE TIMES YET, REPEAT IT
   48.          SETI CODE = CODE + 1
   49.          REMARK *** INCREASE THE CURRENT VALUE OF 'CODE' BY 1
   50.          IF CODE EQ 2 THEN 30
   51.          REMARK *** IF 'CODE' HAS REACHED THE VALUE OF 2 THE PICK-
   52.          REMARK *** AND-PLACE LOOP HAS BEEN COMPLETED 3 TIMES
   53.          REMARK *** THUS TRANSFER PROGRAM CONTROL TO LABEL 30
   54.          TYPE ************* WARNING ***************
   55.          TYPE   THE BASE POSITION IS GOING TO BE CHANGED
   56.          TYPE ***************************************
   57.          TYPE
   58.          PAUSE TYPE 'PROCEED' AND PRESS RETURN TO CONTINUE
   59.          BASE 0.00, 0.00, -100.00, -30.00
   60.          REMARK *** REPOSITION ROBOT WORLD REFERENCE FRAME. NEW VALUES:
   61.          REMARK *** X,Y=0, Z=-100, AND ROTATE OLD FRAME BY -30 DEGREES.
   62.          GOTO 10
   63.          REMARK *** CONTINUE PROGRAM EXECUTION AT LABEL 10
   64.          REMARK *** BEGIN MOTION CONTROL AND RESET COUNTER
   65.     30 BASE 0.000, 0.000, 0.000, 0.000
   66.          REMARK *** THIS IS THE DEFAULT WORLD REFERENCE FRAME
   67.          MOVE START
   68.          REMARK *** RETURN TO THE STARTING POINT IN THE DEFAULT FRAME
   69.          TYPE
   70.          TYPE ************* END OF PROGRAM ****************
   71.          RETURN 0
.END
```

Fig. 8.26 Sample VAL program: the use of the 'BASE' instruction

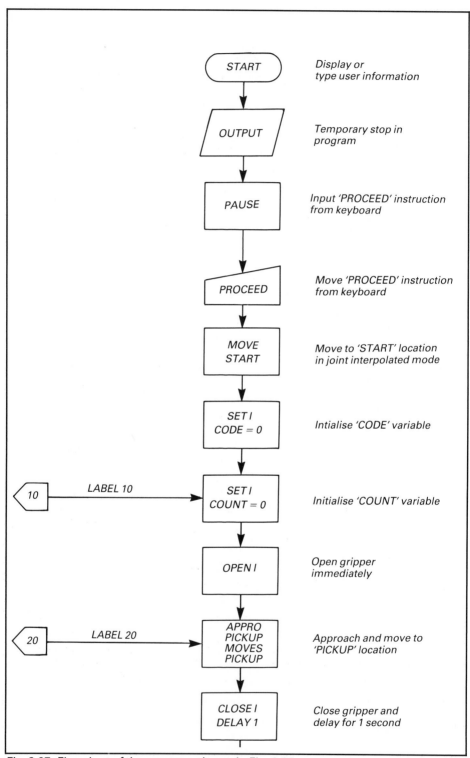

Fig. 8.27 Flowchart of the program shown in Fig. 8.26

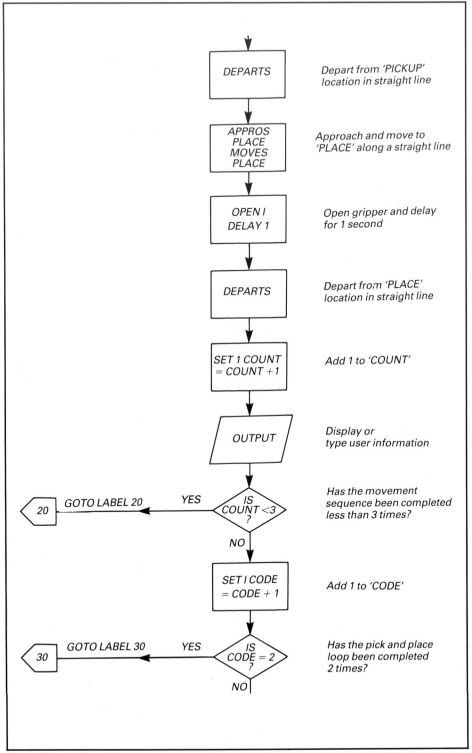

Fig. 8.27 continued

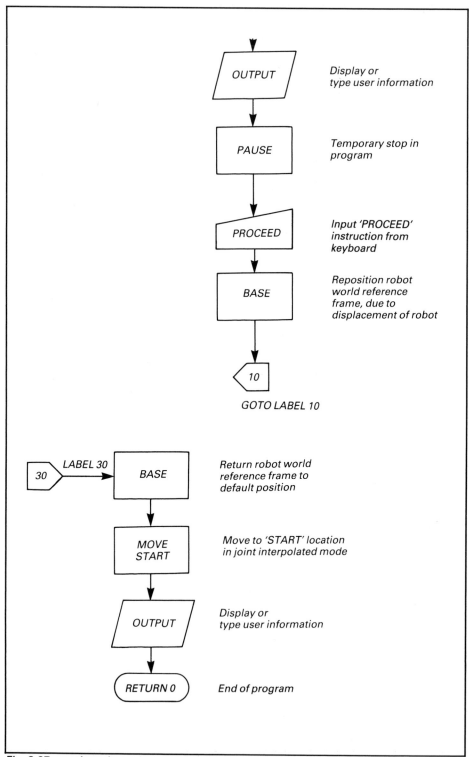

Fig. 8.27 continued

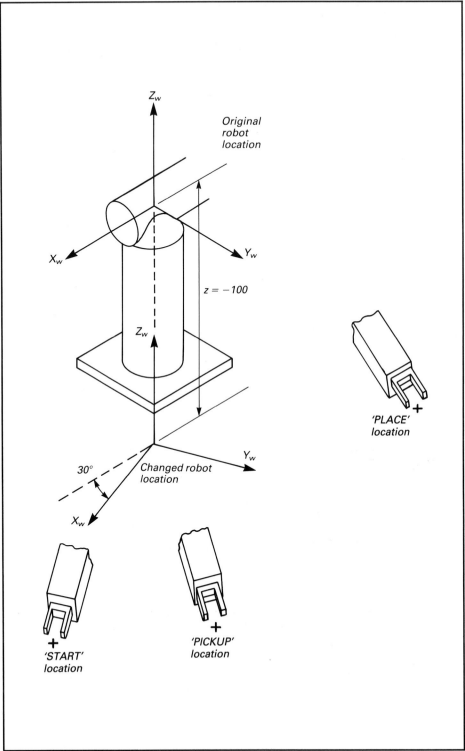

Fig. 8.28 Diagrammatic representation of the program shown in Fig. 8.26

```
.PROGRAM DEMO9
   1.          TYPE   PROGRAM DEMO9 RUNNING...
   2.          TYPE
   3.          TYPE ********************************************************
   4.          TYPE * THIS PROGRAM DEMONSTRATES THE 'TOOL' INSTRUCTION.    *
   5.          TYPE * THIS INSTRUCTION IS VERY USEFUL WHEN THE ROBOT IS    *
   6.          TYPE * USING MORE THAN ONE TOOL, SINCE   DIFFERENT TOOL     *
   7.          TYPE * LENGTH (Z) AND X AND Y DATA CAN BE AUTOMATICALLY     *
   8.          TYPE * COMPENSATED FOR.     (SET SLOW MONITOR SPEED)        *
   9.          TYPE ********************************************************
  10.          TYPE
  11.          PAUSE TYPE 'PROCEED' AND PRESS RETURN TO CONTINUE
  12.          REMARK *** PROGRAM WILL WAIT AS LONG AS REQUIRED
  13.          MOVE START
  14.          REMARK *** START PROGRAM AT LOCATION 'START'
  15.          SETI CODE = 0
  16.          REMARK *** CODE IS USED TO STORE THE TOOL TRANSFORMATION
  17.          REMARK *** STATUS. (IF CODE = 0 THEN NO TRANSFORMATION)
  18.      10 SETI COUNT = 0
  19.          REMARK *** THIS COUNTER SHOWS THE NUMBER OF TIMES THE ROBOT
  20.          REMARK *** HAS PERFORMED THE PICK-AND-PLACE OPERATION
  21.      20 OPENI
  22.          APPRO PICKUP, 120.00
  23.          MOVES PICKUP
  24.          REMARK *** APPROACH AND MOVE TO LOCATION 'PICKUP' WITH AN
  25.          REMARK *** OPEN GRIPPER
  26.          APPROS PLACE, 150.00
  27.          MOVES PLACE
  28.          CLOSEI
  29.          DELAY 1.00
  30.          REMARK *** APPROACH AND MOVE TO LOCATION 'PLACE' AND CLOSE
  31.          REMARK *** THE GRIPPER UPON ARRIVAL
  32.          SETI COUNT = COUNT + 1
  33.          REMARK *** THE FIRST LOOP HAS BEEN COMPLETED
  34.          TYPE
  35.          TYPE NUMBER OF COMPLETED LOOPS
  36.          TYPEI COUNT
  37.          REMARK *** DISPLAY NUMBER OF COMPLETED LOOPS
  38.          IF COUNT LT 3 THEN 20
  39.          REMARK *** DO NOT CONTINUE UNTIL ALL THREE LOOPS HAVE BEEN
  40.          REMARK *** COMPLETED
  41.          SETI CODE = CODE + 1
  42.          REMARK *** CURRENT CODE INCREMENTED BY 1
  43.          IF CODE EQ 2 THEN 30
  44.          REMARK *** IF CODE HAS REACHED THE VALUE OF 2, THEN ALL
  45.          REMARK *** ALL LOOPS HAVE BEEN COMPLETED
  46.          TOOL T1
  47.          REMARK *** TRANSFORMATION T1 REPRESENTS 150 MM DISPLACEMENT
  48.          REMARK *** ALONG THE TOOL 'Z' AXIS. REPOSITION ROBOT TO
  49.          REMARK *** ACCOMODATE THIS NEW TOOL
  50.          GOTO 10
  51.          REMARK *** CONTINUE PROGRAM AT LABEL 10 TO BE ABLE TO
  52.          REMARK *** ACCOMPLISH THREE CYCLES AND TOOL OFFSETS
  53.      30 TOOL T1.INV
  54.          REMARK *** SET TOOL TRANSFORMATION BACK TO NORMAL
  55.          TYPE
  56.          TYPE **************  END OF PROGRAM  ****************
  57.          RETURN 0
.END
```

Fig 8.29 Sample VAL program: the use of the 'TOOL' instruction

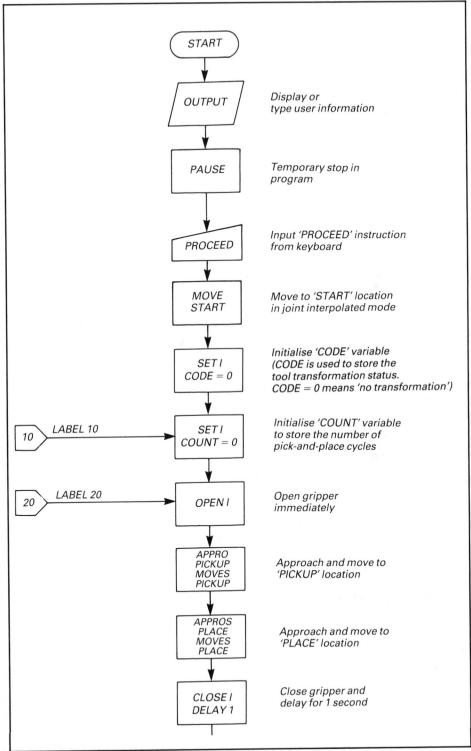

Fig. 8.30 Flowchart of the program shown in Fig. 8.29

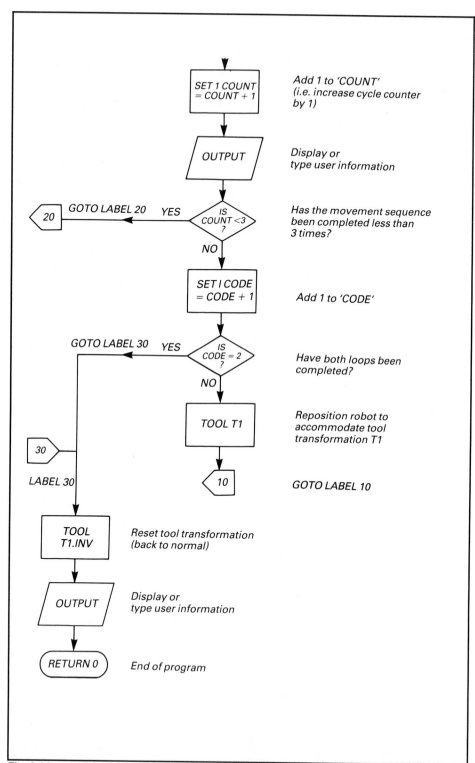

SET 1 COUNT = COUNT + 1 — Add 1 to 'COUNT' (i.e. increase cycle counter by 1)

OUTPUT — Display or type user information

GOTO LABEL 20 — 20 — YES — IS COUNT <3 ? — Has the movement sequence been completed less than 3 times?

NO

SET I CODE = CODE + 1 — Add 1 to 'CODE'

GOTO LABEL 30 — YES — IS CODE = 2 ? — Have both loops been completed?

NO

TOOL T1 — Reposition robot to accommodate tool transformation T1

30 — LABEL 30

10 — GOTO LABEL 10

TOOL T1.INV — Reset tool transformation (back to normal)

OUTPUT — Display or type user information

RETURN 0 — End of program

Fig. 8.30 continued

Tool coordinate transformations

The purpose of the 'TOOL' instruction is to allow the use of several different tools of different length and orientations. (For an application example see Chapter Twelve, where automated robot hand changers (among others) are discussed.)

In the demonstration program (Fig. 8.29) the tool transformation facility is used in the format given in Table 8.2.

Table 8.2 Tool transformation facility format

Variable	X/JT1	Y/JT2	Z/JT3	O/JT4	A/JT5	T/JT6
T1	0.00	0.00	150.00	90.00	−90.00	0.00
T1.INV	0.00	0.00	0.00	90.00	−90.00	0.00

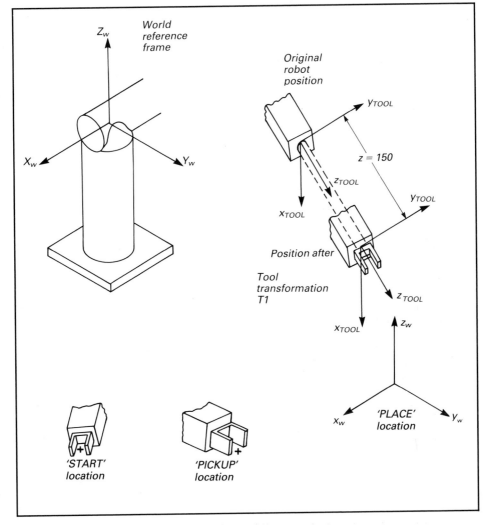

Fig. 8.31 Diagrammatic representation of the program shown in Fig. 8.29

```
.PROGRAM STARTME
    1.        TYPE   ROUTINE STARTME RUNNING...
    2.        TYPE
    3.        TYPE ***************************************************************
    4.        TYPE * THIS PROGRAM DEMONSTRATES A STANDARD STARTUP ROUTINE     *
    5.        TYPE * IT CONTAINS ROBOT CONFIGURATION AND ORIENTATION          *
    6.        TYPE * INSTRUCTIONS TO ENSURE THAT PROPER CONDITIONS ARE SET    *
    7.        TYPE * BEFORE PROGRAM EXECUTION. (SET SLOW MONITOR SPEED)       *
    8.        TYPE ***************************************************************
    9.        TYPE
   10.        RIGHTY
   11.        REMARK ***   ORIENTATE ROBOT LIKE A HUMAN RIGHT ARM
   12.        ABOVE
   13.        REMARK ***   ELBOW POINTED UPWARDS
   14.        NOFLIP
   15.        REMARK ***   JOINT 5 POSITIVE ANGLES ONLY
   16.        APPRO START, 5.000
   17.        MOVE START
   18.        REMARK ***   APPROACH AND MOVE TO LOCATION START
   19.        OPENI
   20.        DELAY 1.00
   21.        REMARK ***   OPEN GRIPPER IMMEDIATELY
   22.        TYPE
   23.        TYPE ************** END OF ROUTINE ****************
   24.        RETURN 0
.END


.PROGRAM PALLET1
    1.        TYPE   ROUTINE PALLET1 RUNNING...
    2.        TYPE
    3.        TYPE ************************************************************
    4.        TYPE * THIS PROGRAM DEMONSTRATES PALLETISING.                  *
    5.        TYPE * PALLET REFERENCE LOCATION, PALREF AND PICKUP            *
    6.        TYPE * LOCATION PICKREF MUST BE TAUGHT PRIOR TO                *
    7.        TYPE * PROGRAM EXECUTION                                       *
    8.        TYPE *----------------------------------------------------------*
    9.        TYPE * LAYOUT:            * PUMA 560 *                          *
   10.        TYPE *                                                         *
   11.        TYPE * (+X) COLUMN +  50  +  50  +  50  +  50  + PALREF        *
   12.        TYPE *            100                                          *
   13.        TYPE *              +      +      +      +      +              *
   14.        TYPE *            100                                          *
   15.        TYPE *              +      +      +      +    + ROW (+Y) *
   16.        TYPE *                                                         *
   17.        TYPE *----------------------------------------------------------*
   18.        TYPE * MOTION SEQUENCE: PICKUP-PALREF-COLUMN LOCATIONS         *
   19.        TYPE * NEXT ROW-COLUMN LOCATIONS-NEXT ROW...UNTIL ROW=3        *
   20.        TYPE *----------------------------------------------------------*
   21.        TYPE * SET SLOW MONITOR SPEED AND OPEN THE GRIPPER BEFORE *
   22.        TYPE * PROGRAM EXECUTION.                                      *
   23.        TYPE ************************************************************
   24.        TYPE
   25.        PAUSE TYPE 'PROCEED' AND PRESS RETURN TO CONTINUE
   26.        REMARK *** PROGRAM WILL WAIT AS LONG AS REQUIRED
   27.        SET PALORIG=PALREF
   28.        REMARK ***   STORE LOCATION PALREF FOR FURTHER USE
   29.        SETI COL=0
   30.        SETI ROW=1
   31.        REMARK ***   INITIALISE COLUMN AND ROW COUNTER VARIABLES
   32.   10   SETI COL=COL+1
   33.        REMARK *** INCREASE CURRENT COLUMN COUNTER BY ONE (IE. ROBOT
   34.        REMARK *** MOVES TO THE NEXT NEXT COLUMN ON THE PALLET)
   35.        TYPE
   36.        TYPE CURRENT ROW
```

Fig. 8.32 Sample VAL program: a general purpose palletising program

```
37.       TYPEI ROW
38.       TYPE
39.       TYPE CURRENT COLUMN
40.       TYPEI COL
41.       REMARK ***  DISPLAY ACTUAL ROW AND COLUMN VALUES
42.       IF COL GT 5 THEN 20
43.       REMARK *** IF ALL COLUMNS COMPLETED TRANSFER PROGRAM CONTROL
44.       REMARK *** TO LABEL 20
45.       REMARK ***  ELSE CONTINUE HEREBELOW...
46.       REMARK *** IF THE NUMBER OF COLUMNS IN YOUR APPLICATION ARE
47.       REMARK *** DIFFERENT, CHANGE THE VALUE 5 HERE...
48.       APPRO PICKREF, 100.00
49.       MOVES PICKREF
50.       REMARK *** APPROACH AND MOVE TO LOCATION 'PICKREF'
51.       CLOSEI
52.       DELAY 1.00
53.       REMARK *** CLOSE GRIPPER IMMEDIATELY
54.       DEPARTS 100.00
55.       REMARK *** DEPART FROM 'PICKREF' ALONG A STRAIGHT LINE
56.       APPRO PALREF, 100.00
57.       MOVES PALREF
58.       REMARK *** APPROACH AND MOVE TO LOCATION 'PALREF'
59.       OPENI
60.       DELAY 1.00
61.       REMARK *** OPEN GRIPPER AND WAIT 1 SECOND
62.       DEPARTS 100.00
63.       SHIFT PALREF BY 50.00, 0.00, 0.00
64.       REMARK ***  FIRSTLY SERVE ALL COLUMN LOCATIONS, THEN INCREMENT
65.       REMARK *** TO THE NEXT ROW
66.       REMARK *** NOTE: IF COLUMN INCREMENT MUST BE CHANGED, REWRITE
67.       REMARK *** 50 TO THE NEW VALUE. (VAL IS NOT CAPABLE OF PASSING
68.       REMARK ***  A VARIABLE NAME HERE)
69.       GOTO 10
70.       REMARK *** REPEAT UNTIL NOT COMPLETED
71.  20   SETI ROW=ROW+1
72.       REMARK *** TAKE NEXT ROW
73.       SETI COL=0
74.       REMARK *** RESET THE COLUMN COUNTER TO 0 AND CONTINUE TO WORK
75.       IF ROW GT 3 THEN 30
76.       REMARK ***  ELSE
77.       REMARK *** NOTE IF THE NUMBER OF ROWS ARE DIFFERENT IN YOUR
78.       REMARK *** CASE, REWRITE 3 TO THE NEW VALUE
79.       SET PALREF=PALORIG
80.       REMARK *** RESET PALREF TO START POSITION TO MAKE SHIFT
81.       REMARK *** STATEMENT SIMPLE
82.       SHIFT PALREF BY 0.00, 100.00, 0.00
83.       REMARK *** STEP RAW INCREMENT OF 100 MM ALONG THE ROWS
84.       REMARK *** REWRITE THIS VALUE AS REQUIRED
85.       GOTO 10
86.  30   APPRO PICKREF, 100.00
87.       MOVES PICKREF
88.       REMARK ***  ROBOT IN PICK-UP LOCATION
89.       SET PALREF=PALORIG
90.       REMARK *** RESET PALREF TRANSFORMATION TO PALORIG
91.       TYPE
92.       TYPE ************** END OF PROGRAM  ***************
93.       RETURN 0
.END
```

Fig. 8.32 continued

```
.PROGRAM GOHOME
    1.          TYPE    ROUTINE GOHOME RUNNING...
    2.          TYPE
    3.          TYPE ************************************************************
    4.          TYPE * THIS PROGRAM DEMONSTRATES A STANDARD 'GO HOME' ROUTINE *
    5.          TYPE * (IT MOVES THE ROBOT TO A SAFE LOCATION)                *
    6.          TYPE * LOCATION 'GHOME' MUST BE TAUGHT PRIOR TO PROGRAM       *
    7.          TYPE * EXECUTION.                                             *
    8.          TYPE ************************************************************
    9.          TYPE
   10.          PAUSE TYPE 'PROCEED' AND PRESS RETURN TO CONTINUE
   11.          REMARK *** PROGRAM WILL WAIT AS LONG AS REQUIRED
   12.          APPRO GHOME,100
   13.          MOVE GHOME
   14.          TYPE
   15.          TYPE *** IN GHOME POSITION ***
   16.          TYPE
   17.          TYPE *************** END OF PROGRAM  ****************
   18.          RETURN 0
.END


.PROGRAM PALLETMAIN
    1.          TYPE    PROGRAM PALLETMAIN (DEMO10) RUNNING...
    2.          TYPE
    3.          TYPE ************************************************************
    4.          TYPE * THIS PROGRAM DEMONSTRATES THE WAY SUBROUTINES CAN   *
    5.          TYPE * BE USED IN A GENERAL PURPOSE PALLETISING PROGRAM.   *
    6.          TYPE ************************************************************
    7.          GOSUB STARTME
    8.          REMARK *** CALL SUBROUTINE 'STARTME' (SEE ABOVE)
    9.          GOSUB PALLET1
   10.          REMARK *** CALL SUBROUTINE 'PALLET1' (SEE ABOVE)
   11.          GOSUB GOHOME
   12.          REMARK *** CALL SUBROUTINE 'GOHOME' (SEE ABOVE)
   13.          TYPE
   14.          TYPE *************** END OF MAIN PROGRAM  ****************
   15.          TYPE
   16.          REMARK *** NOTE THE LACK OF 'RETURN 0'.
   17.          REMARK *** BECAUSE OF THIS IT CANNOT BE USED AS A SUBROUTINE,
   18.          REMARK *** ONLY AS A MAIN PROGRAM CALLING THE INDICATED
   19.          REMARK *** ROUTINES.
.END
```

Fig. 8.32 continued

It should be noted that the transformation for correcting the length of the tool in the Z direction is used and that 'T1.INV' is not the inverse transformation of transformation T1 as generated by the 'INVERSE' command. It merely returns the tool transformation to its normal value. (See also Figs. 8.30 and 8.31).

General-purpose palletising program

Fig. 8.32 illustrates the way a general purpose palletising program could be written. Note the way the routines 'STARTME', 'PALLET1' and 'GOHOME' are organised and called from a main program 'PALLETMAIN'. Also note the pallet row and column size changes advised in the program listings. The operation of the program is explained in Figs. 8.33 and 8.34.

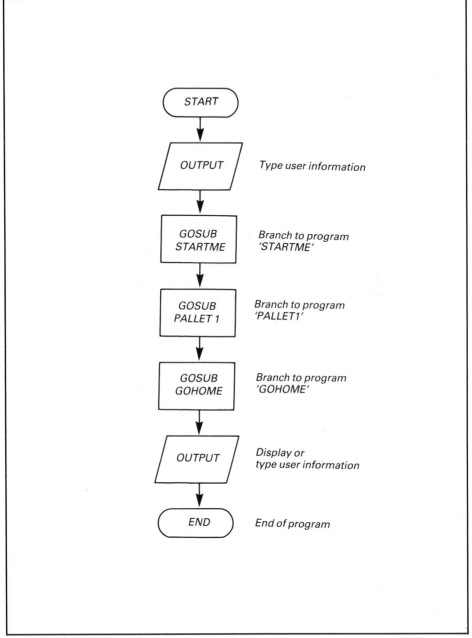

Fig. 8.33(a) Flowchart of the program shown in Fig. 8.32

Although VAL is not a structured language, it is very much advised to use this type of subroutine . . . main program organisation, as well as parametrised programming wherever VAL allows this. (Unfortunately these facilities seem to be very limited in VAL compared to Pascal or AML.)

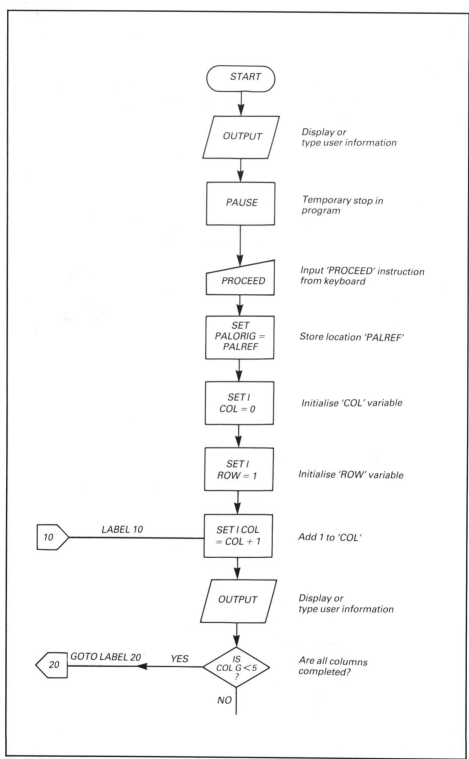

Fig. 8.33(b) Flowchart of the program shown in Fig. 8.32

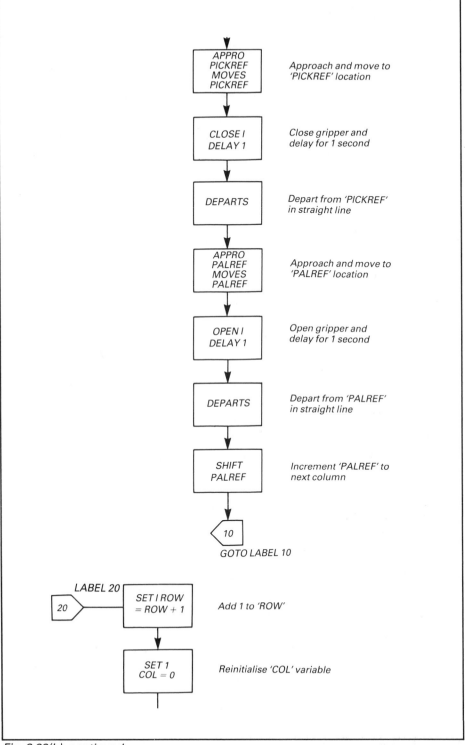

APPRO
PICKREF
MOVES
PICKREF — Approach and move to 'PICKREF' location

CLOSE I
DELAY 1 — Close gripper and delay for 1 second

DEPARTS — Depart from 'PICKREF' in straight line

APPRO
PALREF
MOVES
PALREF — Approach and move to 'PALREF' location

OPEN I
DELAY 1 — Open gripper and delay for 1 second

DEPARTS — Depart from 'PALREF' in straight line

SHIFT
PALREF — Increment 'PALREF' to next column

10
GOTO LABEL 10

LABEL 20
20
SET I ROW
= ROW + 1 — Add 1 to 'ROW'

SET 1
COL = 0 — Reinitialise 'COL' variable

Fig. 8.33(b) continued

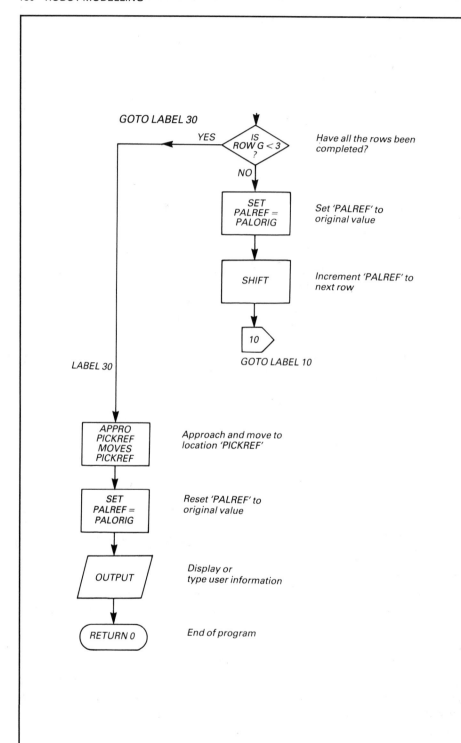

Fig. 8.33(b) continued

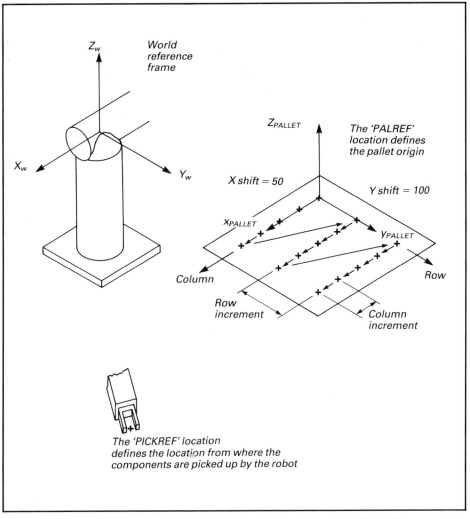

Fig. 8.34 Diagrammatic representation of the program shown in Fig. 8.32

Definition of new work planes

These routines explain the use of the 'FRAME' statement and some limited inverse transformation, or compound transformation, with the 'INVERSE' instruction (Fig. 8.35).

8.6 New developments in robot programming

There is still a great deal of work required in robot programming and control, and more generally in machine and manufacturing cell programming and control. The major problems relating to robot programming include:

● Lack of general understanding of the robot integrated manufacturing cell concept, as outlined in Chapter One.

```
.PROGRAM DEF
   1.        TYPE   PROGRAM DEF (DEMO11) RUNNING...
   2.        TYPE
   3.        TYPE ********************************************************
   4.        TYPE * GENERAL PROCEDURE FOR DEFINING NEW FRAME LOCATIONS *
   5.        TYPE ********************************************************
   6.        TYPE
   7.        PAUSE TYPE 'PROCEED' AND PRESS RETURN TO CONTINUE
   8.        REMARK *** PROGRAM WILL WAIT AS LONG AS REQUIRED
   9.        TYPE
  10.        TYPE POSITION THE ROBOT ARM AT THE FRAME ORIGIN, 'FORIG'
  11.        PAUSE TYPE 'PROCEED' AND PRESS RETURN TO CONTINUE
  12.        HERE FORIG
  13.        TYPE
  14.        TYPE POSITION THE ROBOT ARM AT A LOCATION ON THE FRAME, 'FX'
  15.        TYPE ALONG THE DESIRED 'X' AXIS
  16.        PAUSE TYPE 'PROCEED' AND PRESS RETURN TO CONTINUE
  17.        HERE FX
  18.        TYPE
  19.        TYPE POSITION THE ROBOT ARM AT A LOCATION ON THE FRAME, 'FY'
  20.        TYPE ALONG THE DESIRED 'Y' AXIS
  21.        PAUSE TYPE 'PROCEED' AND PRESS RETURN TO CONTINUE
  22.        HERE FY
  23.        REMARK *** A FRAME MAY NOW BE DECLARED
  24.        TYPE
  25.        TYPE ********   END OF NEW FRAME LOCATION DEFINITION  *********
  26.        RETURN 0
.END


.PROGRAM FRAMEDEF
   1.        TYPE   PROGRAM FRAMEDEF (DEMO12) RUNNING...
   2.        TYPE
   3.        TYPE ********************************************************
   4.        TYPE * A GENERAL PROCEDURE FOR DEFINING A NEW WORK PLANE   *
   5.        TYPE * WITH THE 'FRAME' INSTRUCTION.                       *
   6.        TYPE * IF THIS ROUTINE IS EXECUTED BEFORE ANY OTHER PROGRAM *
   7.        TYPE * THE INSTRUCTIONS AND LOCATIONS WILL BE INTERPRETED  *
   8.        TYPE * ON THIS NEW 'WORKPLANE' AS DEFINED HERE, RATHER THAN *
   9.        TYPE * THE DEFAULT WORLD CO-ORDINATE SYSTEM.               *
  10.        TYPE ********************************************************
  11.        TYPE
  12.        PAUSE TYPE 'PROCEED' AND PRESS RETURN TO CONTINUE
  13.        REMARK *** PROGRAM WILL WAIT AS LONG AS REQUIRED
  14.        FRAME WORKPLANE = FORIG, FX, FY
  15.        REMARK *** IT IS OUR AIM TO PALETTISE ON A NEW FRAME,
  16.        REMARK *** NAMED 'WORKPLANE'
  17.        INVERSE INV.WORKPLANE = WORKPLANE
  18.        REMARK *** THE ROBOT HAS NOW DEFINED THE NEW WORK PLANE
  19.        REMARK ***
  20.        SET PALSTART = INV.WORKPLANE:FORIG
  21.        REMARK *** THE PALLET STARTING POINT HAS BEEN DEFINED IN
  22.        REMARK *** PALLET X,Y CO-ORDINATES
  23.        SET PALREF = PALSTART
  24.        REMARK *** NEW PALLET REFERENCE POINT IS DEFINED
  25.        SET PLACE = WORKPLANE:PALREF
  26.        REMARK *** LOCATION 'PALREF' IS DEFINED IN THE ROBOT
  27.        REMARK *** CO-ORDINATE FRAME.
  28.        REMARK *** THIS INSTRUCTION WILL ALLOW LOCATION 'PLACE'
  29.        REMARK *** TO BE UPDATED BY SHIFTING LOCATION 'PALREF'.
  30.        TYPE
  31.        TYPE *********   END OF PROGRAM   **********
  32.        RETURN 0
.END
```

*Fig. 8.35 Sample VAL program: explanation of the 'FRAME' and 'INVERSE'
instructions*

- Slow rate of introduction of powerful computer based machine and robot controllers, which have similar operating system and interfacing capacities as the current 16-32 bit mini and microcomputers.
- Lack of intelligent high-level standard machine control languages, providing access not only at a high level but also at a robot system programming level.
- High cost of three-dimensional vision systems which could provide data for an image database interfaced with the robot control system and a currently commercially unavailable suitable expert system.
- Slow speed and high cost of commercial automated robot hand changing and part transfer systems.
- Slow rate and high cost of the introduction of solid-modelling techniques to be used in component and manufacturing process design (CAD/CAM), and the lack of systems offering solid-model-based generation and animation of machine-independent part programs.

Obviously the above points need several years of research and development work. There is a lot of research already done and there are new developments in all fields outlined above, but since it is often hard to justify economically the need for such improvements and new products in industry very few of them are commercially available.

A robot-independent programming exercise

To illustrate one aspect of the concept that users and robot programmers must be provided with a robot and/or robot language independent, task-orientated, interactive, user-friendly operator interface and a language generator which is robot dependent, a simple rule-based language module (MARTI) is introduced. This was developed by one of the authors at Trent Polytechnic, Nottingham as a research exercise[4].

This task-orientated module was developed because it was often felt in practice, particularly when programming assembly tasks or relatively complex palletising, routing, glueing, drilling, etc. jobs, that neither the teach-mode nor the supplied off-line language provide a sufficient amount of programming efficiency.

The concept is to provide a robot language independent, interactive, user-friendly operator interface, and a language generator which is robot dependent. Following this concept, task-orientated programming modules could be used for different applications, capable of helping the operator in generating programs for palletising, routing, assembly and drilling tasks, and motions where circular and linear interpolation is required. The module to be demonstrated, as well as the robot language dependent software interface, have been written in portable Pascal using a run-time operating system capable of running on most 16-bit microcomputers.

The major benefit of this concept is that although it is off-line, interactive and high level, it combines the benefits of the off-line, the run-time interpretative and the teach-mode type of robot programming, by offering a robot-dependent interface as well as the high-level user-interface. In other words, the operator can write and link any robot dependent modules in the robot dependent language, for example VAL, to the previously off-line generated programs.

The other benefit is the possibility of standardising robot programming in a workshop where different types of robots are to be used. The program is capable of being interfaced directly with the particular robot, provided a DNC link can be established between the computer and the particular robot controller.

The interactive 'fill-in-the-screen' type of input, provided in the robot-independent module of the program, needs no special training or knowledge of a particular robot language and can be used to generate different robot programs using the same input language. This feature improves the reliability and the efficiency of robot programming, reflected in programming economics as well.

Having access both to the robot-independent, intelligent user-interface, as well as to the robot-dependent language, it is more than likely that this type of modular and multi-level programming will have a great future in real-time machine and robot control, particularly in those areas where sensors are developing faster than the high-level languages which should handle them. (The current level of CNC machine tool intelligence and the rigid APT languages are good examples.)

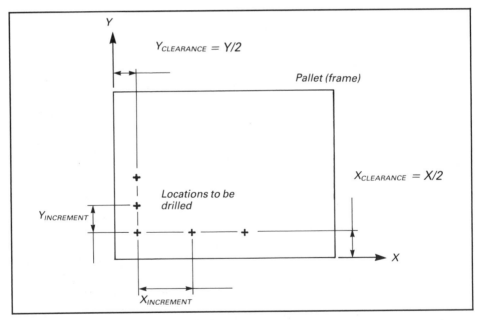

Fig. 8.36 *Input data in the task orientated robot programming modules is provided by the operator, by simply filling in a series of user-friendly screens. (All input is run-time checked for range errors providing a safe man-machine interface.) The input screen in this example contains the following data: 12 drilled holes along the 'X' coordinate axis incrementing by 22.00mm centre-to-centre; 5 drilled holes along the 'Y' coordinate axis incrementing by 31.00mm centre-to-centre; the drill approach speed is = 60 (robot speed value); the drilling speed (robot feed motion) is equal to 50% of the approach speed; the distance of the drill above the surface is 15mm; the depth the drill enters into the material is 12mm*

To illustrate the MARTI off-line robot program generator, Fig. 8.36 shows the input data collection and Fig. 8.37 the generated program for the calculated pattern to be drilled with the PUMA 560 robot. In this program the following options can be selected:

DRILLING ORDER	THESE ARE THE INTERSECTION POINTS	
	X-COORDINATES	Y-COORDINATES
1(0, 0)	1.69	5.17
2(0, 1)	1.69	10.33
3(0, 2)	1.69	15.50
4(0, 3)	1.69	20.67
5(0, 4)	1.69	25.83
6(1, 0)	3.38	5.17
7(1, 1)	3.38	10.33
8(1, 2)	3.38	15.50
9(1, 3)	3.38	20.67
10(1, 4)	3.38	25.83
11(2, 0)	5.08	5.17
12(2, 1)	5.08	10.33
13(2, 2)	5.08	15.50
14(2, 3)	5.08	20.67
15(2, 4)	5.08	25.83
16(3, 0)	6.77	5.17
17(3, 1)	6.77	10.33
18(3, 2)	6.77	15.50
19(3, 3)	6.77	20.67
20(3, 4)	6.77	25.83
21(4, 0)	8.46	5.17
22(4, 1)	8.46	10.33
23(4, 2)	8.46	15.50
24(4, 3)	8.46	20.67
25(4, 4)	8.46	25.83
26(5, 0)	10.15	5.17
27(5, 1)	10.15	10.33
28(5, 2)	10.15	15.50
29(5, 3)	10.15	20.67
30(5, 4)	10.15	25.83
31(6, 0)	11.85	5.17
32(6, 1)	11.85	10.33
33(6, 2)	11.85	15.50
34(6, 3)	11.85	20.67
35(6, 4)	11.85	25.83
36(7, 0)	13.54	5.17
37(7, 1)	13.54	10.33
38(7, 2)	13.54	15.50
39(7, 3)	13.54	20.67
40(7, 4)	13.54	25.83
41(8, 0)	15.23	5.17
42(8, 1)	15.23	10.33
43(8, 2)	15.23	15.50
44(8, 3)	15.23	20.67
45(8, 4)	15.23	25.83
46(9, 0)	16.92	5.17
47(9, 1)	16.92	10.33
48(9, 2)	16.92	15.50
49(9, 3)	16.92	20.67
50(9, 4)	16.92	25.83
51(10, 0)	18.62	5.17
52(10, 1)	18.62	10.33
53(10, 2)	18.62	15.50
54(10, 3)	18.62	20.67
55(10, 4)	18.62	25.83
56(11, 0)	20.31	5.17
57(11, 1)	20.31	10.33
58(11, 2)	20.31	15.50
59(11, 3)	20.31	20.67
60(11, 4)	20.31	25.83

Fig. 8.37 Calculated X, Y coordinate values and the order in which the drilling procedure will be executed

- Hole centre coordinates and distances (i.e. increments) are defined by the user.
- Hole centre coordinates are calculated by the program based on user-defined grid dimensions and the number of holes in the X/Y direction. In this case there will be a clearance of half the hole centre distance all around the edge of the grid (e.g. pallet). (Note that this option has been selected for demonstration in Fig. 8.38.)

To achieve the illustrated results, the data was input simply by filling in a screen offered by the computer and by teaching two points on the PUMA robot to define the frame in the three-dimensional space. (It is important to emphasise that the pattern can be drilled in *any* three-dimensional plane, the only limits being the trajectory limits of the actual robot.)

Within this module, input data can also be tested by means of a graphic display output, indicating all important drilling data as well as the layout of the specified drill pattern.

Graphical robot programming and simulation

There is currently a great deal of research and development in integrating CAD/CAM systems and generating robot programs in a similar way to NC machine tools. Compared to machine tools, where in most cases the number of controlled motions are less or equal to three, the task is often more complex in robotics where six or more controlled axes are common.

Three-dimensional models can be generated with so-called wire frames, made up of interconnected lines. They are easy to create and provide accurate definition of surface discontinuities, but they do not contain information about the surfaces themselves. Through hidden line removal these models often appear to be solids, but in fact they represent only a shell of the component.

Some of the tasks that can be solved, or at least simulated with computer graphics, include:

- Collision free trajectory generation.
- Motion and work cell layout and cycle time optimisation.
- Robot motion and program demonstration by computer animation in 3D.

To illustrate the point, one of the growing number of impressive systems, McAuto's PLACE package, is introduced. This system allows the comparison of different robot layouts and programmed motions in different cells in interaction with feeders, conveyors and other types of part input/output devices. The system is capable of automatically calculating desired 'go-to locations' from the graphics screen (Fig. 8.39).

More advanced systems make use of solid-modelling techniques and solid-model animation. Solid modelling incorporates the design and analysis of virtual parts and assemblies created from primitives of solids stored in an image database. Solid modellers not only contain sufficient information to produce hidden surface drawings, but are also 'clever' enough to prevent accidental creation of unrealistic combinations of such primitives.

```
    PROGRAM GRID
    REM ************************************************
    REM * THIS VAL PROGRAM WAS CREATED BY THE         *
    REM * MARTI OFF-LINE ROBOT PROGRAMMING SYSTEM     *
    REM ************************************************
    REM ***   THIS PROGRAM USES GRID DIMENSIONS AND THE NUMBER
    REM ***   OF HOLES IN THE X/Y DIRECTION,
    REM ***   BEFORE OPERATION THE WORK FRAME MUST BE DEFINED
    GOSUB PLANE
    REM ***   THIS SUBROUTINE IS USED TO DEFINE THE WORKING PLANE
    REM ***   THE ROBOT IS TO EXECUTE THE MACHINING OPERATIONS
    SPEED 60 ALWAYS
    MOVE F1
    SHIFT DOWN BY  0.85, 2.58 ,0.0
    REM ***   MOVE TOOL FROM CORNER OF GRID TO FRIST HOLE
    REM ***   CENTRE.
    SET DRILLING =GRID:DOWN
    REM ***   DEFINE DRILLING WRT FRAME (GRID)
    MOVE DRILLING
    REM ***   MOVE FROM GRID CORNER I.E. FIRST HOLE CENTER
    SETI ROW=0
    SETI COL=0
    REM ***   THE X/Y COUNTERS SET TO ZERO
    GOSUB DRILL
    REM ***   VERTICAL OPERATION OF THE ARM FOR DRILLING
 10 SETI ROW=ROW+1
    IF ROW GT 4 THEN 20
    REM ***   WHEN LOGIC IS TRUE GOTO LINE 20
    REM ***   ELSE
    TYPEI ROW
    TYPEI COL
    REM ***   IF ROW VALUE NOT GREATER THAN 4 THEN
    REM ***   INCREMENT IN THE Y-DIRECTION BY 1.
    SHIFT DOWN BY 0, 5.17 ,0
    SET DRILLING=GRID:DOWN
    MOVE DRILLING
    DELAY 3
    GOSUB DRILL
    REM ***   VERTICAL OPERATION OF THE ARM FOR DRILLING
    REM ***   THIS PROCEDURE NOW COMPLETED,RETURN TO LINE 10
    GOTO 10
 20 SETI COL=COL+1
    REM *** ROW NUMBER NOW EXCEEDS INPUT, THEREFORE INCREMENT COL
    IF COL GT 11 THEN 30
    REM *** IF COL VALUE NOT GREATER THAN 11 THEN
    REM *** INCREMENT THE X-DIRECTION BY 1
    SHIFT DOWN BY   1.69,- 25.83,0.0
    REM *** ELSE GOTO LINE NUMBER 30
    SET DRILLING=GRID:DOWN
    MOVE DRILLING
    GOSUB DRILL
    REM *** DRILL HOLE INCREMENTED IN THE X-DIRECTION
    REM *** RE-SET ROW COUNTER FOR RE-RUN OF PROCEDURE
    SETI ROW=0
    GOTO 10

 30 TYPE ===== THIS IS THE END OF PROGRAM GRID =====


    PROGRAM DRILL
    REM *********************************************
    REM * THIS SUBROUTINE IS CONCERNED WITH THE     *
    REM * ACTUAL DRILLING OPERATION.                *
    REM *********************************************
    REM
    SHIFT DOWN BY 0,0,  15.00
    REM *** MOVE  15.00MM IN THE Z-AXIS
    SET DRILLING=GRID:DOWN
    MOVE DRILLING
```

Fig. 8.38 Sample robot program generated by one of the modules of the MARTI system for the PUMA 560 robot. This VAL program drills holes into a plane in the specified three-dimensional space, following the off-line calculated pattern interactively defined by the user as described in the text

```
REM *** DOWNWARD MOVEMENT (ZD) BEFORE ACTUAL DRILLING.
REM *** REDUCE SPEED FOR ACTUAL DRILLING INTO MATERIAL
REM *** A DISTANCE   12.00
SHIFT DOWN BY 0,0,   12.00
SET DRILLING=GRID:DOWN
SPEED 50
REM *** SPEED NOW 50 % OF NORMAL
MOVE DRILLING
REM *** MOVE TO LOCATION
SPEED 100
REM *** SPEED BACK TO NORMAL(60).
DELAY 3
SHIFT DOWN BY 0,0, -27.00
REM *** MOVE ARM UPWARDS A DISTANCE   27.00
SET DRILLING=GRID:DOWN
MOVE DRILLING
RETURN 0
REM *** RETURN TO MAIN PROGRAM AT LINE AFTER
REM *** BRANCHING COMMAND


PROGRAM PLANE
REM *********************************************
REM * THIS SUBROUTINE IS USED TO DETERMINE THE *
REM * PLANE OF OPERATION FOR THE DRILLING      *
REM *********************************************
REM
TYPE POSITION THE ROBOT AT THE CORNER OF THE GRID (F1)
TYPE AT THE CORRECT HEIGHT (TOOL Z=AXIS)
GOSUB REPLY
HERE F1
TYPE POSITION THE ROBOT AT A POINT ON THE
TYPE FRAMES +VE X-AXIS(F2)
GOSUB REPLY
HERE F2
TYPE POSITION THE ROBOT AT A POINT IN THE
TYPE X/Y PLANE (F3)
GOSUB REPLY
HERE F3
FRAME GRID=F1,F2,F3
REM *** NEW FRAME DEFINED BY GRID
INVERSE DRILL=GRID
REM *** ROBOT IS NOW DEFINED WITH RESPECT TO FRAME GRID
REM *** NOW DEFINE DRILL START POINT IN GRID COORDINATES
SET HOLE=DRILL:F1
REM *** ORIGIN IS NOW SET
SET DOWN=HOLE
SET DRILLING=GRID:DOWN
REM *** NEW REFERENCE FRAME FULLY DEFINED
RETURN 0
REM *** RETURN TO MAIN PROGRAM AT LINE AFTER
REM *** BRANCHING COMMAND


PROGRAM REPLY
REM *************************************************
REM * THIS SUBROUTINE ALLOWS THE ROBOT TO BE MOVED  *
REM * TO THE DESIRED POSITION BEFORE DEFINING IT    *
REM *************************************************
REM
PAUSE *** PLEASE TYPE PROCEED [CR] TO CONTINUE
RETURN 0
```

Fig. 8.38 continued

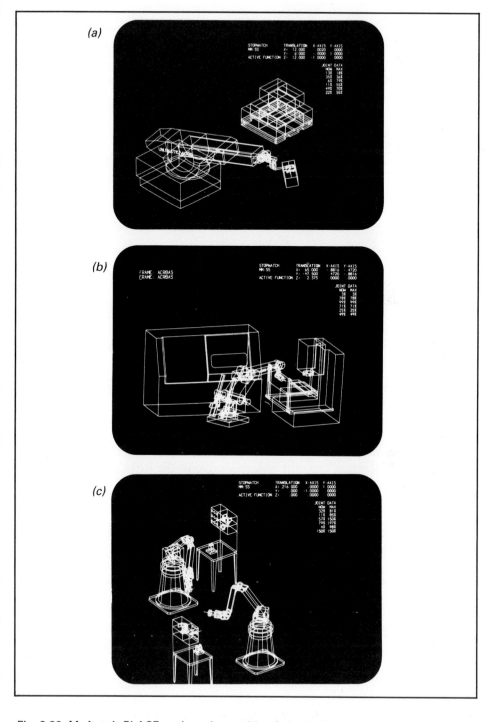

Fig. 8.39 McAuto's PLACE package is capable of simulating robot layouts and programs dynamically: (a) Unimate 4000 robot simulation, (b) ASEA IRb60 robot simulation, and (c) Cincinnati T^3 robot simulation. (Courtesy of McDonnel Douglas McAuto Corp., St. Louis, Missouri)

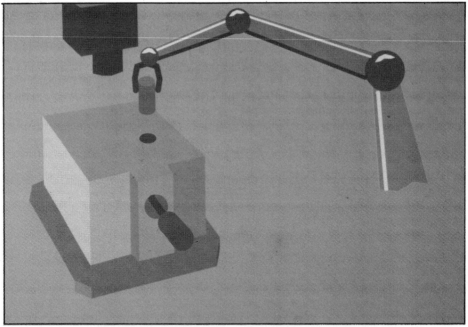

Fig. 8.40 Solid model simulation of an industrial robot assembly operation

This feature can be of use when simulating robotised assembly using solid objects, primitives and subassemblies, stored in a geometrical database (Fig. 8.40).

These and other tasks need CAD/CAM systems which are capable of learning from the data they have processed. This additional knowledge is not fed back into the design process at present (although application programs contain a certain amount of knowledge in the form of algorithms and formulae), and the decisions made in most CAD/CAM systems are deterministic (i.e. choices which are static) do not represent intelligence and can be calculated.

In the authors' view a CAD/CAM system is intelligent if it can self-determine choices in its decisions, based upon the experience gained in the past both from failures and successful solutions which are stored in the form of rules in the system's knowledge base. Because present CAD/CAM users have to take too many subjective decisions on the basis of primitive deterministic (in artificial intelligence terms) programming methods, databases and graphics systems, the expert system should act as an adviser.

Of all the representation modes of solids, Such as the sweep representation, the cell decomposition, constructive solid geometry, boundary representation, template instancing and spatial occupancy enumeration, only three are to be discussed:

● The translational sweep representation, where a solid is represented as the volume swept by a 2D image when it is translated along a line. Translation may also include rotation, when a 2D image is rotated around a line which serves as an axis.

- Constructive solid geometry handles primitives of solids, which are bounded intersections of closed half-spaces, defined by planes or shapes (i.e. $F(x,y,z) \geqslant 0$, where F is an analytical function. More complex solids can be built by composition and decomposition using set operations, such as union, intersection and difference of solid bodies, or extrusion. Extrusion operators allow the creation of faces which are then usually swept to create solids or holes in solids. The Boolean and the extrusion operations are generally intermixed during the design process to provide maximum flexibility.

- Boundary representations allow the definition of solids by means of their enclosing surfaces (i.e. boundaries). This representation handles surfaces, curves, lines and points, which can then be used to create boundaries.

Solid modellers use raster graphics to produce solid images. They follow the scan-conversion principles, by which software using an imaginary light source and a defined viewing plane as a guide, scans the model to determine the colour and the intensity of each pixel on the screen. This means that each of these images contains millions of different picture elements. To create realistic (i.e. hidden surface) images, ray-tracing algorithms are used, where each ray is evaluated for intersection with the bounding surfaces of the solid model, thus allowing creation of true images, as well as shades and reflective images.

Although graphics simulation in general can be a very useful tool in machine and robot programming, major drawbacks include:

- Unless already stored in the CAD database, the graphic images of the robot(s), the parts and the other components of the cell must be designed by the user. This can be often be time consuming.
- Using three-dimensional wire-frame models, collision can often be only detected by visual rather than fully computerised methods. Thus it is not reliable.
- At present, computer graphic simulation is not capable of automatically detecting either compliance or dynamic problems of the arm.

In summary, it is the authors' view that as soon as all parts and system components are to be designed in solid-modelling systems, and as soon as such systems are available at a low cost, graphics simulation will be an important tool not only for sales engineers but also for the production engineer in his every day robot cell layout and programming work.

References

[1] Ránky, P. G. 1982. *The Design and Operation of FMS.* IFS (Publications) Ltd, Bedford, UK.
[2] Coiffet, P. 1983. *Robot Technology,* Vol. 2, *Interaction with the Environment.* Kogan Page, London.
[3] Koren, Y. 1983. *Computer Control of Manufacturing Systems.* McGraw-Hill, New York.
[4] Ránky, P. G. 1984. Programming industrial robots in FMS. *Robotica,* 2: 87–92.

Chapter Nine

TASK PLANNING AND COMPLIANCE

A S ROBOTICS technology becomes more complex, the strategies involved in designing robotics systems should become increasingly sophisticated. Presently, robotic manipulators are usually designed with a particular type of task in mind, with the appropriate functions specified in the construction of the hardware and low-level software, and the robot placed in service until either the task or the robot becomes obsolete. Various problems can result from this type of set-up. Foremost is when an expensive piece of machinery has been dedicated to either a short-term or an overly simple task, and upgrading the robot's capabilities may prove too costly or impossible. Adding performance criteria such as speed or accuracy might require altering the manner in which a manipulator performs its tasks, while small changes in the conditions for a work environment which have not been predefined can have the effect of causing the desired operation to fail.

Instead, some new ideas and considerations in assessing robotics systems are evolving. Robots should be general purpose, in that they should be able to perform a number of different tasks in various work environments without requiring extensive modification of software or hardware. Robots should provide generalised structures and capabilities, invoking meaningful descriptions of the work environment and its associated tasks, rather than highly detailed task definitions guaranteed to be precise, often awkward to develop, and ultimately confusing for those granted the favour of updating someone else's design and code. Robots should be forgiving, so that slight misalignments of objects in the environment result in slightly altering the robot's response, rather than aborting the execution of the task. Acknowledging these concerns should result in systems which are more versatile, more powerful, and more expensive in the short run at least, but hopefully easier to use and design.

The example command 'PLACE BLOCK IN BLUE BOX' seems much

friendlier to use than a dialect requiring the user to specify the position and orientation of both objects, how to grasp the block, what path to take to the box, and how to drop the block in. Surely these basic actions must be specified in great detail somewhere along the way, but the end-users should only have to deal with a set of primitives on a much higher level.

Unfortunately, absolute models are not achievable. By trying to specify complex tasks to such a degree that success is guaranteed, too much information is required to evaluate decisions, and the system performance becomes slow and expensive. Instead, fuzzy maps of a system are developed to allow for specified levels of uncertainty in the model of a plan, whereby possible deviations from a desired state are noted and compensated for in such a way that the achievement of a prescribed goal can be verified[1].

The solutions for task planning are non-trivial, and present methods for developing task plans have verifiable flaws in one area or another. In working towards near future potential applications, say where many manipulators are active within a cramped space, closely coupled synchronisation would be most essential, and parallel plans would be developed to account for interactions.

9.1 Robot planners

The majority of present day robot applications involve fairly simple activities, such as inserting screws, painting surfaces or drilling holes. Pick-and-place operations, pulling parts from a bin and inserting them into certain slots or holes, account for a third of industrial robot tasks. Even as tasks become complicated, most can be described in terms of grabbing an object, moving to a new setting, and applying the object to the setting in some relatively simple manner. Problems generally entail an unspecified collision between two objects or between one object and the manipulator, or a failure to achieve the desired result (goal state) due to misalignments of some nature. These types of situations, such as in Fig. 9.1[2], are conducive to the development of generalised terminologies for describing them.

In providing the higher level task environment, plans for future actions by the robot are required. The robot plan is the program which transforms task-level descriptors into manipulator-level actions. The robot planner (or task planner) is responsible for specifying what actions are needed to accomplish a task, and producing a program to implement those actions. The resultant program is then submitted to the plan checker to determine the feasibility of the plan. The approved plan is then executed by the robot controller using whatever manipulators and sensors are at its disposal[3].

The robot planner is in general not available at execution time. Instead, it develops a plan using a description of the system's environment and the appropriate tasks at hand (the use of the term 'system' in this chapter implies an array of one or more coordinated manipulators, while the 'environment' refers to the system and the various objects and obstacles residing within the system's workspace). Included in the description are the initial and goal states, motion operators on those states, plus a set of constraints defining which motions are allowed. Much of the environment is

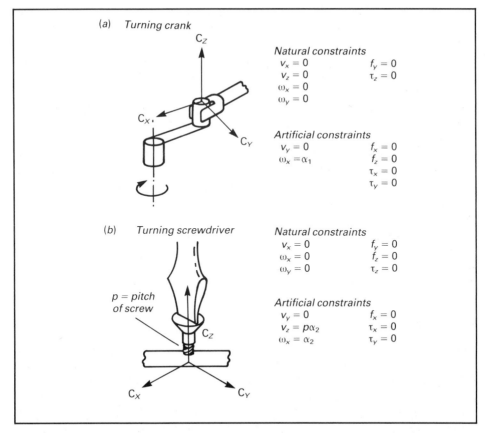

(a) Turning crank

Natural constraints

$$v_x = 0 \qquad\qquad f_y = 0$$
$$v_z = 0 \qquad\qquad \tau_z = 0$$
$$\omega_x = 0$$
$$\omega_y = 0$$

Artificial constraints

$$v_y = 0 \qquad\qquad f_x = 0$$
$$\omega_x = \alpha_1 \qquad\qquad f_z = 0$$
$$\qquad\qquad\qquad \tau_x = 0$$
$$\qquad\qquad\qquad \tau_y = 0$$

(b) Turning screwdriver

Natural constraints

$$v_x = 0 \qquad\qquad f_y = 0$$
$$\omega_x = 0 \qquad\qquad f_z = 0$$
$$\omega_y = 0 \qquad\qquad \tau_z = 0$$

p = pitch
of screw

Artificial constraints

$$v_y = 0 \qquad\qquad f_x = 0$$
$$v_z = p\alpha_2 \qquad\qquad \tau_x = 0$$
$$\omega_x = \alpha_2 \qquad\qquad \tau_y = 0$$

Fig. 9.1 Examples of force control tasks showing the constraint frame $\{C\}$, natural constraints, and artificial constraints. Here, $[v_x,v_y,v_z,\omega_x,\omega_y,\omega_z]^T$ is the hand's velocity vector, three translational and three angular components, given in $\{C\}$. $(f_x,f_y,f_z,\tau_x,\tau_y,\tau_z)^T$ is the force vector acting on the hand, three forces and three torques, also given in $\{C\}$. The α's are constraints. (a) Turning a crank at a constant rate, α_1; and (b) turning a screw with constant rate, α_2. Note that screw is frictionless

in a 'don't care' state at any given time, and it is often advantageous to task efficiency to limit the available information in a task description to the data which is essential to the task, or at least prune the set of possibilities to a more manageable number. Uncertainties in task and environment descriptions can make optimal decisions difficult, such that the planner must balance its purpose between achieving efficiency and assuring the success of the operation.

Lozano-Perez[4] divides the robot planner's function into three parts (conceptually distinct but somewhat overlapping in basic functions): world modelling, task specification and manipulator program synthesis. The modelling aspect attempts to derive the physical information from the environment using mathematical and relational descriptions. Task specification designates appropriate checkpoint states or motions to ensure the completion of a task. Program synthesis prescribes the behaviour of the manipulator in relation to task goals and sensor inputs (see Fig. 9.2).

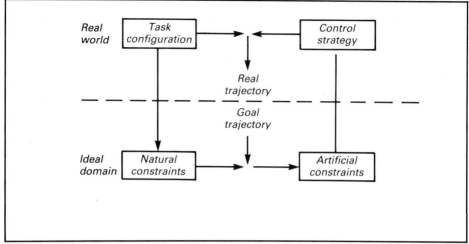

Fig. 9.2 *Overview of control strategy synthesis*

9.2 World modelling

In the world modelling state of task planning, various world parameters are obtained so as to define objects and their behaviours. These parameters can be split into five groups:

● A geometric description of the robots and objects in the task environment.
● A physical description of all objects, e.g. mass and inertia.
● Kinematic descriptions of all linkages present in the environment.
● Descriptions of various robot characteristics, such as joint limits and acceleration bounds.
● The object and linkage configurations, and their uncertainties.

The geometric model provides the bulk of the information concerning the environment's status. Geometric descriptions can be derived using CAD or machine vision representations or other techniques, and they are usually based on combinations of line, surface or solid primitives. Obviously, not all objects are well suited for definition by something as simple as perhaps a wire frame or a combination of circles and squares, so that solid modelling should often be introduced into the modelling process. The interpretation of three-dimensional information from two-dimensional data is one source for ambiguity. Tactile sensing can be used to discern surface changes on a more localised basis, and can aid in enhancing low-precision descriptions.

A physical model of an environment can take into account such basic qualities as mass, inertia, colouring, texture, and material malleability. In certain applications, deformation of materials for either the manipulator or the object during the execution of a task is a large source for uncertainty, but for most situations modelling a deformation is more difficult than correcting for its perceived impact. In many pattern recognition tasks, colour and texture data are acquired at the same time as the geometric description, and stored with the geometry as an attribute. For example, a light green 'smooth' area next to a dark green 'rough' area in an aerial photograph might be recognised as a field next to the woods, where the only information stored is the two boundaries matched to the designations 'field' and 'woods'. Similarly, differentiating the head and thread ends of a screw, determining where the holes are for putting on a wheel, or identifying the kind of soda from can or bottle colour and size are all activities appropriate for assembly lines. In fact product inspection for quality control and arc welding of metal joints are two areas requiring as much recognition capability as robotics potential.

Kinematic equations are used to describe the linkage characteristics of such mechanisms as springs, gears and joints, or where objects are interacting due to friction and part insertions or connections. The changing interactions require the kinematic description to be updated with events, i.e. new object and linkage configurations. As an example, the characteristics of the robot are defined through the various arm and wrist kinematic and dynamics descriptions which seldom change during a task, but when the robot picks up a new tool, the manipulator description must be altered to allow for the extension from the hand. However, many of the robot characteristics, such as sensing capabilities or the position, velocity and acceleration bounds, most likely are modelled only once. Placing bounds on the extent of uncertainties in the configuration is required as part of setting up the compliance mechanism, such that limits for the margin of error are defined prior to task specification.

9.3 Task specification

Closely allied with the world modelling phase is the task specification process, whereby a task is usually defined as a sequence of model states or operations. In the interest of concurrency or efficiency, the sequence called for might allow for some freedom of choice in the order of execution of subtasks, subject to various conditions or events occurring in the task environment. For now, we will assume tasks to be simpler. As a minimum, for a primitive operation, designation of initial and final task states is required. More likely, a group of intermediate steps and relevant information concerning them will be necessary.

The world (task environment) configuration can be specified in a number of ways, such as by an explicit model state, a CAD system model, a robot designated model or symbolic representation. Two of the concerns expressed in choosing a particular specification method are accuracy, in both orientation and position, and the ease with which the model can be interpreted and modified. An explicit model, defining all pertinent parameters as precisely as possible so as to solve a set of associated equations, is quite accurate but inflexible and difficult to set-up. Large tasks turn into unwieldy sets of delimiting equations. Inherent accuracy and interpretation problems also exist for robot- and CAD-based models. Symbolic representation appears to satisfy the noted concerns somewhat more adequately, though cost factors might warrant using a less sophisticated approach.

By providing for symbolic spatial relationships between features, a more convenient mode for referencing tasks or subtasks is allowed. Families of related configuration can be expressed with a high degree of accuracy. In these families, alternating computational methods would allow important constraints to dominate certain configurations, while permitting the use of less complex models where appropriate. For example, choosing a particular site to tap a hole might call for a relatively imprecise operation, as long as a screw can be precisely inserted later. In this case, the conditions for the screw insertion would be much more carefully defined than the parameters for tapping.

The operation sequence is object orientated, and the goal states require the confirmation of various relationships between objects, as well as final placement data, so as to avoid events such as collision. Relational information can also be used to specify more efficient state sequences. A classic human example – the husband parking behind the wife's car in the driveway when she is about to drive somewhere – is also applicable to robot systems. By altering task execution in ways which do not directly affect the immediate goal, events inhibiting the execution of later tasks might be avoided. In a cluttered workspace, this type of congestion might be unavoidable, no matter how much forethought has been given, but specific moves might be seen as creating higher probabilities of future interference. Parallel planning allows for the distinction between subtask processes by recognising interactions between branches, by correcting harmful interactions that keep the plan from accomplishing its overall goal, and by taking advantage of helpful interactions on parallel branches so as not to produce inefficient plans[5].

Symbolic representations must somehow be mapped onto the configuration parameters of objects, along with their associated constraints. This process is bound to introduce uncertainties into the model, possibly of great significance. Reusing the 'PLACE BLUE BLOCK IN BOX' example, specifying a constraint of the type 'WITH BLOCK AGAINST NORTH WALL' is intrinsically easier to comprehend than the designation of the specific coordinates and orientation for the box. Converting the semantics of the term 'AGAINST' into a verifiable condition, in this case a relation between a block and a specific wall, introduces a large set of equally valid goal configurations into the database. One strategy, pushing the block towards the supposed location of the wall until the block stops rotating or moving translationally, would probably have a better chance for success than attempting to specify an absolute motion in aligning the face with a certain segment of the wall. The latter method could produce an erroneous goal state, possibly unverified, with even small miscalculations in the equations describing the plane for either the block or the wall segment, so that some way smoothing out discrepancies is needed.

In choosing criteria for compensating for possible errors, one should assess each error's significance. A compliance or error-correcting scheme should be robust, but reasonable limits should be set on how much uncertainty is tolerated. Always allowing for the worst possible error could be as expensive and wasteful as, say, designing all cars to withstand 100mph collisions. Instead, consideration should be paid to which inaccuracies require the most attention, and what the probabilities and consequences are for the occurrence of various possible 'disasters'.

9.4 Manipulator program synthesis

Any particular manipulator activity can probably be described in terms of picking up a new part, moving to a new location, and affixing the part in some way to that site. Thus, three areas of concern are defined for manipulator action: grasp planning, motion planning, and error detection. As a result, the manipulator program synthesis step takes on the duties of developing feasible strategies for grasping and movement activity and determining to what extent the implementation of those functions has deviated from the planned action.

Grasp commands should satisfy a few basic guidelines in their behaviour. The grasp should be safe for the forces applied, resulting in no damage to the object or the gripper, while allowing no inadvertent collisions to result from either the grasp itself or the motion involved with reaching the grasp configuration. The grasp state should be both stable, i.e. controlled, and reachable by the manipulator. Also, the certainty in the representation of the configuration should be increased, or at least not decreased, as a result of the grasp action.

Motion commands are defined to be either free motion, guarded motion or compliant motion. The guarded motion is a closely controlled motion used when the possibility of collision is high, whereas a compliant motion refers to the fine adjustments required to reach the goal configuration from a proximal configuration. Both of these motions are in general sensor based, whereas in free motion, for moving between configurations, the major responsibility is obstacle avoidance, and usually entails low enough precision to be a non-feedback servo operation.

Implicit in the grasp and motion commands is the job of error detection and correction. Deviations from the underlying models and specifications must be monitored and compensated for at various stages during the execution of a task in order to prevent the errors from accumulating and creating gross system malfunctions.

There are three sources of uncertainty at the planning level: the manipulator itself, the objects being manipulated, and the process of introducing those objects into the environment. One common cause of error is when the robot drifts while repeating a task, such that the manipulator never quite resets to the same position on each repetition. Inaccurate machining, high object design tolerances, bin feeders and conveyor belts are all possible sources of error. If one is confident that the task has been specified loosely enough, the effect of uncertainties might be ignored (a robotically controlled wrecking ball comes to mind). An estimation of inaccuracy might be used to determine whether the task implementation is feasible and whether to abort the mission.

By computing the effects of uncertainties in a symbolic fashion, other options are available. Significant uncertainties could be identified and reduced, with the plan constrained in some manner so as to guarantee success. Posting constraints refers to the practice of delaying decisions until later constraints are made available. Multiple approaches for performing the same function may be allowed, by sorting through and examining particular strategies until a feasible or efficient one is found. Instead of a particular implementation, a virtual plan would actually offer a group of existing possible subplans, or planning islands, which when used together as a pool to choose from, ensure the completion of a task. Through combining plans to form new semantic levels of abstraction, more symbolic representations for robot behaviour can be introduced.

9.5 Task definitions

The definition of a task involves reducing the task to its pertinent subcomponents, generating a feasible strategy for the subcomponents, and insuring their execution. The common basis for these operations is that they require a knowledge base to work from. The assimilation of data from the environment into a relational, interpretable form is an imprecise art. McDermott and Davis [6] use the idea of a fuzzy map to relate multiple frames of reference as a single instantiation. The entities of 'frame' and 'object' are combined into a hybrid 'frob' such that (a) these coordinate frames have position, orientation and scale with respect to other frames, and (b) between any two frames there must be an easy-to-compute frame path of frames between them, ensuring efficiency and dispelling ambiguity. In addition to the intrinsics of the former, objects would also have shape plus extrinsic parameters such as length, colour, etc.

In order to define the function of a task, we must first come up with methods for describing its composite structures, as in determining the appropriate algorithms on which to base it. The work is divided into four basic areas: symbolic spatial relationships, path description, grasp planning, and motion strategies.

Symbolic spatial relationships

A process is needed for mapping the symbolic level of spatial relationships between objects onto an appropriate set of descriptive equations. A coordinate system is set-up for the objects and the object features in the world view and used for setting-up equations for the relationships between features, based on the various object configuration parameters. These equations are then combined and solved for each object's configuration parameters. A complete task can be specified using a set of target spatial relationships along with a group of partially constrained motions required to reach the target state.

Solving a complete set of equations for every task is both time consuming and somewhat unnecessary. One alternative is to provide standard sets of relationships, such that larger sets of equations can be rewritten and reduced to a combination of the appropriate standard sets. Inequality constraints can be bounded by applying linear programming or symbolic simplification methods.

The treatment by Popplestone et al.[7] concerns the coordinate transformations involved in describing workspace assemblies. The various configuration features are defined by assigning them a unique coordinate frame with its base axes. Parameters of the feature would be defined in relation to the base frame origin. A simple instance would be a unit cube, whose base coordinate frame might be taken as a corner of the cube with the X, Y, Z axes initially defined along the cube edges. Using the knowledge that each face is orthogonal to the adjacent faces, the cube can be represented using the origin and the far corners of two adjacent faces, such as the points (0,0,0), (1,0,1) and (0,1,1). These points would be specified in 4×4 matrix form in allowing traditional positional and rotational matrix transformations. The translational operator 'trans' would transform the cube's coordinate frame with respect to the world frame. The rotational operator 'twix' spins the cube about its origin such that it would return new locations for the two far corners. From there, the location of the six faces can be ascertained in their new configuration. In achieving goals, the calculations might be evaluated in reverse order, such that the desired final cube configuration is known, and a combination of twix and trans operations are evaluated to meet the interim and goal configuration constraints (i.e. FACE 1 flush against FACE 2) and the subsisting equations are solved (see Fig. 9.3).

More irregular polygons would obviously require a more explicit description of shape, but in all but the most precise action specifications, could be represented by a smaller set of features. A mirror ball, an approximate sphere with many facets, could be designated by its origin and two other defining points, along with a stored complex description in terms of the defining points. At the conclusion of a group of transformations on the three points, the position/orientation of particular facets would be calculated rather than evaluating each transformation for every vertex of the polyhedron. Possibly more advantageous would be the declaration of each facet as a feature while solving the constraint equations only for the facet(s) significant to the task. The convenience of particular strategies would depend on the types of motion involved and the problems associated with a specific configuration for the task environment.

Path definition

Path definition essentially involves finding a proper trajectory to avoid collisions with objects in the world. Though various schemes are available, most robotics systems at this time are designed stringently enough that the chance of an inadvertent collision during free motion is slim and not taken into account. The simplest form of planning is the 'hypothesise and test' method, whereby a path is chosen almost at random, checked for success, and used or discarded accordingly. This trial and error method suffers in a crowded environment. Slightly missing an allowed path might precede the testing of a large number of failures before a successful path is finally encountered. If the probability of conflict is high, a more sophisticated path determination procedure is needed. Penalty functions can be used to assign a weighting to a path, based on whether the path is a collision course (given

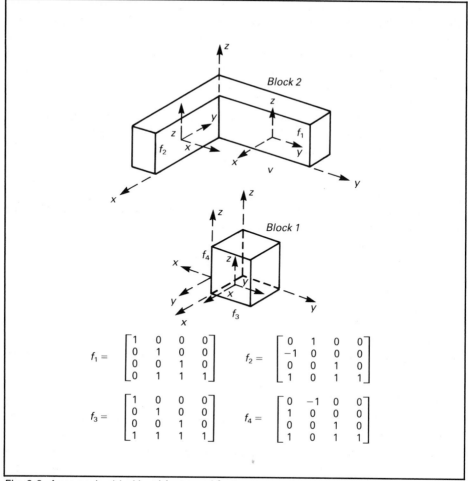

Fig. 9.3 Axes embedded in objects and features

a weight of infinity), close to the shortest path to the goal (given a low weighting), or somewhere 'out in left-field'. Sets of potential paths can be pruned quickly from a decision tree based on their penalty functions, so that more efficient or more likely paths can be searched for. The problem here is that the assignment of penalties is complicated in itself, especially for a manipulator with a large number of links. Paths cannot be determined solely on the hand trajectory, as a good hand path might run the arm into walls or require impossible motions from the other links (the perils of eating at a crowded table should provide enough examples). Much of the problem stems from relying on localised information to assign the penalties, such that using paths which seem optimal might result in deadlock or require the manipulator to backtrack to find a feasible path.

Brooks[3] and Lozano-Perez[8] have attempted to develop a somewhat more efficient representation for coarse motion. By designating specific regions of free space, or allowed paths, path selection amounts to determining whether a given path resides within that safe region. Here, the arm and hand assemblies are approximated by describing their swept volume for the task. The hand and payload can usually be described as a fixed volume on the end of the upper arm, while the other link configurations vary, the exceptions being when the payload is much larger than the hand or the workspace is exceptionally cramped (constraints on wrist orientation for painting a contour or handling open containers might require even more complexity). Due to the nature of joint interaction, the problem cannot be decomposed into separate joint equations, but must be treated as coupled trajectories forming the swept volume for a boom. Constraints on legal positions and orientations are propagated along the joints, and search techniques can be used to narrow the choices.

The free space solution to the find-path problem offers a greater chance for success in the case of crowded environments, as long as a path exists in the subset of free space described for the task. To ensure success, the task can be broken into smaller portions at greater expense, such as for approach motions. For efficiency's sake, the class of target solutions would be restricted to provide a more practical path planner.

Path control as presented by Paul and Shimano[9] is effected through pure position control. In the interest of efficient motion, velocity functions are interpolated to provide smooth velocity changes at intermediate points. High gain servos are used to prevent overshoot. Deviations by the manipulator from its defined path are represented as a separate 'compliance joint' coupled to the end of the ideal non-compliance effector.

Grasp planning

Initiating a grasp motion requires information concerning the state of the target object, the manipulator gripping surfaces, and the initial and final grasp configurations. As mentioned above, the grasp plan should be feasible, stable in motion, non-damaging, and provide more knowledge concerning the configuration. The difference between grasp operations and the path planning is essentially that the grasp is a lower level, more precise action. We are concerned with only one configuration, as opposed to a path or set of configurations, and its interaction with subsequent operations. In this, more detailed information concerning the interactions between the manipulator's and the target objects' shapes and surfaces is required.

In finding a safe path, a set of potential configurations is chosen, based on the geometries and forces involved as well as the stability and uncertainty reduction principles. The unreachable configurations and those resulting in collision are then removed from the list of potential choices. An appropriate plan is then selected from the remainder according to greater stability, less uncertainty or other considerations. This is a rather simplified methodology, yet its underlying descriptions are often imprecise. Motion during the actual grasping action, more diverse or generalised geometries and surfaces, and more complex manipulator grippers are properties of grasp planning which have not been modelled properly.

Trade-offs between stability and incidental forces are required. Visual and tactile sensing capabilities are often demanded, though they can provide misleading information if not powerful enough (e.g. the handle of a coffee cup could go unnoticed if the cup were viewed or touched from the other side). Disambiguation information, such as gradient and slippage, or general facts, such as texture, hardness and temperature, can be provided through the placement of appropriate gripper sensors. Some of the newer finger-based grippers provide excellent dexterity without ever addressing issues concerning safety and reachability issues. Traditional grippers are more like automated pliers. What will be needed are more innovative and unified approaches in assuring controlled handling. One approach presented in 1977 is the finger system proposed in Fig. 9.4[10].

Sensor-based motion strategies

In a free motion situation, it is assumed that the path has been chosen so that manipulator deviations from the specified trajectory will not result in collisions, and the motion can be executed with the greatest speed without further input from the manipulator's environment. The other situation of concern is that of fine-motion, be it with configuration approach, grasping, or compliance in parts mating. In these cases, the interaction between the manipulator and objects is expected, and the assurance of a particular behaviour must be provided. That assurance is usually implemented using sensor information in conjunction with a descriptive model designating the response.

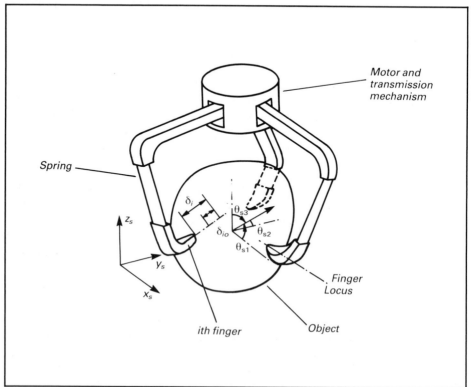

Fig. 9.4 Schematic diagram of stable prehension system

The model is developed according to geometric and kinematic descriptions of the manipulator and the objects in its environment, a specification for legal configurations which lie outside of the configuration space obstacles, and the determination of a path preventing undesired contact between objects along with the appropriate control theory for desired contact. Sensor-based strategies are in a sense a product of and an extension to the earlier work in mechanical compliance devices for error compensation, and a presentation of that evolution is appropriate here.

9.6 Compliance

By its original definition in the mechanical sense, a compliance is a multi-axis spring which deforms translationally and rotationally in the presence of contact forces and moments, and reverts to its original orientation when those forces are removed. The present robotics definition is more complex. A passive compliance would designate a non-sensor driven mechanism which responds to applied forces by realigning closer to a local goal state, as with a spring-loaded wrist that, when inserting a peg into a hole, deflects from any surface contact in such a fashion as to shove the peg further into the hole. If the tolerances for the peg fit are fairly loose, compliance might not be required at all, but in many conceivable parts mating tasks, where high precision in placement specification is needed, accuracy is either unavailable or too slow and expensive to implement. Thus came the development of alternate methods to achieve the goal.

Much of the literature is based on the two-dimensional model for peg insertion developed by Drake[11] in 1977. Here, the effects of different insertion strategies were analysed to see the differences between straight insertion (peg axis parallel to hole axis), angling the peg with a wrist capable of responding to the contact forces encountered, and chamfering the edges of the peg or the hole so that the peg would be forced closer to alignment with the hole. The geometrical relationships were then examined to see which situations would result in success, jamming or wedging, or outright failure.

It was found that the angling and chamfering scenarios provided similar advantages in that they resulted in the expansion of the region of success for a parts mating. In the straight insertion, if the peg hits the lip surrounding the hole, the result is a force orthogonal to the hole without any torque, and therefore failure. In an angling scheme, surface contact in the proximity of the hole produces forces and torques which tend to push the end of the peg further into the opening, and contact with the far wall of the hole realigns the peg shaft along the hole axis (i.e. corrects out-of-square alignment errors).

Using chamfers on the peg tip or bevelling the hole opening provides the necessary deflection at initial parts contact, and combined with a search technique for sliding along the surface near the hole, this concept has proved to be practical and very effective in industrial applications (see Fig. 9.5).

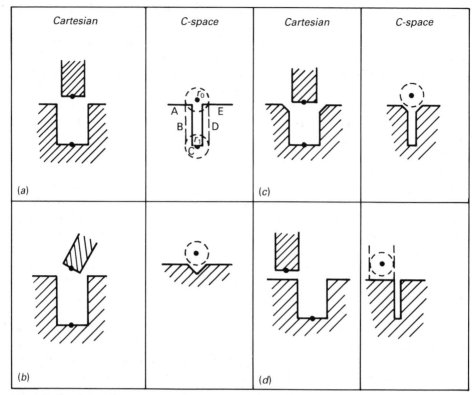

Cartesian	C-space	Cartesian	C-space
(a)		(c)	
(b)		(d)	

Fig. 9.5 Configuration space illustration of peg-in-hole insertion strategies

After examination of the force/torque graphs for chamfers, the usual questions develop for how the mechanism could be optimised, and how it could be used in the modelling of particular situations. Adjusting the shape, and especially the angles, for a chamfer produces different degrees of success for different strategies. As important as the parts shape is the behaviour of the supporting device, be it a wrist mechanism, a gripper or an equivalent assembly. All supports will have at least some compliance due to materials deformation, float in linkages, and the like, but those supports which are compliant to an extent of being modelled and applied usefully are termed remote centre compliances (RCCs).

In specifying the RCC parameters, specification of the centre of compliance location and values for lateral and rotational response is required. The centre of compliance is the point of pure response, where forces produce only latitudinal displacements and torques produce only rotational deflections. Generally, locating the centre of compliance near the point of initial parts contact minimises insertion and contact forces as well as the chance of jamming. A stiff RCC would require greater forces to create responses, but would be less susceptible to vibrational and gravitational effects.

In the peg-in-hole problem, the centre of compliance would most likely be near the tip of the peg. Rotational compliance about the insertion axis would be necessary if the peg needed to be aligned in a certain way in the hole (as with a square peg in a square hole). Also, it would be forbidden where rotational alignment is critical or torque is applied around the insertion axis (such as a cotter pin or a threaded screw), and would be unimportant in other cases (such as for the insertion of a radially symmetric rod). Many of the constraints and concerns are dependent on the specific application, and make RCC's inappropriate for handling dissimilar tasks. One way to alleviate this problem is the use of mechanical locks to adjust the values of compliance along the compliance frame's coordinate axes according to the task at hand. Parts geometries and the friction present between surfaces also affect the evaluation of the RCC.

Whitney[1] presents an analysis of the peg insertion to determine which regions guarantee success when one or both parts are compliant. He breaks the parts assembly down into four stages: approach, chamfer crossing, one-point contact, and two-point contact. The part mating event is described by the paths of both the peg and its support (say the wrist or another source of compliance), as well as the compliant forces/torques applied to the peg, both by the support compliance as the two paths deviate, and by the hole or its surrounding area due to surface contact and friction.

Other objects can be substituted for the peg and the hole in this scenario. What is important is the detection of failures. Assuming initial parts contact occurs somewhere within the chamfered edge of the hole, the two possible catastrophes would be jamming and wedging. Jamming occurs when the compliant forces/torques applied through the support are in the wrong proportions, and can be alleviated be reproportioning those forces. Wedging results from geometrical mismatching or deformed parts, and sometimes can be alleviated only by damaging or deforming the involved structures. In Fig. 9.6, events on or inside the parallelogram represent

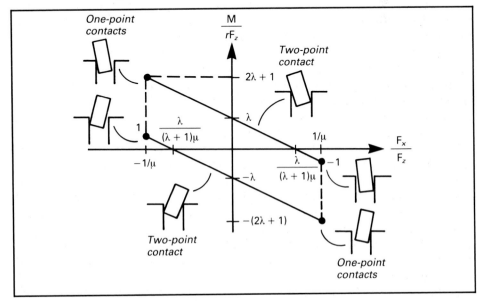

Fig. 9.6 The jamming diagram for the peg-in-hole problem

successful insertion, whereas events outside result in jamming. M and F represent the moments and forces, μ the friction, r the peg radius, and λ a resulting function in terms of friction and geometry. The wedging situation can be similarly represented.

9.7 Hybrid control

Force feedback control is a method for providing active compliance. By comparing a force sensor signal with the force specified, motor output torques can be measured and governed. In industrial automation control, this has been the most commonly used monitor for regulating machines. On the other hand, traditional manipulator motion strategies have been position controlled. At present, robotics research is most concerned with a combination of the two forms, termed hybrid control, whereby the execution of fine motion is defined under constraints provided by force/torque and position/orientation (hereinafter referred to simply as force and position) models and sensor data.

A hand immobile in a particular direction has no positional freedom (and as a result no control) along that axis, whereas its force control is absolute. Similarly, a hand in free space would warrant the use of position control but not force control. By specifying a surface in the configuration space such that positional freedom exists tangent to the surface and force freedom exists normal to the surface, we define what Mason[12] terms a C-surface in the task configuration. Natural constraints are specified through the interacting surfaces, while artificial constraints are added to comply with desired motion or force behaviour.

Control strategies can then be developed with respect to the defined compliance frame. Strategies are created through higher level function descriptions acting on sensor-provided data to specify the appropriate input to the robot actuators. An example would be the guarded move, whereby a slow approach would be terminated upon contact with a surface (i.e. when a force sensor produces a signal). At this primitive level, there are three basic criteria for control strategies: that their functions be relevant to the specified task, that the manipulator behaviour be conceptually simple, and that they execute quickly. These factors help enable the primitives to be combined into more useful, albeit more complicated, tokens. Paul and Shimano[9] express the concern that the matrix equations defining their path control primitives can be solved fast enough so as to provide a moderate refresh rate of 40Hz.

9.8 Fine-motion strategies

In Fig. 9.5 a configuration space representation of various peg approaches was depicted, illustrating the differences in their regions of success. This concept is expanded by Lozano-Perez et al.[13] to handle cases where the initial configurations are uncertain. A pre-image defines the regions from which the goal can be reached in a single motion under a specified velocity, denoted in a reachability graph. Fine motion strategies are construed as a sequence of guarded motions defined under the C-space. Recursive

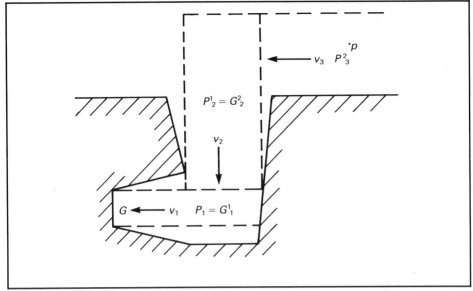

Fig. 9.7 Backward chaining of pre-images

computations would backtrack from the goal to the initial configuration set to ensure that all possible initial points are satisfied by a successful motion.

Fig. 9.7 shows a sequence of guarded moves satisfying pre-images P, goals G and velocity vectors V. A hybrid position/force pre-image would allow compliant motions in the success set. Thus, the C-space fine motion representation would include a set of possible initial and final configurations, a geometric spatial description, a kinematic surface description, and a set of correctness conditions for a class of synthesis algorithms. Correctness would assess successful motions under termination predicates specifying position, force and time conditions.

The fine-motion strategy has yet to be implemented, and its authors will most likely encounter many computational and decision based problems in doing so. The important aspect is that this approach, combined with algorithms for 'find-path' and 'grasp' tasks, falls under a more generalised methodology for robot planning. Sensor-based robot programming has been rather ill-defined, and any progress towards unifying the theories of task planning should have the results of reducing redundant effort, expanding robot capabilities, and clarifying our goals and methodologies. Naturally, few managers are willing to bear the costs of untried technologies on their projects, and the 'quick-and-dirty' approaches will continue to thrive where successful, but the eventual blossoming of robotics applications will result from innovation in the synthesis of ideas.

References

[1] Whitney, D. E. 1982. Quasi-static assembly of compliantly supported rigid parts. *J. Dyn. Syst. Measurement Control,* 104: 65–77.
[2] Raibert, M. H. and Craig, J. J. 1981. Hybrid position/force control of manipulators. *J. Dyn. Syst. Measurement Control,* 102: 126–133.

[3] Brooks, R. A. 1982. Symbolic error analysis and robot planning. *Int. J. Robotics Research,* 1(4): 29–68.

[4] Lozano-Perez, T. 1983. Task planning. In, *Robot Motion* (Eds. Brady, M. et al), pp. 473–566. MIT Press, Cambridge, MA, USA.

[5] Wilkins, D. E. 1984. Domain-independent planning: Representation and plan generation. *Artificial Intelligence,* 22(3): 269–301.

[6] McDermott, D. and Davis, E. 1984. Planning routes through uncertain territory. *Artificial Intelligence,* 22(3): 269–301.

[7] Popplestone, R., Ambler, A. and Bellos, I. 1980. An interpreter for a language describing assemblies. *Artificial Intelligence,* 14(1): 79–107.

[8] Lozano-Perez, T. 1983. Spatial planning: A configuration space approach. *IEEE Trans. Comput.* C–32(2): 108–120.

[9] Paul, R. P. and Shimano, B. 1976. Compliance and control. In, *Proc. Joint Automatic Control Conf.,* pp. 694–699. ASME, New York.

[10] Hanafusa, H. and Asada, H. 1977. A robot hand with elastic fingers and its application to assembly process. In, *IFAC Symp. on Information Control, Problems in Manufacturing Technology,* pp. 127–138.

[11] Drake, S. H. 1977. *Using Compliance in lieu of Sensory Feedback for Automatic Assembly.* Ph.D. Thesis, MIT, Cambridge, MA, USA.

[12] Mason, M. T. 1981. Compliance and force control for computer controlled manipulators. *IEEE Trans. Syst. Man Cybern.,* SMC–11(6): 418–432.

[13] Lozano-Perez, T., Mason, M. and Taylor, R. 1984. Automatic synthesis of fine-motion strategies for robots. *Int. J. Robotics Research,* 3 (1): 3–24.

Chapter Ten

SYSTEMATIC PLANNING OF ROBOT PROJECTS AND ROBOTISED SYSTEM INSTALLATIONS

IT IS a fact in a wide variety of industrial robot and high-technology installations that the most valuable experience can be gathered by making a series of mistakes. A more acceptable alternative to this is to learn from other peoples' experiences and mistakes. The effect of this latter method may be less dramatic, but it is less expensive and often faster, and thus advisable to all potential robot users.

A few rules of thumb, based on the authors' and other peoples' past experience are now listed. Each item is very true and is worth remembering before starting any project, and in particular any advanced manufacturing system or robot project.

- The carelessly planned project will take three times longer than expected; carefully planned projects will take only twice as long.
- No major project is ever installed on time, within budgets, with the same staff that started it.
- Projects progress quickly until they become 90% complete.
- One advantage of fuzzy project objectives is that they let you avoid the embarrassment of estimating the corresponding costs.
- When things are going well, something will go wrong. When things seemingly just can't get any worse, they will. When things appear to be going better, something has been overlooked.
- If the project content is allowed to change freely, the rate of change will exceed the rate of progress.
- No system is ever completely debugged. Attempts to debug a system inevitably introduce new bugs that are even harder to find.

A summary of points and tasks considered essential in achieving a successful robot installation include:

- Careful robot project planning.
- Identification of the application, or in other words, the specification of the goals.
- Selection of the right robot manufacturer and/or vendor.
- Selection of the necessary devices and the assurance of the possibility of their integration.
- Design, manufacturing and integration of the components (i.e. robot tools, feeders and part orientating devices, etc.).
- Analysis and preparation of the workplace.
- Training.
- Maintenance.

This chapter concentrates on two aspects of this list: project planning and robot selection criteria. The other items are discussed elsewhere in the book.

10.1 The robot project

When beginning the design of a robotised system installation in industry one should consider the following main points:

- Does the responsible team or person have sufficient knowledge in robotics and related sciences to be able to guide and control the development?

    If not, basic research and study must be done on the available robot technology, on data processing and control equipment, software, sensors, grippers, robot test methods, etc.

    Many companies employ a fully autonomous in-house project team which works independently from other departments. Another approach is to set up a project coordination team using the existing management structure.

    Most likely, the best solution is when there is a fully autonomous in-house project team headed by a highly reputable and knowledgeable system and production engineer, with team members being specialists in production, control, electronic, mechanical and software engineering.

- An important step is the identification of the application, or in other words the specification of the goals and the economic justification of the installation.

    Many industrial robot and other projects did not fulfil the expectations of the management because the goals were not specified properly, thus nobody really knew what to expect of the robot installation.

- Before discussing the robotics system development project with the upper management and/or the interested parties, one should carefully consider the following questions and points of economic evaluation factors:

    –What is the required minimum rate of return on the investment?
    –Should the machine, the robot, the tools, the feeders, etc. be purchased or produced in-house?

−Are there any labour savings? (What are the social implications?)

−What are the installation costs, including power supply, electrical alterations and other associated costs?

−What are the operating expenses, including production staff, maintenance, programmers, electrical and other energy consumption, etc.?

−What are the educational and training expenses, including courses, fees, travelling expenses, etc.?

−Are there any tax implications?

Since the key to any successful project is the careful planning, the commonly used techniques for project planning are now discussed with relevant examples.

10.2 Robot project planning and scheduling

The robotised system development project can easily contain a hundred or more important activities, thus computer-assisted project planning and scheduling methods are advisable.

In practice, bar charts, or Gantt charts, the critical path method (CPM) and the PERT method (program evaluation and review technique) are widely used. New methods, such as the IDEF model and related computer programs proved to be useful in the case of very large computer-integrated manufacturing system design projects, however they seem to be hard to handle and follow by the middle management in industry, because of its complexity and often time-consuming input data procedures [1,2].

To illustrate different project scheduling methods as used in a robotics project, an industrial case study is presented. The key to both the CPM and PERT methods is the project network. The basic idea is to break the project into logical steps or activities and then to order these activities into a possible and logical order. This is usually done manually because of the several management and engineering rules and limitations one should consider whilst doing this job.

In this case study, the following procedure for the construction of the robot project network was suggested:

● Determine the project objectives.
● Determine the required activities.
● Determine the key events.
● Use the CPM and/or PERT programs to analyse the critical path (i.e. shortest processing route) of the network.

A flowchart outlining the systematic approach for consideration of robot application is given in Fig. 10.1. Since most managers are familiar with Gantt charts, the first step was to indicate all the important activities (Fig. 10.2). A detailed list of these activities was also prepared of which only a summary is shown in Fig. 10.3.

Once all activities and their relationships were clear, the precedence relationship and eventually the network to be used in the CPM and PERT

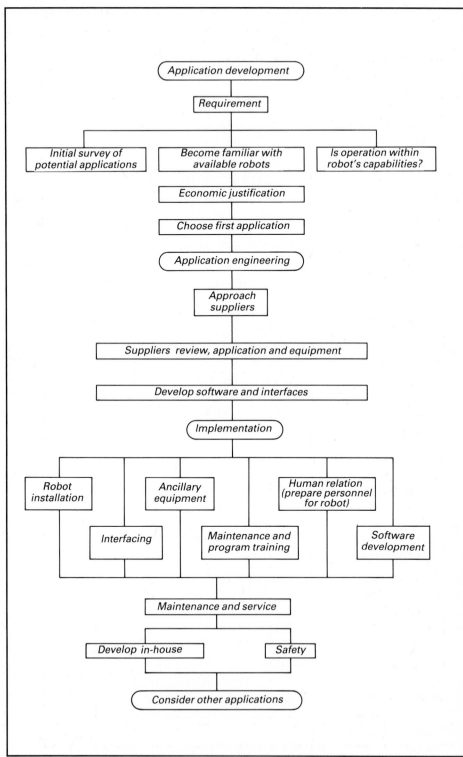

Fig. 10.1 Systematic approach for consideration of robot application

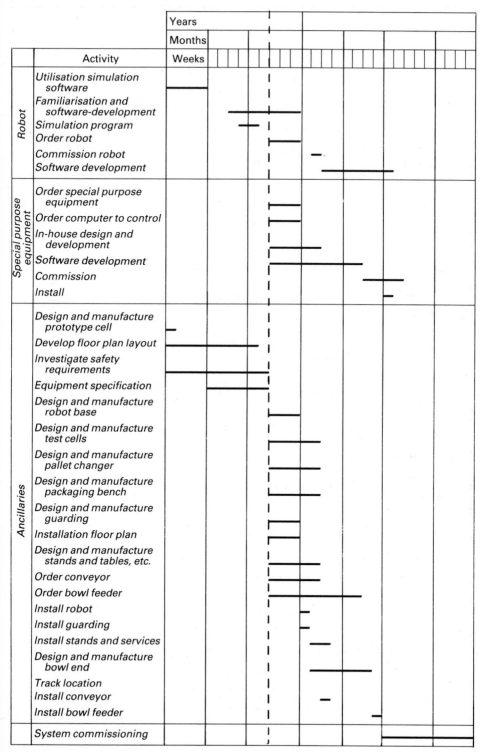

Fig. 10.2 Gantt chart for a typical robot application and implementation plan

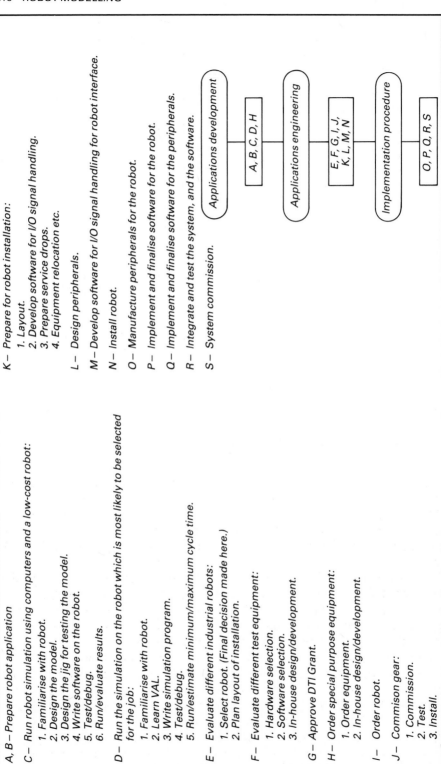

A, B – Prepare robot application

C – Run robot simulation using computers and a low-cost robot:
1. Familiarise with robot.
2. Design the model.
3. Design the jig for testing the model.
4. Write software on the robot.
5. Test/debug.
6. Run/evaluate results.

D – Run the simulation on the robot which is most likely to be selected for the job:
1. Familiarise with robot.
2. Learn VAL.
3. Write simulation program.
4. Test/debug.
5. Run/estimate minimum/maximum cycle time.

E – Evaluate different industrial robots:
1. Select robot. (Final decision made here.)
2. Plan layout of installation.

F – Evaluate different test equipment:
1. Hardware selection.
2. Software selection.
3. In-house design/development.

G – Approve DTI Grant.

H – Order special purpose equipment:
1. Order equipment.
2. In-house design/development.

I – Order robot.

J – Commison gear:
1. Commission.
2. Test.
3. Install.

K – Prepare for robot installation:
1. Layout.
2. Develop software for I/O signal handling.
3. Prepare service drops.
4. Equipment relocation etc.

L – Design peripherals.

M – Develop software for I/O signal handling for robot interface.

N – Install robot.

O – Manufacture peripherals for the robot.

P – Implement and finalise software for the robot.

Q – Implement and finalise software for the peripherals.

R – Integrate and test the system, and the software.

S – System commission.

Applications development: A, B, C, D, H

Applications engineering: E, F, G, I, J, K, L, M, N

Implementation procedure: O, P, Q, R, S

Fig. 10.3 Summary of activities

Fig. 10.4 Precedence
relationships

Activity	Precedence relation
A	–
B	–
C	B
D	C
E	C
F	C
G	A
H	F, D, E, G
I	D, E, G
J	H
K	I
L	I
M	I
N	K
O	L
P	N, M, J
Q	M, P
R	Q
S	R

analysis were established (Figs. 10.4 and 10.5). To provide a full analysis, both the CPM and the PERT programs were used (Figs. 10.6 to 10.9).

The outputs of these programs show the critical activities and also the slack for those activities which are not critical. (It must be emphasised that if any of the critical activities are late, then the whole project will also be late.

The early start times and the latest finish times can be used by the project manager for determining the resources required during the timespan of the project. The output of the PERT program shows the estimates of the expected duration and the standard deviation for each activity in the project. This data collection can be used for determining the standard deviation of the critical path and the expected project completion time.

The information obtained from the computer programs enabled the creation of a modified Gantt chart (Fig. 10.10). This can be used to follow the status of the critical activities, and use together with the network effective coordination of the project was possible. The chart would also help in optimising resource allocation during the project. This can be done by adjusting the start times of those activities which have slack time so that minimum resources are used at any one time.

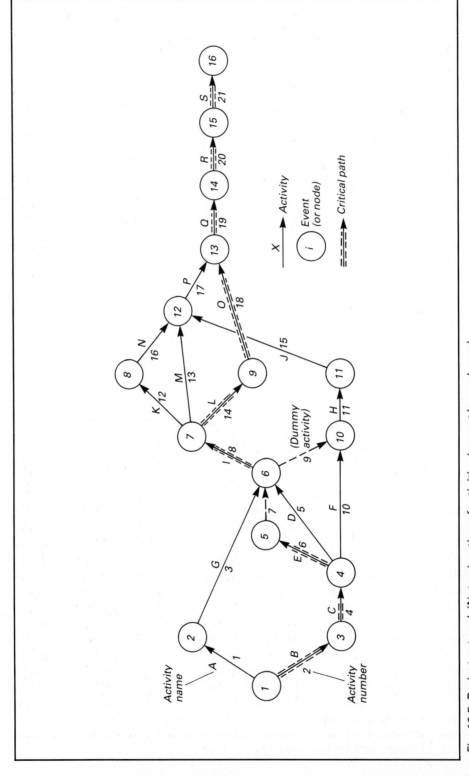

Fig. 10.5 Project network (Note, duration of activities has not been shown)

ACTIVITY No	NAME	START NODE	END NODE	DURATION	COST
1:	A	1	2	12.000	0.000
2:	B	1	3	14.000	0.000
3:	G	2	6	15.000	0.000
4:	C	3	4	5.000	0.000
5:	D	4	6	5.000	0.000
6:	E	4	5	10.000	0.000
7:	DUMMY	5	6	0.000	0.000
8:	I	6	7	6.000	0.000
9:	DUMMY	6	10	0.000	0.000
10:	F	4	10	8.000	0.000
11:	H	10	11	4.000	0.000
12:	K	7	8	8.000	0.000
13:	M	7	12	8.000	0.000
14:	L	7	9	10.000	0.000
15:	J	11	12	5.000	0.000
16:	N	8	12	5.000	0.000
17:	P	12	13	4.000	0.000
18:	O	9	13	8.000	0.000
19:	Q	13	14	4.000	0.000
20:	R	14	15	3.000	0.000
21:	S	15	16	3.000	0.000

Fig. 10.6 Input data using the CPM program

ACTIVITY	START NODE	END NODE	EARLY START	LATE FINISH	DURATION	SLACK	COST
1	1	2	0.000	14.000	12.000	2.000	0.000
2	1	3	0.000	14.000	14.000	CRITICAL	0.000
3	2	6	12.000	29.000	15.000	2.000	0.000
4	3	4	14.000	19.000	5.000	CRITICAL	0.000
5	4	6	19.000	29.000	5.000	5.000	0.000
6	4	5	19.000	29.000	10.000	CRITICAL	0.000
7	5	6	29.000	29.000	0.000	CRITICAL	0.000
8	6	7	29.000	35.000	6.000	CRITICAL	0.000
9	6	10	29.000	40.000	0.000	11.000	0.000
10	4	10	19.000	40.000	8.000	13.000	0.000
11	10	11	29.000	44.000	4.000	11.000	0.000
12	7	8	35.000	44.000	8.000	1.000	0.000
13	7	12	35.000	49.000	8.000	6.000	0.000
14	7	9	35.000	45.000	10.000	CRITICAL	0.000
15	11	12	33.000	49.000	5.000	11.000	0.000
16	8	12	43.000	49.000	5.000	1.000	0.000
17	12	13	48.000	53.000	4.000	1.000	0.000
18	9	13	45.000	53.000	8.000	CRITICAL	0.000
19	13	14	53.000	57.000	4.000	CRITICAL	0.000
20	14	15	57.000	60.000	3.000	CRITICAL	0.000
21	15	16	60.000	63.000	3.000	CRITICAL	0.000

THE CRITICAL PATH LENGTH IS: 63.000

Fig. 10.7 Output using the CPM program

NO	ACTIVITY NAME	START NODE	END NODE	OPTIMISTIC TIME	MOST LIKELY TIME	PESSIMISTIC TIME
1:	A	1	2	10.000	12.000	16.000
2:	B	1	3	12.000	14.000	16.000
3:	G	2	6	12.000	15.000	18.000
4:	C	3	4	3.000	5.000	7.000
5:	D	4	6	3.000	5.000	8.000
6:	E	4	5	8.000	10.000	12.000
7:	DUMMY	5	6	0.000	0.000	0.000
8:	I	6	7	4.000	6.000	8.000
9:	DUMMY	6	10	0.000	0.000	0.000
10:	F	4	10	5.000	8.000	12.000
11:	H	10	11	3.000	4.000	5.000
12:	K	7	8	6.000	8.000	10.000
13:	M	7	12	6.000	8.000	10.000
14:	L	7	9	8.000	10.000	12.000
15:	J	11	12	3.000	5.000	8.000
16:	N	8	12	3.000	5.000	8.000
17:	P	12	13	2.000	4.000	6.000
18:	O	9	13	6.000	8.000	10.000
19:	Q	13	14	2.000	4.000	6.000
20:	R	14	15	2.000	3.000	4.000
21:	S	15	16	2.000	3.000	5.000

Fig. 10.8 Input data using the PERT program

10.3 Selection of the robot manufacturer and/or vendor

Unfortunately there are many vendors in this field who are aiming to make large profits without great understanding of the product they are selling. Because of the high level of systems, control and software engineering skills required to support, develop and install individual robots, as well as systems utilising many robots and other devices working together in a distributed processing environment (e.g. in a flexible assembly system or in a flexible spot welding line), this aspect will be even more important in the future. For a user-company the solution is to develop an in-house team.

10.4 Selection of necessary devices and ensuring integration

A clear understanding of the related technology is essential before selecting any device. Because up-to-date robots incorporate computers and micro-processors, knowledge of data processing cannot be ignored.

The variety and differences of terminology used by different manu-facturers must also be emphasised. To offer some guidance within this field, besides the already defined and/or explained robot terminology used within this book, some commonly used robot and computer terms which are important when selecting industrial robots are given below:

● *Accuracy* is a measurement of deviation from a straight line or a par-ticular taught point in space. Accuracy deviations are attributed to calcu-lation errors, arm geometry and manufacturing errors, and poor location alignment due to control system adjustment or functional faults. (Note,

ACTIVITY 1: A
(NODE 1 TO NODE 2) IS A
NON-CRITICAL EVENT.
EXPECTED DURATION: 12.333 STD DEVIATION: 1.000
EARLY START: 0.000   LATE START: 1.667
EARLY FINISH: 12.333   LATE FINISH: 14.000
SLACK TIME: 1.667

ACTIVITY 2: B
(NODE 1 TO NODE 3) IS A
CRITICAL EVENT.
EXPECTED DURATION: 14.000 STD DEVIATION: 0.667
START NO LATER THAN: 0.000
MUST BE COMPLETED BY: 14.000

ACTIVITY 3: G
(NODE 2 TO NODE 6) IS A
NON-CRITICAL EVENT.
EXPECTED DURATION: 15.000 STD DEVIATION: 1.000
EARLY START: 12.333   LATE START: 14.000
EARLY FINISH: 27.333   LATE FINISH: 29.000
SLACK TIME: 1.667

ACTIVITY 4: C
(NODE 3 TO NODE 4) IS A
NON-CRITICAL EVENT.
EXPECTED DURATION: 5.000 STD DEVIATION: 0.667
START NO LATER THAN: 14.000
MUST BE COMPLETED BY: 19.000

ACTIVITY 5: D
(NODE 4 TO NODE 6) IS A
NON-CRITICAL EVENT.
EXPECTED DURATION: 5.167 STD DEVIATION: 0.833
EARLY START: 19.000   LATE START: 23.833
EARLY FINISH: 24.167   LATE FINISH: 29.000
SLACK TIME: 4.833

ACTIVITY 6: E
(NODE 4 TO NODE 5) IS A
CRITICAL EVENT.
EXPECTED DURATION: 10.000 STD DEVIATION: 0.667
START NO LATER THAN: 19.000
MUST BE COMPLETED BY: 29.000

ACTIVITY 7: DUMMY
(NODE 5 TO NODE 6) IS A
CRITICAL EVENT.
EXPECTED DURATION: 0.000 STD DEVIATION: 0.000
START NO LATER THAN: 29.000
MUST BE COMPLETED BY: 29.000

ACTIVITY 8: I
(NODE 6 TO NODE 7) IS A
CRITICAL EVENT.
EXPECTED DURATION: 6.000 STD DEVIATION: 0.667
START NO LATER THAN: 29.000
MUST BE COMPLETED BY: 35.000

ACTIVITY 9: DUMMY
(NODE 6 TO NODE 10) IS A
NON-CRITICAL EVENT.
EXPECTED DURATION: 0.000 STD DEVIATION: 0.000
EARLY START: 29.000   LATE START: 39.833
EARLY FINISH: 29.000   LATE FINISH: 39.833
SLACK TIME: 10.833

ACTIVITY 10: F
(NODE 4 TO NODE 10) IS A
NON-CRITICAL EVENT.
EXPECTED DURATION: 8.167 STD DEVIATION: 1.167
EARLY START: 19.000   LATE START: 31.667
EARLY FINISH: 27.167   LATE FINISH: 39.833
SLACK TIME: 12.667

ACTIVITY 11: H
(NODE 10 TO NODE 11) IS A
NON-CRITICAL EVENT.
EXPECTED DURATION: 4.000 STD DEVIATION: 0.333
EARLY START: 29.000   LATE START: 39.833
EARLY FINISH: 33.000   LATE FINISH: 43.833
SLACK TIME: 10.833

ACTIVITY 12: K
(NODE 7 TO NODE 8) IS A
NON-CRITICAL EVENT.
EXPECTED DURATION: 4.000 STD DEVIATION: 0.667
EARLY START: 35.000   LATE START: 35.833
EARLY FINISH: 43.000   LATE FINISH: 43.833
SLACK TIME: 0.833

ACTIVITY 13: M
(NODE 7 TO NODE 12) IS A
NON-CRITICAL EVENT.
EXPECTED DURATION: 8.000 STD DEVIATION: 0.667
EARLY START: 35.000   LATE START: 41.000
EARLY FINISH: 43.000   LATE FINISH: 49.000
SLACK TIME: 6.000

ACTIVITY 14: L
(NODE 7 TO NODE 9) IS A
CRITICAL EVENT.
EXPECTED DURATION: 10.000 STD DEVIATION: 0.667
START NO LATER THAN: 35.00
MUST BE COMPLETED BY: 45.000

ACTIVITY 15: J
(NODE 11 TO NODE 12) IS A
NON-CRITICAL EVENT.
EXPECTED DURATION: 5.167 STD DEVIATION: 0.833
EARLY START: 33.000   LATE START: 43.833
EARLY FINISH: 38.167   LATE FINISH: 49.000
SLACK TIME: 10.833

ACTIVITY 16: N
(NODE 8 TO NODE 12) IS A
CRITICAL EVENT.
EXPECTED DURATION: 5.167 STD DEVIATION: 0.833
START NO LATER THAN: 43.000
MUST BE COMPLETED BY: 49.000

ACTIVITY 17: P
(NODE 12 TO NODE 13) IS A
NON-CRITICAL EVENT.
EXPECTED DURATION: 4.000 STD DEVIATION: 0.667
EARLY START: 48.167   LATE START: 49.000
EARLY FINISH: 52.167   LATE FINISH: 53.000
SLACK TIME: 0.833

ACTIVITY 18: O
(NODE 9 TO NODE 13) IS A
CRITICAL EVENT.
EXPECTED DURATION: 8.000 STD DEVIATION: 0.667
START NO LATER THAN: 45.000
MUST BE COMPLETED BY: 53.000

ACTIVITY 19: Q
(NODE 13 TO NODE 14) IS A
CRITICAL EVENT.
EXPECTED DURATION: 4.000 STD DEVIATION: 0.667
START NO LATER THAN: 53.000
MUST BE COMPLETED BY: 57.000

ACTIVITY 20: R
(NODE 14 TO NODE 15) IS A
CRITICAL EVENT.
EXPECTED DURATION: 3.000 STD DEVIATION: 0.333
START NO LATER THAN: 57.000
MUST BE COMPLETED BY: 60.000

ACTIVITY 21: S
(NODE 15 TO NODE 16) IS A
CRITICAL EVENT.
EXPECTED DURATION: 3.167 STD DEVIATION: 0.500
START NO LATER THAN: 60.000
MUST BE COMPLETED BY: 63.167

THE CRITICAL PATH LENGTH IS: 63.167

PLUS OR MINUS: 2.041

Fig. 10.9 Output using the PERT program

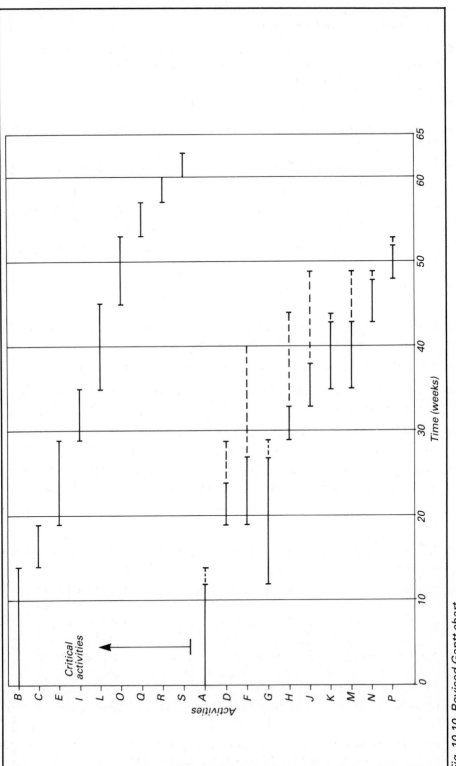

Fig. 10.10 Revised Gantt chart

that in the authors' view, robot position accuracy should be measured in 3D, as described in Chapter Eleven).

● *Adaptive control* is a control method in which control parameters are continuously adjusted using sensors, in response to the monitored process to achieve better performance.

● *Axes* are the directions of possible movements.

● *Coding* is the preparation of a computer/robot machine control program to solve a problem.

● *Cycle time* is the period of time from starting one machine operation to starting another.

● *Drift* is the tendency of the robot to gradually move away from the desired response.

● *Duty cycle* is the time required for a system to become active and operate at full power and accuracy.

● *Encoder* is a transducer used to convert position data into electrical signals.

● *End-point rigidity* is the resistance of the hand, robot tool or end-point of the manipulator arm to motion under applied force.

● *EPROM*, erasable programmable read only memory.

● *Gripper* is a device by which the robot is capable of grasping and holding one or more parts.

● *Hardware* is the mechanical, magnetic, electrical and electronic devices from which a computer and the robot controllers are built.

● *Image processing* is a computational phase prior to the feature extraction phase when analysing an image by machine.

● *Interface* is a mechanical, electrical, pneumatic or hydraulic connection between two devices.

● *Joint* is a rotational or translational degree of freedom in a manipulator system.

● *Memory* is a device or media used to store information in a form that can be understood by the computer hardware. Memory size in computing is given in bytes (1 byte = 8 bits), or K bytes (1K = 1024 bytes).

In robotics, and also in NC/CNC technology, the memory size is sometimes given by the maximum number of possible stored points. This is confusing and does not define the available memory at all, if one assumes that robots and other machines can be programmed using high-level languages and can execute subroutines and other program structures.

● *Microcomputer* is a computer which is using a single integrated circuit (i.e. a microprocessor) based hardware architecture.

- *Multiprocessor* is a computer or a network of computers capable of processing more different tasks concurrently (i.e. in the same time).

- *PROM* stands for programmable ROM. A read only memory that can be initialised once by the user.

- *Real-time* pertains to the actual time during which a physical process transpires.

- *Repeatability,* as opposed to accuracy (see above), is a measurement of the deviation between a taught location point (i.e. a transformation) and the played back location. Under identical conditions of load and velocity this deviation will be smaller than the tolerance given by accuracy.

- *Sensor* is a transducer whose input is a physical phenomenon and whose output is a quantitative measure of the phenomenon.

- *Sensory control* of the robot arm or its motion means that its control is based on sensor readings.

- *Software* is the programs that control the machine.

- *Subroutine (procedure, routine)* is a series of machine control or program instructions describing a certain sub-task that is written only once, but is likely to be required more than once during program execution.

- *Tactile sensor* is a transducer that is sensitive to touch (i.e. pressure).

- *Working envelope* is the set of points representing the maximum extent or reach of the robot hand (or arm) in all permitted directions.

As mentioned previously, most current robots lack proper interfaces, intelligent sensors and programming languages. Hence, the integration aspect should not be overlooked, even if initially the device is going to be standalone.

10.5 Safety aspects when designing robotised systems

Since robots can change their configuration under software control and can produce unpredictable combinations of motions, safety aspects must not be overlooked when selecting industrial robots[1]. Although in most countries standards are available to guide users in this matter, it is important to check the following features of the robot to be purchased:

- Do the controls follow the fail-safe design concept? (i.e. is there enough space between different controls; is the system user-friendly from the safety point-of-view?).
- What kind of emergency stops are provided? Do they operate via software, or directly via the hardware, or in both ways?
- What controls are available for programming, for occasions when the operator must stay close to the end-effector, particularly in welding and assembly applications? Can the robot be stopped quickly without losing the currently edited and/or written program?

Safeguarding is important, but unfortunately strict standards often limit the access of even those programmers, systems and maintenance people who need to stay close to the machine to perform their task. There are many different possible guarding methods, including fixed mechanical guards, interlocked mechanical guards, control system safeguards, pressure sensitive mats, and non-contact or photoelectric guards, etc. Guards should be interlocked with the robot control system, allowing maintenance and programming staff to access the robot safely.

Probably the most important safeguard of all is training and proper system documentation, and these aspects must be not forgotten, even if training costs are high.

References

[1] Bonney, M. C. and Yong, Y. F. (Eds.) 1985. *Robot Safety*. IFS (Publications) Ltd, Bedford, UK.

Chapter Eleven

COMPUTER-ASSISTED TESTING OF THREE-DIMENSIONAL ROBOT HAND POSITIONING AND ORIENTATION ERRORS

TO SELECT the most suitable device, both in technical and economic terms, before purchasing a robot to perform different tasks with it, one should carefully analyse its specification. Unfortunately robot manufacturers often do not provide sufficiently accurate and/or precise technical data of their robots, of their tools, hand and other accessories and their peripherals; thus the laboratory test at the user or at the vendor is often unavoidable.

A carefully designed robot test always depends on the design characteristics of the robot and its planned applications, but generally a full test includes the following main areas:

● Since the controller is the heart of any robot system, one should put a lot of emphasis on its power supply and noise values, its operating system, the programming language(s) it uses, its interfaces, (e.g. simple I/O port, analogue, standard digital interface), its communications functions and interfaces to other computers, and the way the system peripherals (e.g. display, keyboard, teach-pendant, disk store, printer) can be accessed from the programming language and/or operating system.

● Geometrical value measurements should incorporate workspace test (e.g. stroke, angle of rotation), static positioning errors, reproduction of the smallest step, overshoot, path accuracy, orientation errors of the hand, compliance errors.

It should be emphasised that there are several robots which incorporate a tool and/or hand (e.g. IBM's RS2 Manufacturing System), thus in such cases the geometrical error tests should include the tool as well. If the robot does not have an integrated gripper it is always a wise idea to test it with those grippers/tools which are going to be utilised and with the maximum load it will carry.

	Contacting		Non-contact			
Description	Touch-trigger probe	3-D measuring head	3-D measuring heads with 3 sensors	3-D measuring heads with 6 sensors	2-D measuring head with 2 sensors	2-D measuring head with 4 sensors
Diagrammatic representation	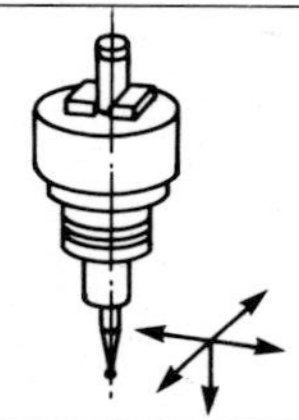	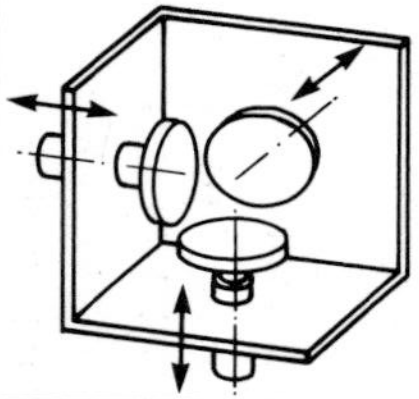	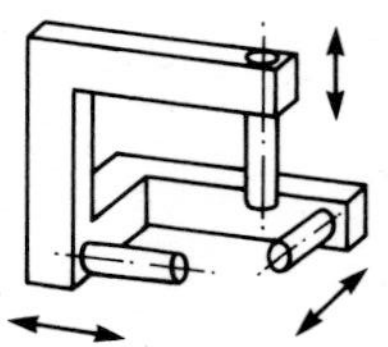	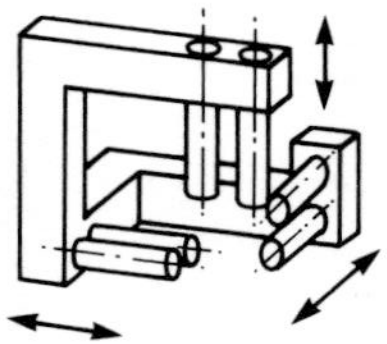	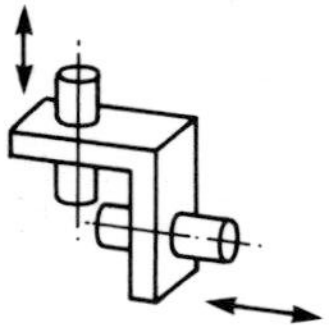	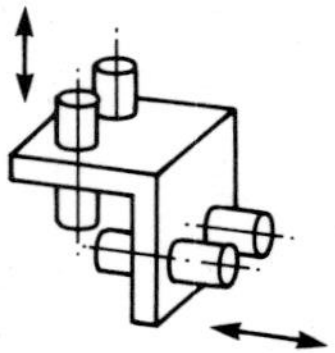
Direction of contact	Distributed arbitrarily over hemisphere	X, Y, Z	X, Y, Z	X, Y, Z	Y, Z	Y, Z
Contact force	1.5 N (at contact) 2.5 N (over-travel)	1 ÷ 3N	–	–	–	–
Stroke or measuring distance	7mm over-travel	20mm	10mm	10mm	10mm	10mm
Weight	1kg	1kg	1kg	1.2kg	0.5kg	1kg
Reference Body — Description	Cube with notch	Sphere	Cube		Straight edge	
Diagrammatic representation	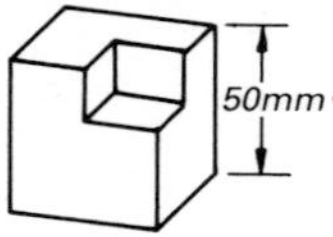 50mm	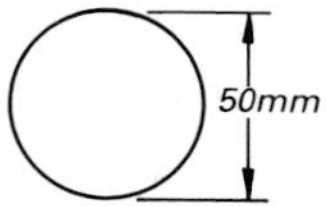 50mm	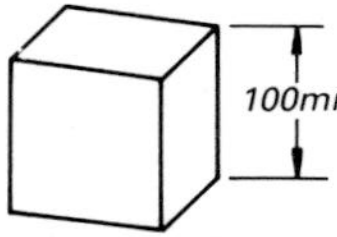 100mm		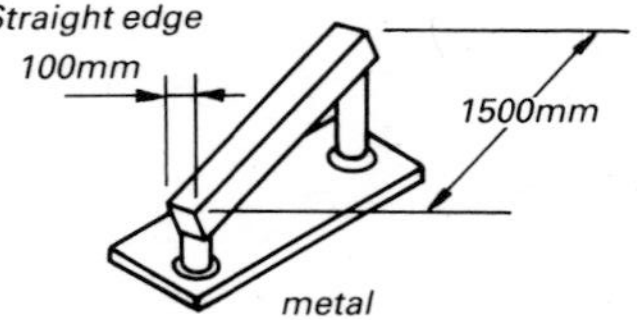 100mm / 1500mm	
Material	any	any	metal		metal	
Mounting	Probe on 3D measuring instrument Reference body on robot	Head on 3-D measuring instrument, reference body on robot	Head on 3-D measuring instrument Reference body on robot		Head on robot Ruler on measuring plate	
Contact	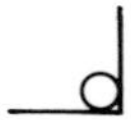		Distance a 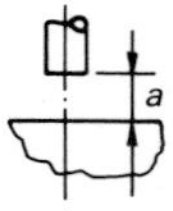	Distance $a \sim \dfrac{a_1 + a_2}{2}$ Angle $\alpha \sim \dfrac{a_1 - a_2}{b}$ 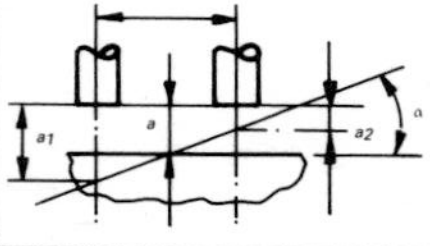	Distance a 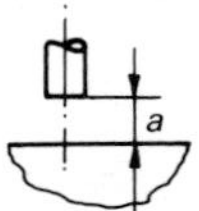	Distance $a \sim \dfrac{a_1 + a_2}{2}$ Angle $\alpha \sim \dfrac{a_1 - a_2}{b}$ 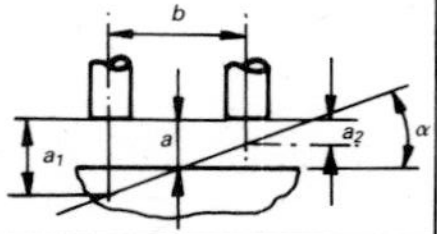

Fig. 11.1 Overview on some robot test models and test-rigs offering robot hand position error, but no hand orientation error analysis and no test-rig self-diagnosis features[1]

- The kinematic test usually includes the measurement of the maximum, minimum, average and the maximum variation of the specified speed and acceleration values for each linear, rotary and combined motions the robot can perform.

- The dynamic test of a robot is usually complex and requires sensory based test equipment. The following tests are advised: gripping force, maximum load capacity, rigidity, vibration, dynamic repeatability, dynamic compensation (if available).

- Robot safety aspects and their assurance is guided by standards in most countries. In general, test for safety should cover the environmental aspects, such as temperature, humidity, electronic sensitivity, and a variety of emergency situations and devices, such as software and hardware emergency stop, physical guarding, the effect of the loss of power supply, unauthorised access, human errors and their possible effects, pneumatic, electronic and hydraulic faults and their effect on the operator and the environment.

11.1 Repeatability and positioning error analysis

Test methods based on scalar data currently used by most robot manufacturers often do not provide sufficient and precise information about the robot's positioning and orientation errors. As an increased number of robots are going to be used by the precision assembly industry in the very near future this was a problem which had to be solved.

One must realise that repeatability and positioning error analysis methods adapted from the machine tool industry, and currently widely used by robot manufacturers, should not be used in robotics without a great deal of modification. Because of the infinite number of robot arm orientation possibilities, compared to the fixed axes of most machine tools, it is advised to adapt the three-dimensional error analysis method.

Most robot position error measurement methods make use of a special purpose measuring head offering the possibility of measuring a point, or a series of points in the three-dimensional space, reached by a distinguished point of the robot hand or tool. A comprehensive comparison of such measuring heads and the utilised test methods is given in Fig. 11.1.

The proposed method and software developed by one of the authors and described in this chapter utilises vector analysis and approaches to introduce a new standard in measuring positioning and orientation errors of industrial robots. The difference compared to other methods is that it not only offers the three-dimensional robot position (repeatibility) test facility, but also provides the orientation errors, and the self-diagnosis of the measured data based on statistical error analysis.

As in other methods it requires an accurate test-cube which is loaded and unloaded by the robot into a test-rig, but utilises displacement data received from nine test probes, rather than three, offering the possibility of the orientation error analysis as well as the test-rig self-diagnosis by calculating three normal vectors independently and by checking the test-cube/

Fig. 11.2 The Unimation PUMA 560 robot loading the test-cube into the 'Ranky-type' test-rig

test-rig perpendicularity errors in the software after each load/unload operation. This test method and the software are universally applicable, thus can be utilised for any industrial robot (Fig. 11.2).

This chapter provides the description and the results of the TEST-ROBOT program, a module of 'The FMS Software Library' capable of performing the necessary robot positioning and orientation error calculations, based on measured data and a subsequent statistical error analysis.

The described test method, the robot test-rig and software can be utilised as an off-line test facility and the software can also be integrated into robot controllers and executed in real-time in unmanned computer-integrated flexible assembly systems where increased accuracy and consequently higher loading reliability are essential requirements.

The results of the robot error analysis can be useful to robot manufacturers, test engineers and production engineers, who wish to test the most suitable assembly and/or inspection robots and to test their robots in action. The benefits of this package can be realised in general by all those designers and software engineers who are involved in designing unmanned assembly, welding, part handling, inspection, etc. cells for FMS and Computer Integrated Manufacturing (CIM), not only mechanically but also from the real-time control and sensory feedback processing points of view.

The real benefits can be achieved if the program is integrated into powerful robot controllers and executed in real-time. In this case the method will not only contribute to a software-corrected increased robot position and orientation accuracy, but also will help to improve the reliability of unmanned computer-controlled assembly, inspection, welding, machine loading, etc. robots and pick-and-place devices.

By integrating the program with an expert system, typical faults can be learnt and analysed more easily, and more effective real-time action can be taken.

11.2 Mathematical model of the TESTROBOT program

In general, robot hand alignment errors occur because each robot carries different joint, arm and tool manufacturing errors and because there can be varying errors in the robot control system.

The true position of the robot hand (i.e. tool) can be defined by a vector $\bar{r}$ pointing to any known point of the test component in the test-rig coordinate system (in our case this point is the reference point of a test-cube for simplicity), and the robot hand orientation vector $\bar{N}$ defined by $\bar{n}1$, $\bar{n}2$, $\bar{n}3$ unit normal vectors, which define its orientation in the test-rig coordinate system.

If a sufficient number of measurements are carried out, then the analysis of a series of such measurements indicate the statistical average error of the position and the orientation of the robot hand.

The core of the calculation method is based on the following mathematical model. Consider face 1 of the test-cube (Fig. 11.3 (a,b)). Having measured the indicated points in the test-rig coordinate system, we can define $\bar{p}1$, $\bar{p}2$ and $\bar{p}3$ vectors pointing to points P1, P2 and P3 on face 1, respectively.

$$\bar{p}1 = \begin{matrix} x1 \\ y1 \\ z1 \end{matrix} \qquad \bar{p}2 = \begin{matrix} x2 \\ y2 \\ z2 \end{matrix} \qquad \bar{p}3 = \begin{matrix} x3 \\ y3 \\ z3 \end{matrix}$$

The orientation of face 1 can be determined by calculating the normal (unit) vector ($\bar{n}1$) to the face.

The $\bar{p}2\bar{p}1$ and $\bar{p}2\bar{p}3$ vectors can be defined as:

$$\bar{p}2\bar{p}1 = \bar{p}1 - \bar{p}2 \quad \text{and} \quad \bar{p}2\bar{p}3 = \bar{p}3 - \bar{p}2$$

Both of these vectors lie in the plane of face 1.
A vector normal to face 1 will be $\bar{p}2\bar{p}3 \times \bar{p}2\bar{p}1$, since

$$\bar{p}2\bar{p}1 = (x1-x2)\bar{i} + (y1-y2)\bar{j} + (z1-z2)\bar{k}$$

and

$$\bar{p}2\bar{p}3 = (x3-x2)\bar{i} + (y3-y2)\bar{j} + (z3-z2)\bar{k}$$

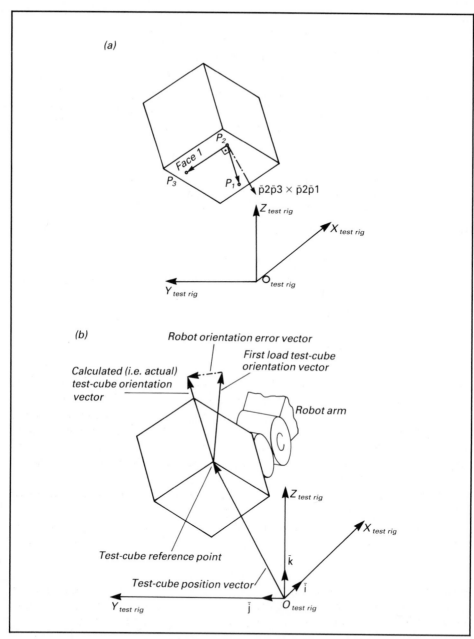

Fig. 11.3(a) Definition of the p̄2p̄1 and p̄2p̄3 vectors on the selected face 1 of the test-cube. It is important to select the indicated P1, P2 and P3 points so that the p̄2p̄3 x p̄2p̄1 vector points outwards from the test-cube surface. This condition should apply for the other two surfaces as well, otherwise the interpretation of the calculated orientation error vector will be different to the calculated orientation. (b) General purpose explanation of the TESTROBOT program results. It should be noted that the first load test-cube orientation does not have to be perfectly aligned with the test-rig coordinate system, the first load test-cube orientation and position are taken as the base for the statistical error analysis, and the drawn vectors indicate a general case and do not relate to the actual results (given in Fig. 11.13)

then

$$\begin{array}{ccc} \bar{i} & \bar{j} & \bar{k} \end{array}$$

$$\begin{aligned} \bar{p}2\bar{p}3 \times \bar{p}2\bar{p}1 &= \begin{array}{ccc} x3-x2 & y3-y2 & z3-z2 \end{array} \\ &= \begin{array}{ccc} x1-x2 & y1-y2 & z1-z2 \end{array} \\ &= [(y3-y2)(z1-z2) - (y1-y2)(z3-z2)]\bar{i} \\ &\quad +[(x1-x2)(z3-z2) - (x3-x2)(z1-z2)]\bar{j} \\ &\quad +[(x3-x2)(y1-y2) - (x1-x2)(y3-y2)]\bar{k} \end{aligned}$$

For simplicity, if

$$\begin{aligned} C1 &= (y3-y2)(z1-y2) - (y1-y2)(z3-z2) \\ C2 &= (x1-x2)(z3-z2) - (x3-x2)(z1-z2) \\ C3 &= (x3-x2)(y1-y2) - (x1-x2)(y3-y2) \end{aligned}$$

then

$$\bar{p}2\bar{p}3 \times \bar{p}2\bar{p}1 = C1\,\bar{i} + C2\,\bar{j} + C3\,\bar{k}$$

where C1, C2 and C3 contain only sensory feedback (i.e. measured) data and are easily computed.

The unit normal vector to face 1 is

$$\bar{n}1 = \frac{C1\,\bar{i} + C2\,\bar{j} + C3\,\bar{k}}{C1^2 + C2^2 + C3^2}$$

The other two unit vectors n2 and n3 can be found correspondingly on face 2 and face 3, after having measured points P4, P5, P6, P7, P8 and P9.

It is important to realise that by calculating and measuring any of the above three normal vectors, the robot test-rig and test-cube perpendicularity errors can also be detected. (This is a useful self-calibrating and diagnosis feature.) A sample display of this feature is shown in Fig. 11.4.

One should assume of course, that the test-cubes themselves are correctly manufactured and fully tested before the robot alignment error measurements and analysis are carried out.

Having found ñ1, ñ2 and ñ3 normal unit vectors independently, the orientation vector of the robot hand can be defined:

$$\bar{N} = \bar{n}1 + \bar{n}2 + \bar{n}3$$

Assuming that $\bar{r}_{i+1}$ is the subsequent position of $\bar{r}_i$, generally the relative position error of the robot (i.e. the position error between two subsequent test-cube loading and unloading operations) can be calculated as:

$$\bar{E}_i = \bar{r}_{i+1} - \bar{r}_i$$

The absolute position and orientation error calculation is based on the first test-cube load position and orientation. In other words, the first load (or set) position and orientation provides the base data (i.e. error free data) for the calculations.

To measure the robot alignment errors one can use the in-cycle gauging technique, appropriate robot integrated non-contact sensors or a specially designed robot-test rig (Figs. 11.5 and 11.6).

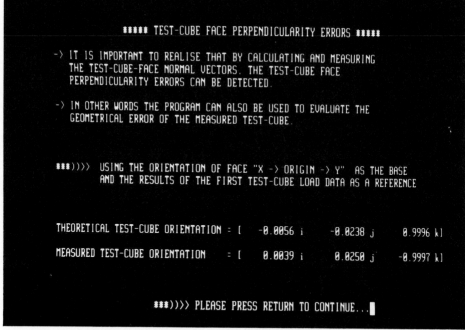

***** TEST-CUBE FACE PERPENDICULARITY ERRORS *****

-> IT IS IMPORTANT TO REALISE THAT BY CALCULATING AND MEASURING
   THE TEST-CUBE-FACE NORMAL VECTORS. THE TEST-CUBE FACE
   PERPENDICULARITY ERRORS CAN BE DETECTED.

-> IN OTHER WORDS THE PROGRAM CAN ALSO BE USED TO EVALUATE THE
   GEOMETRICAL ERROR OF THE MEASURED TEST-CUBE.

***))>>   USING THE ORIENTATION OF FACE "X -> ORIGIN -> Y" AS THE BASE
           AND THE RESULTS OF THE FIRST TEST-CUBE LOAD DATA AS A REFERENCE

THEORETICAL TEST-CUBE ORIENTATION = [-0.0056 i   -0.0238 j   0.9996 k]

MEASURED TEST-CUBE ORIENTATION   = [0.0039 i   0.0250 j   -0.9997 k]

***))>> PLEASE PRESS RETURN TO CONTINUE...█

Fig. 11.4 The program is capable of analysing test-cube and test-rig perpendicularity errors, thus providing a self-diagnosis feature. (Results shown are in millimetres and represent an error within the tolerance of the test-rig and test-cube, utilising dial gauges rather than non-contact type displacement measuring sensors.)

After a sufficient number of robot loading and unloading operations, a statistical analysis of the position and orientation error is made, and on request the measured as well as the calculated data are displayed and/or printed. For user-convenience and for experimentation, the input data table can be edited (partially or fully, i.e. for each test-cube face and/or point) and the program can be re-run as many times as requested.

11.3 Features of the TESTROBOT program

The main features of the TESTROBOT program are as follows:

● Industrial robot hand position and orientation error analysis and real-time error correction facility by means of software. The TESTROBOT package is capable of statistically analysing measured data and determining the robot orientation and position error.
● User-friendly and run-time-checked screen management system is incorporated to ensure a robust user-interfact (Figs. 11.7 and 11.8).
● The program can be used as a simulation tool for evaluating different solutions by using the 'What if?' input data edit command.
● Results can be saved in a user-specified text file on disk. (This file can also be used in user-written programs for further processing.)
● The input data collection and the calculated results can be downloaded via any valid port into any piece of equipment capable of receiving this data (in ASCII format) for further processing.

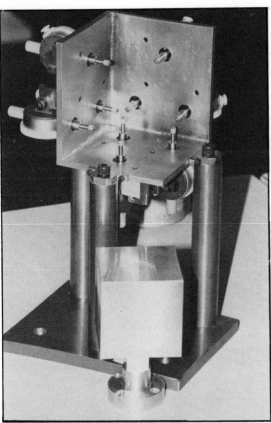

Fig. 11.5 The robot test-rig using low-cost dial gauges for displace measurement. A more advanced (and expensive) solution could incorporate digital sensors interfaced on-line with the robot controller, allowing real-time accuracy checks after a certain number of misloads or compliance errors

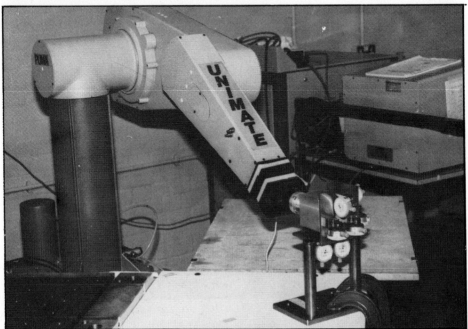

Fig. 11.6 The Unimation PUMA 560 robot under test

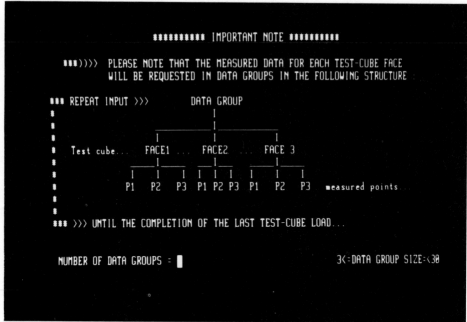

Fig. 11.7 *The input data structure of the program. Since there are three measured points on each surface of the test-cube a data group will contain nine measured points. (A 16-bit microcomputer with 256K memory will be able to analyse about 30 data groups. In most cases, ten data groups provide sufficient results.)*

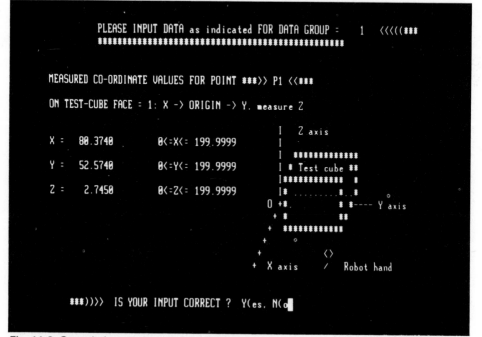

Fig. 11.8 *Sample input screen of the TESTROBOT program. (Each input data is run-time checked and each screen can be re-edited as many times as required in this user-friendly screen management system.)*

● Formatted sheets can be printed with reasonably large margins, user-defined page heading and sequential page numbering.
● Run-time input line editing and disk input/output error handling is also incorporated.
● Capability of evaluating the perpendicularity errors of the utilised robot test-rig.
● Compatibility with the other programs of 'The FMS Software Library'.
● Portable Pascal code, capable of running on over 30 different computers, including the IBM PC/XT, ACT-SIRIUS/VICTOR, SAGE, ACT-APRICOT.

11.4 Program menu

The main menu of the program (Fig. 11.9) includes the following commands:

P (rint input and calculated results

This will put all input data and the calculated data for each cell to the PRINTER: port. If a printer is attached to this port a print of this set of data can be obtained; if another piece of equipment capable of receiving this data is available, then downloading is possible.

N (ew run with new input data

Clears the memory for a new run.

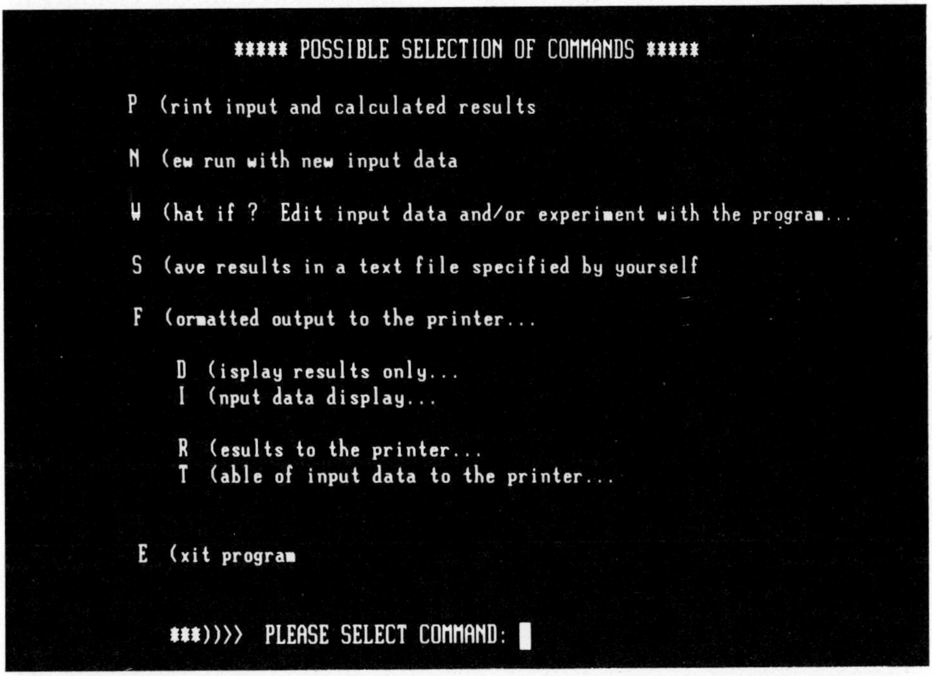

```
      ***** POSSIBLE SELECTION OF COMMANDS *****

  P  (rint input and calculated results

  N  (ew run with new input data

  W  (hat if ?  Edit input data and/or experiment with the program...

  S  (ave results in a text file specified by yourself

  F  (ormatted output to the printer...

          D  (isplay results only...
          I  (nput data display...

          R  (esults to the printer...
          T  (able of input data to the printer...


  E  (xit program


      ***)))>  PLEASE SELECT COMMAND: █
```

Fig. 11.9 The main menu of the program as based on single command characters

```
    DATA GROUP:   4 <<<(####         ####)>>> X CO-ORDINATES OF...

    >>> MEASURED POINTS...                P1          P2          P3
                          FACE 1       80.3740     31.1540     31.1740
                          FACE 2        2.7000      2.3300      1.4500
                          FACE 3       80.6740     50.2340     31.1740

                                   ####)>>> Y CO-ORDINATES OF...

    >>> MEASURED POINTS...                P1          P2          P3
                          FACE 1       52.5740     30.9140     81.3740
                          FACE 2       31.5940     59.5540     81.4940
                          FACE 3        1.3000      2.2000      0.7000

                                   ####)>>> Z CO-ORDINATES OF...

    >>> MEASURED POINTS...                P1          P2          P3
                          FACE 1        2.5400      3.3600      2.0000
                          FACE 2       30.8540     80.2940     29.3940
                          FACE 3       29.6400     80.6340     30.6340



                      ###)>>> PLEASE PRESS RETURN TO CONTINUE...█
```

Fig. 11.10 Sample display of a measured data collection. (Note that the display also shows the test-rig calibration values.)

S (ave results in a specified text file

This command allows the user to transfer information from the memory to a specified disk file. It will take care of mistakes and all possible (i.e. detectable) input/output errors, which might occur during the data transmission process between the machine and the selected disk drive.

F (ormatted output to the printer

This feature allows the operator to make formatted prints of the input and the calculated results. The user can specify the page heading and it will take care of page numbering and adjust the page size.

D (isplay results only . . .

This command will transfer the current results (i.e. calculated data) to the display.

I (nput data display . . .

This command will allow the user to have a look into the current input data collection, by displaying a data group at any one time. (A sample input data group is shown in Fig. 11.10.)

T (able of input data to the printer . . .

This command transfers the current input data collection to the PRINTER: port. (If necessary with this command one can download the data to another machine, similar to the P(rint . . . command.)

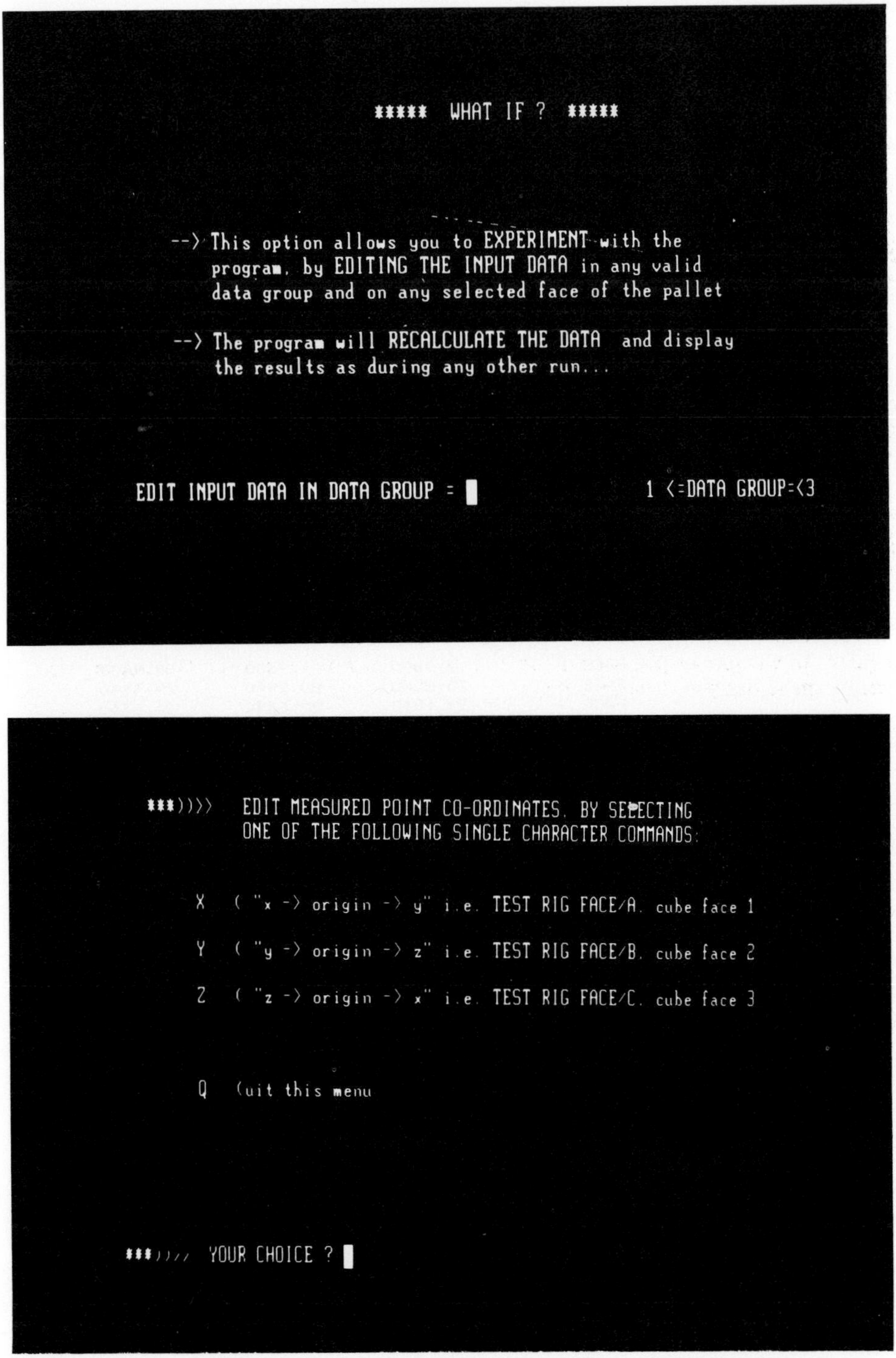

```
               *****  WHAT IF ?  *****



     -->'This option allows you to EXPERIMENT with the
        program. by EDITING THE INPUT DATA in any valid
        data group and on any selected face of the pallet

     --> The program will RECALCULATE THE DATA  and display
        the results as during any other run...




EDIT INPUT DATA IN DATA GROUP = █          1 <=DATA GROUP=<3
```

```
   ***))))    EDIT MEASURED POINT CO-ORDINATES, BY SELECTING
             ONE OF THE FOLLOWING SINGLE CHARACTER COMMANDS:


       X   ( "x -> origin -> y" i.e. TEST RIG FACE/A. cube face 1

       Y   ( "y -> origin -> z" i.e. TEST RIG FACE/B. cube face 2

       Z   ( "z -> origin -> x" i.e. TEST RIG FACE/C. cube face 3


       Q   (uit this menu




   ***////  YOUR CHOICE ? █
```

Fig. 11.11 The 'What if?' feature of the program allows input data editing and experimentation with the program. This feature can also be used for simulating different measured data and their effect on the position and orientation errors, when designing industrial robots

```
**************
INPUT DATA ECHO
**************


DATA GROUP:   1          ****))>> X CO-ORDINATES OF...
**********

>>>>>  MEASURED POINTS...          P1          P2          P3

                  FACE 1      80.3740      31.1540      31.1740
                  FACE 2       2.9000       1.3500       2.3900
                  FACE 3      80.6740      50.2340      31.1740

                  ****))>> Y CO-ORDINATES OF...

>>>>>  MEASURED POINTS...          P1          P2          P3

                  FACE 1      52.5740      30.9140      81.3740
                  FACE 2      31.5940      59.5540      81.4940
                  FACE 3       1.4200       2.9400       1.8900

                  ****))>> Z CO-ORDINATES OF...

>>>>>  MEASURED POINTS...          P1          P2          P3

                  FACE 1       2.9600       1.9600       0.9400
                  FACE 2      30.8540      80.2940      29.3940
                  FACE 3      29.6940      80.6340      30.6340
- - - - - - - - - - - - - - - - - - - - - - - - - - - - - - - - -
DATA GROUP:   2          ****))>> X CO-ORDINATES OF...
**********

>>>>>  MEASURED POINTS...          P1          P2          P3

                  FACE 1      80.3740      31.1540      31.1740
                  FACE 2       3.2500       1.2100       2.3700
                  FACE 3      80.6740      50.2340      31.1740

                  ****))>> Y CO-ORDINATES OF...

>>>>>  MEASURED POINTS...          P1          P2          P3

                  FACE 1      52.5740      30.9140      81.3740
                  FACE 2      31.5940      59.5540      81.4940
                  FACE 3       1.4400       2.9500       1.9000

                  ****))>> Z CO-ORDINATES OF...




>>>>>  MEASURED POINTS...          P1          P2          P3

                  FACE 1       2.9700       1.9600       0.9300
                  FACE 2      30.8540      80.2940      29.3940
                  FACE 3      29.6940      80.6340      30.6340
- - - - - - - - - - - - - - - - - - - - - - - - - - - - - - - - -
```

Fig. 11.12 The data sheet of the input (i.e. measured displacement data)

```
DATA GROUP:  3          ****))>> X CO-ORDINATES OF...
***********

>>>>>  MEASURED POINTS...          P1          P2          P3

                FACE 1      80.3740     31.1540     31.1740
                FACE 2       3.3500      2.1800      3.1400
                FACE 3      80.6740     50.2340     31.1740

                ****))>> Y CO-ORDINATES OF...

>>>>>  MEASURED POINTS...          P1          P2          P3

                FACE 1      52.5740     30.9140     81.3740
                FACE 2      31.5940     59.5540     81.4940
                FACE 3       1.4400      2.8400      2.0900

                ****))>> Z CO-ORDINATES OF...

>>>>>  MEASURED POINTS...          P1          P2          P3

                FACE 1       2.9700      1.9600      0.9300
                FACE 2      30.8540     80.2940     29.3940
                FACE 3      29.6940     80.6340     30.6340
- - - - - - - - - - - - - - - - - - - - - - - - - - - - - - - -
DATA GROUP:  4          ****))>> X CO-ORDINATES OF...
***********

>>>>>  MEASURED POINTS...          P1          P2          P3

                FACE 1      80.3740     31.1540     31.1740
                FACE 2       3.4600      1.9300      2.9200
                FACE 3      80.6740     50.2340     31.1740

                ****))>> Y CO-ORDINATES OF...

>>>>>  MEASURED POINTS...          P1          P2          P3

                FACE 1      52.5740     30.9140     81.3740
                FACE 2      31.5940     59.5540     81.4940
                FACE 3       1.4500      2.8300      1.9200

                ****))>> Z CO-ORDINATES OF...

>>>>>  MEASURED POINTS...          P1          P2          P3

                FACE 1       2.9800      1.9500      0.9200
                FACE 2      30.8540     80.2940     29.3940
                FACE 3      29.6940     80.6340     30.6340
- - - - - - - - - - - - - - - - - - - - - - - - - - - - - - - -
```

Fig. 11.12 continued

```
******************
CALCULATED RESULTS
******************


ROBOT POSITION ERROR
********************

                                    X               Y               Z

STANDARD DEVIATION =              0.3722          0.1768          0.1129


ROBOT ORIENTATION ERROR
***********************

                                    i               j               k

MEAN VALUE          =             0.0005          0.0007         -0.0020
VARIANCE            =             0.0000          0.0000          0.0000
STANDARD DEVIATION  =             0.0020          0.0056          0.0026



CALCULATED (i.e. ACTUAL) ROBOT ALIGNMENT
*****************************************


TEST-CUBE POSITION VECTOR    = [   -1.3973 i      5.3248 j       2.1436 k]


TEST-CUBE ORIENTATION VECTOR = [    0.9795 i      1.0308 j      -1.0001 k]


**>> NOTE: BOTH VECTORS ARE GIVEN IN THE TEST-RIG CO-ORDINATE SYSTEM
           AND RELATE TO THE PREDEFINED TEST-CUBE REFERENCE POINT



TEST-CUBE FACE PERPENDICULARITY ERRORS
****************************************

THEORETICAL TEST-CUBE ORIENTATION = [    0.0254 i     -0.0247 j    0.9992 k]

MEASURED TEST-CUBE ORIENTATION    = [   -0.0292 i      0.0202 j   -0.9994 k]
```

Fig. 11.13 Partial computer output giving the measured input data and the statistically evaluated results based on the three-dimensional analysis

R (esults to the printer . . .

This command puts the output data table to the PRINTER: port. (If necessary other suitable equipment can receive the results via this port.)

W (hat if? Edit input data and/or experiment with the program

This is a useful feature if at least one valid data collection has already been obtained and if one wishes to experiment with it by editing any measured point on any measured faces of the test component, using the test-rig (Fig. 11.11).

E (xit program

By selecting this command one can exit the package.

11.5 Sample robot test procedure and analysis

Several different industrial robots have been tested successfully with the described method and software. Fig. 11.12 shows a measured data collection; the results of the robot alignment error calculation program are given in Fig. 11.13. The output shows the measured data as well as the calculated position and orientation error vectors together with their standard deviation and variance values.

The results (in millimetres) show that this robot carries reasonably small position and orientation errors, and that the test-rig perpendicularity errors are ignorable.

Chapter Twelve

END-EFFECTORS, SENSORS AND AUTOMATED ROBOT HAND CHANGERS

THE PURPOSE of this chapter is to give an international overview of some interesting end-effectors (i.e. grippers, robot tools), automated robot hand changers (ARHC) and sensors integrated into robot tools, as well as their real-time software.

Robot tools, or in more general terms, end-effectors are general purpose, programmable or task-orientated devices connected between the robot wrist and the object or load to be manipulated and/or processed by the robot. They can offer and/or limit the versatility of grasping and/or processing of different compontents, sensing their characteristics and working together with the robot control system to provide a reliable 'service' throughout the component manipulation cycle.

Usually the robot software takes approximately 30–40% of the total development cost. The design and test of purpose built end-of-arm tooling can cost even more, and may even take 40-60% of the time spent on the development particularly if the tool contains integrated sensors. In the case of handling delicate components, the necessary number of slip, touch, force and other sensors applied in the robot tool can be as high as 120 or more. One can imagine the huge real-time computing tasks in such cases.

Because of the importance of the automated robot hand changers (ARHC), particularly in flexible assembly and when designing other flexible robot cells and FMS, this area is also discussed with examples. Some known systems are demonstrated as well as the 'Ranky-type' ARHC developed at Trent Polytechnic, Nottingham for automated assembly and other operations.

Manipulator and robot tools can be classified in several different ways [1]. Any of the listed aspects could identify different groups of grippers and robot tools for classification purposes:

Kinematic aspects
  – Linkage type (Fig. 12.1)
  – Gear and rack type (Fig. 12.2)
  – Cam type (Fig. 12.3)
  – Screw type (Fig. 12.4)
  – Rope and pulley type (see Fig. 12.44)
  – Miscellaneous

Finger type
  – Single finger
  – Multiple finger (see Fig. 12.58)
  – Special (see Fig. 12.48)
  – Flexible (see Fig. 12.45)

Motion control
  – None
  – Cam mechanism (see Fig. 12.3)
  – Gears (see Fig. 12.4)
  – Electronic (see Fig. 12.4)

Power supply
  – Pneumatic (see Fig. 12.13)
  – Hydraulic (see Fig. 12.18)
  – Electric (see Fig. 12.24)
  – Electromagnetic
  – Mechanical coupling, etc.

Type of grip provided
  – Constant force
  – Distance dependent gripping force
  – Surface dependent gripping force
  – Component material dependent gripping force, etc.

Sensors applied in the tool
  – None
  – Slip (see Fig. 12.52)
  – Force
  – Inductive (see Fig. 12.32)
  – Capacitive (see Fig. 12.32)
  – Optical (see Fig. 12.32)
  – Mechanical
  – Heat
  – Colour
  – Smell (Fig. 12.28)
  – Pneumatic pressure, etc.

The manipulated object
  – Gripping diameter and its tolerance
  – Gripping surface geometry
  – Gripping surface quality (e.g. grease, oil, dust, corrosive material)
  – Gravity centre point

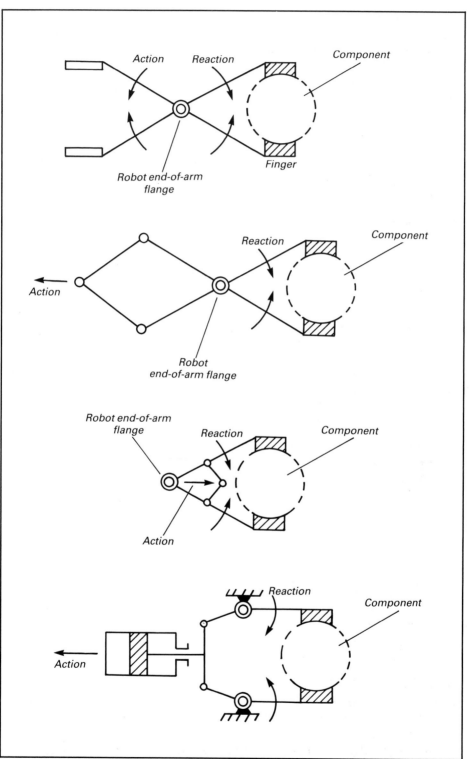

Fig. 12.1 Examples of a linkage type gripper mechanism

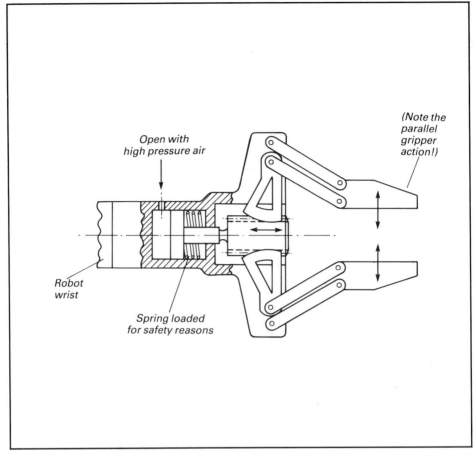

Fig. 12.2 Example of a gear-and-rack type gripper mechanism

- Weight
- Stiffness
- Temperature
- Shock sensitivity

Miscellaneous
- Speed requirements
- Acceleration requirements
- Geometrical requirements
- Loading/unloading safety aspects
- Loading/unloading path limitations
- Positioning accuracy
- Orientation accuracy
- Stiffness
- Wear resistance
- Heat resistance
- Corrosive resistance
- Safety aspects of the robot tool or gripper

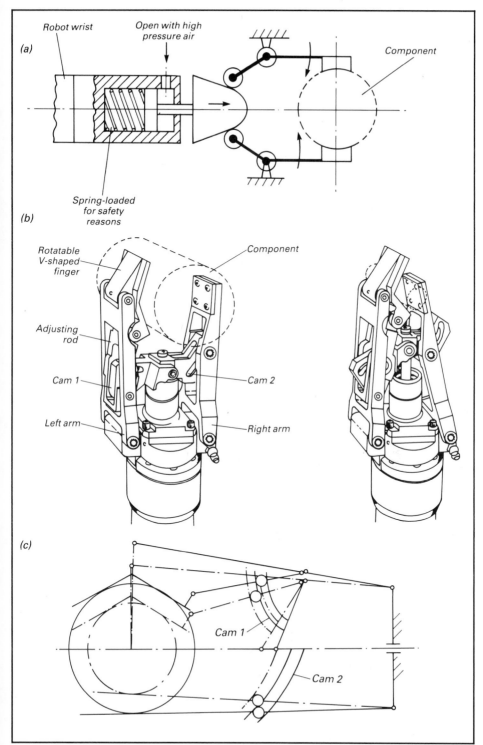

Fig. 12.3(a) Example of a cam type gripper mechanism; (b) mechanism using cams for the concentric gripping of cylindrical components; (c) principle of operation

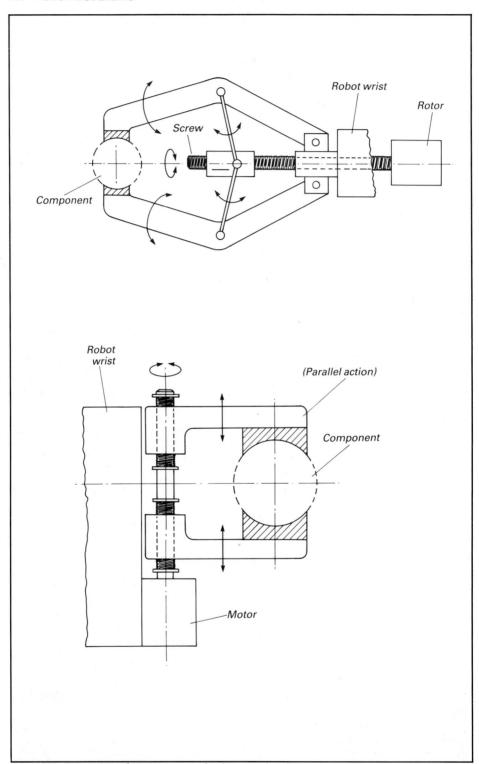

Fig. 12.4 Examples of screw type gripping mechanisms

12.1 Tool changing end-effectors for machine tools

The first machine integrated manipulators were probably used for automated tool changing (ATC) in NC milling centres (Fig. 12.5). Today on the basis of the success of different ATC systems, components of various sizes, type, shape, material, etc. are loaded and unloaded onto a large variety of different machines including milling centres, lathes, plastic moulding machines, and sheet metal manufacturing machines.

To understand the function of the required grippers on such two to three axis point-to-point controlled manipulators, a typical tool changing process in NC or CNC machine tools is briefly summarised:

Step 1. The tool changer arm rotates (or translates) and engages the tools both in the magazine pickup position and the main spindle simultaneously.

Step 2. The arm grips the tools mechanically, then moves forward to remove the old tool from the main spindle and the new tool from the tool magazine socket.

Step 3. The arm changes the two tools, usually by rotating 180°.

Fig. 12.5 Automated tool changing manipulators on the Csepel precision centre (courtesy of Csepel Machine Tool Co., Budapest)

Fig. 12.6 The Hertel Flexible Tooling system block tool magazine (a) and manipulator (b)

Step 4. The arm retracts to put the old tool into the tool magazine, then moves to a safe position.

Step 5. The new tool is firmly located in the machine's main spindle, ready for operation.

Note that machines equipped with some kind of flexible block tool system (such as the Sandvik-Coromant or the Hertel flexible tooling system) pick up the tool from the tool magazine and load it into the nest of the block-tool shank in separate steps. Fig. 12.6 illustrates this kind of hydraulically operated gripper attached to the tool manipulator of the machine tool.

The main problems encountered with tool loading/unloading grippers include:

● The weight and size of certain large tools.
● The accuracy and speed required for loading/unloading tools (e.g. the usual tool change time is 4-7s).
● The possibility of chipping or dirt accumulating on the surface of the tool shank, causing unexpected damage to the part, to the tool and/or to the machine.

12.2 Pneumatically and hydraulically operated mechanical grippers

Since a pneumatic power supply is available in almost every machine shop, pneumatic grippers and actuators are widely used on industrial robots for a large variety of different tasks.

Fig. 12.7 illustrates an actuator principle of PHD Inc. Here, the air pressure forces the piston and cam-bars in the direction of the arrows causing slide bars, which are riding against the inclined cam-bars, to rotate the jaws about the pivot pins which in turn closes the fingers. Reversing the air pressure causes the piston and the cam-bars to move in the opposite direction, and the pivoting fingers to the open position. The cam principle assures uniform gripping force regardless of finger position or variation in grasped component size. The cam angles and materials have been designed to lock fingers against movement by external forces. Fingers may be moved only by internal air pressure, spring or optional manual override buttons. Fig. 12.8 illustrates three typical applications using the PHD gripper.

The 'ERACON Mini-Hands' represent a wide range of pneumatic actuators used for part loading and mechanised assembly operations. The devices consist of two pivoted fingers operated by a single or double action pneumatic cylinder. The single acting system relies upon a spring return to release the product and the double acting system can be positively actuated to achieve release.

The operating principle of the ERACON Mini-Hand shown in Fig. 12.9 is as follows:

● *Grasping action:* with the pressure given to the air through the 'IN' port the piston will move backward to close the fingers and to let the nail hold the external diameter line.

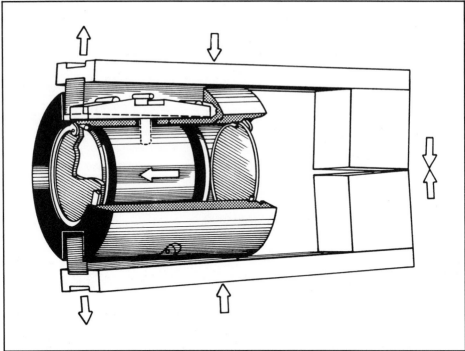

Fig. 12.7 PHD gripper and actuator principle (courtesy of PHD Inc., Fort Wayne, Indiana)

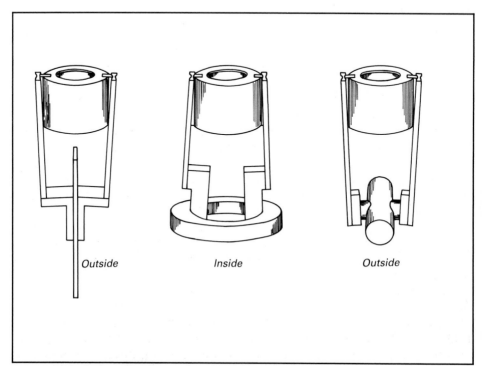

Fig. 12.8 Typical applications using the PHD gripper

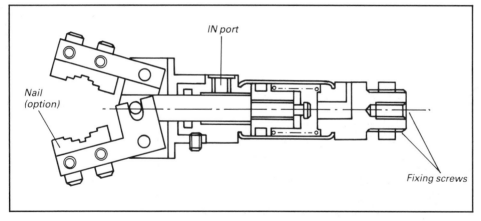

Fig. 12.9 The pneumatic ERACON Mini-Hand (courtesy of Haynes and Fordham Ltd, Leeds)

● *Releasing action:* when the air is discharged from the 'IN' port the resetting power of the reset spring causes the piston to open the fingers and lets the nail release the component.

Figs. 12.10 to 12.12 illustrate a wide variety of different gripper attachments of Air Technical Industries. These demonstrate some useful application areas for pneumatically, hydraulically and electro-hydraulically operated actuators.

12.3 Suction-cup pneumatic grippers

With high pressure air it is simple to generate a vacuum – the essential requirement for suction-cup tools. As discussed previously, if either the components to be handled have an odd gripping surface, or because of part orientation or functional reasons, the simplest solution is often to lift them up by means of vacuum.

Pneumatic grippers can be adapted to accomodate many different (i.e. acceptable or odd) shapes. They provide an easily controllable lifting and holding force and do not deform or scratch the components they manipulate.

On the negative side, pneumatic suction-cup heads cannot compete with the accuracy of mechanical fingers and are limited to relatively light loads and temperatures below 95°C in the case of rubber parts.

Fig. 12.13 illustrates suction-cup holders as used on a Dainichi-Sykes industrial robot for automated assembly. Suction-cup holders are generally used where the workpiece has a smooth surface allowing the cup to maintain the vacuum.

Suction-cup holders can be operated by natural vacuum when pressed onto a dry and clean surface, or by a vacuum pumping system. If the manipulated object is held by natural vacuum, the maximum holding force is about 140N and the part must be released by a pressure impulse [2]. If a vacuum pump is utilised the holding force depends on the surface quality of the object as well as on the power of the pump (see Fig. 12.14).

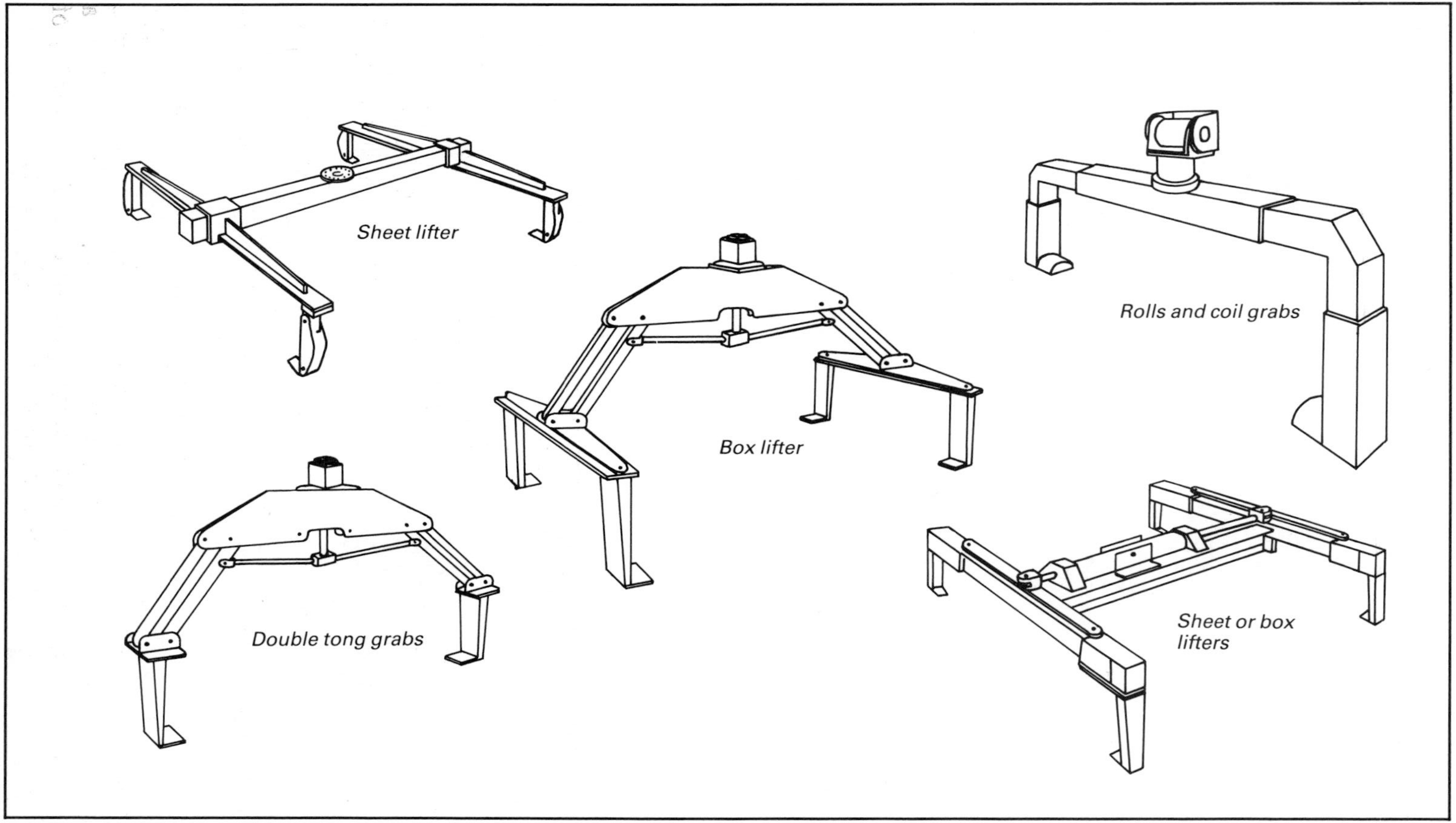

Rolls and coil grabs

Sheet or box lifters

Box lifter

Sheet lifter

Double tong grabs

Fig. 12.10 Pneumatic and hydraulic lifters for sheet metal, paper rolls, boxes, etc., handling (courtesy Air Technical Industries, Mentor, Ohio). Facing page: Pneumatic actuators and lifters used by Fanuc

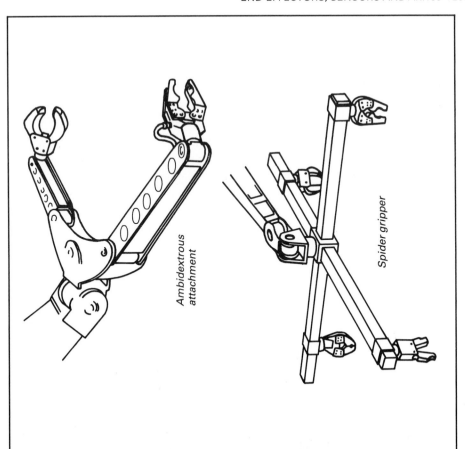

Ambidextrous attachment

Spider gripper

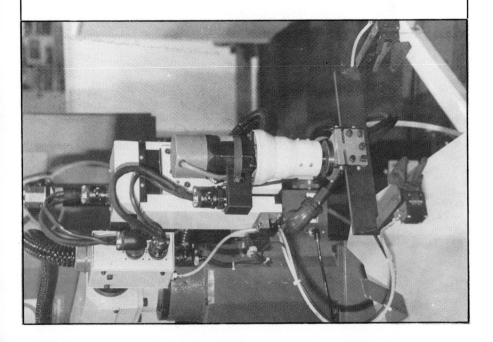

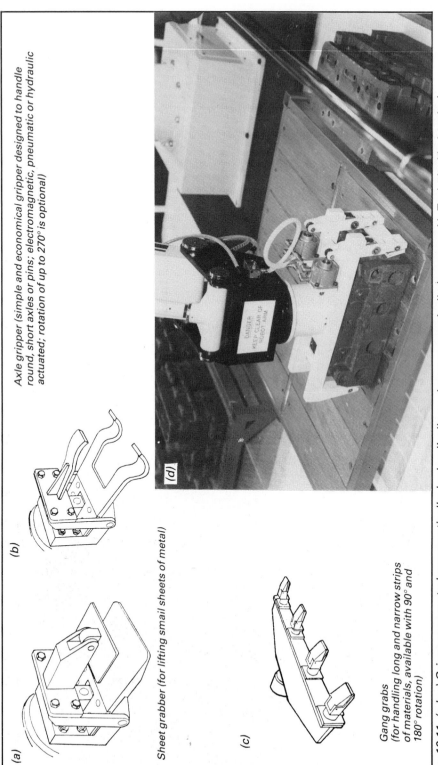

(a)

(b)

Axle gripper (simple and economical gripper designed to handle round, short axles or pins; electromagnetic, pneumatic or hydraulic actuated; rotation of up to 270° is optional)

Sheet grabber (for lifting small sheets of metal)

(c)

Gang grabs
(for handling long and narrow strips of materials, available with 90° and 180° rotation)

(d)

Fig. 12.11 (a,b,c) Grippers operated pneumatically, hydraulically or electromagnetically (courtesy Air Technical Industries), (d) pneumatically operated parallel gripper for machine loading/unloading

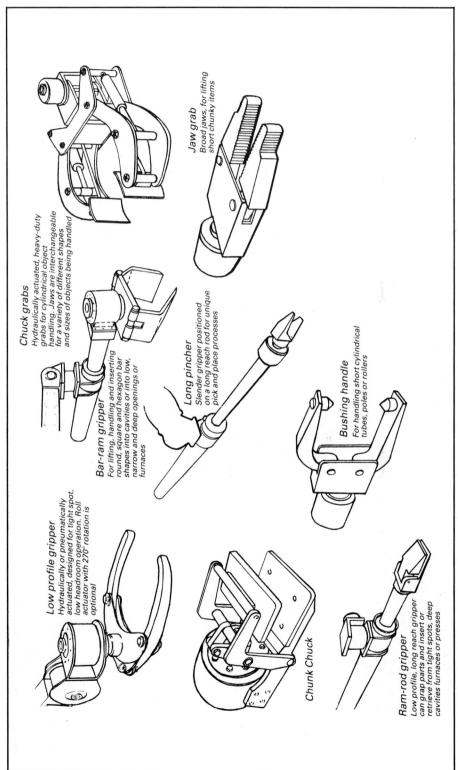

Jaw grab
Broad jaws, for lifting short chunky items

Chuck grabs
Hydraulically actuated, heavy-duty grabs for cylindrical object handling. Jaws are interchangeable for a variety of different shapes and sizes of objects being handled

Bar-ram gripper
For lifting, handling and inserting round, square and hexagon bar shapes into cavities or into low, narrow and deep openings or furnaces

Long pincher
Slender gripper positioned on a long reach rod for unique pick and place processes

Bushing handle
For handling short cylindrical tubes, poles or rollers

Low profile gripper
Hydraulically or pneumatically actuated, designed for tight spot, low headroom operation. Roll actuator with 270° rotation is optional

Chunk Chuck

Ram-rod gripper
Low profile, long reach gripper can grap parts and insert or retrieve from tight spots, deep cavities furnaces or presses

Fig. 12.12 Grippers which can 'grab' components and material (courtesy of Air Technical Industries)

The holding force can be distributed by several suction-cups to ensure a solid and stable grip, as in the case of most sheet metal, plastic, glass, and paper handling operations (see Fig. 12.15).

12.4 Modular and universal end-effectors

Any up-to-date design, whether a gripper or a computer program, should be modular. The reason why this section has been included is the impressive modularity and universality shown by different robot end-effector manufacturers.

The tools offered by BASE Robotic Systems show the concept of the modular system design very well. In the example shown in Fig. 12.16, the fingers can be exchanged as required. This is a widely applied method to accomodate different component shapes, material, and stiffness characteristics without the need for modification on the base of the tool.

Of course, the best solution is when not only the end-effector but also the robot wrist and arm module are of modular design. This feature can be of extreme importance, particularly in the case of mechanised assembly, when a certain location must be reached in a difficult position and arm orientation. Fig. 12.17 illustrates this modular design approach.

12.5 Workpiece loading robot end-effectors

Robotised part loading and unloading operations are widespread on up-to-date machines (e.g. on CNC machine tools and plastic moulding machines) since they provide the unmanned solution to the FMS and FAS cell orientated materials handling tasks.

Fig. 12.13 Suction-cup holders on a Dainichi-Sykes robot

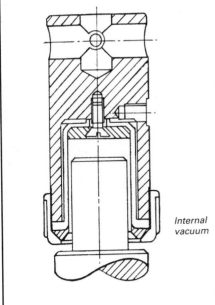

Diam. range (mm)	Holding force (N) at	
	min. diam.	max. diam.
5–11	50	50
11–20	95	120
20–34	180	220
34–43	260	330
43–54	240	380
54–59	220	180

Internal vacuum

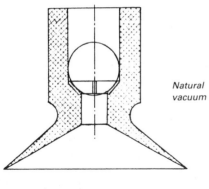

Cup diam. (mm)	Holding force (N)	Release pressure (bar)
6.3	3.0	1.0–1.4
10	4.5	1.0–1.4
16	10.0	1.0–1.4
25	30.0	1.0–1.4
40	60.0	1.0–1.4
63	140.0	1.0–1.4

Natural vacuum

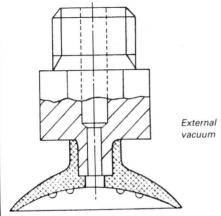

Cup diam. (mm)	Holding force at 25% vacuum (N)
20	7.8
30	35.0
50	43.0
80	160.0
100	173.0

External vacuum

Fig. 12.14 Suction-cup design/selection data

Fig. 12.15 Sheet metal manufacturing all loaded and unloaded by pneumatic manipulator utilising suction-cups for lifting the sheet metal components

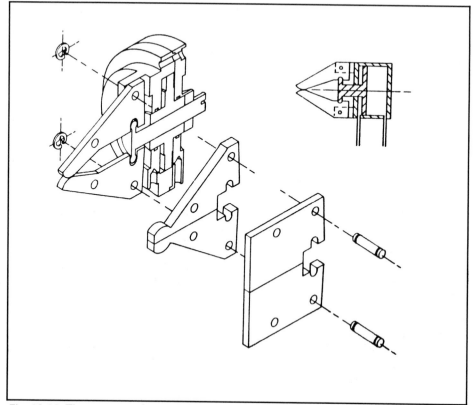

Fig. 12.16 The concept of a modular gripper/finger design

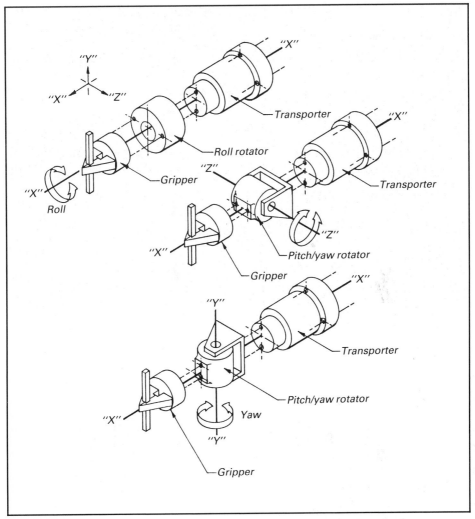

Fig. 12.17 Modular robot wrist/hand design (courtesy of BASE Robotics Corporation)

To perform automated part loading and unloading, the following conditions have to be checked or ensured:

- Is the machine itself capable of receiving components automatically? (Is it fitted with a hydraulic chuck, programmable fixture, or other workholding device controlled by the CNC controller linked to the robot controller?)
- Can the control system of the machine communicate with the robot controller?
- Is there a sufficient workpiece magazine for component storage and orientation?
- Can sensors be integrated both into the machine and the robot gripper to monitor each crucial step of the load/unload operation to avoid any risk of dropping or misloading the component?

Fig. 12.18 OKUMA turning centre loaded and unloaded by a manipulator (1). (Note the dual action gripper (2), as well as the part magazine (3)

To illustrate the above listed features, Fig. 12.18 shows the OKUMA part loading manipulator servicing an integrated turning cell, and Fig. 12.19 demonstrates a part loading sequence and the grippers used on the Yamazaki turning cell.

To fully automate the workpiece loading and unloading process in unmanned manufacturing, robots need part storage and part orientating devices as peripherals. In addition to the workpiece changer, Figs. 12.18 and 12.19 also introduce in-process workpiece storage magazines. The part storage unit can also be put on the top of an automated guided vehicle, thus providing a more flexible solution than the gantry type.

As can be seen in both cases dual action grippers were used. These offer several advantages in part changing operations:

● The minimisation of machine downtime (i.e. 50% reduction can be achieved in loading time compared to single action grippers) by removing the finished component with one gripper and quickly loading the raw part with the other.

● Two completely different grippers may have to be used because of the large part size and/or shape variation manufactured in an alternating way on the machine.

● High production rates would often otherwise require two manipulators or robots, increasing the cost of the manufacturing cell.

On the negative side one should also consider that because of the dual size gripper the robot will carry double weight, and will require larger clearance distances and space in general. Typical dual gripper loading/ unloading sequences are illustrated in Fig. 12.20, and the control requirements are analysed in Table 12.1.

Table 12.1 Actions identified on examination of control requirements at the machine and at the robot during the automated part changing sequence

Machine control system	Machine	Part loading/unloading (manipulator)	Conditions to examine/check at machine control	Manipulator gripper conditions
Executing CNC part program	Machining	Prepare part changing operation	CNC program supervision	Hand 1 open? Hand 2 loaded with raw part?
Part ready Stop spindle Coolant off Open Cover	Spindle stops Coolant off Cover opens		Acknowledge actions (Clear way to chuck?)	
Change part Cycle start	Position for part changing			
	Blow air (or coolant) on finished part to remove chips			
	Grasp part			Hand 1 closed?
	Open cover		Open?	
		Unload finished component		
		Clean swarf again		
		Load new part (Dual action gripper)		
	Close chuck		Closed?	
				Hand 2 open?
	Close cover		Closed?	
End of cycle			No errors?	No errors?
Continue the execution of the CNC part program				

It should be noted that each important action is acknowledged between the CNC and the robot controller. Because most CNC controllers and robot controllers lack proper communications requirements this communication can simply be performed using the I/O port of the robot.

Grippers of machine loading manipulators often have slip sensing and force sensing capabilities to avoid the risk of dropping or misloading a component. Such sensors are also utilised when handling delicate components. When designing such systems one should not overlook the vast amount of control and number of software tasks relating to handling the sensors.

To give some further variants for machine loading grippers, Fig. 12.21 illustrates the Fanuc dual action robot gripper as integrated into a CNC turning centre for changing rotational components.

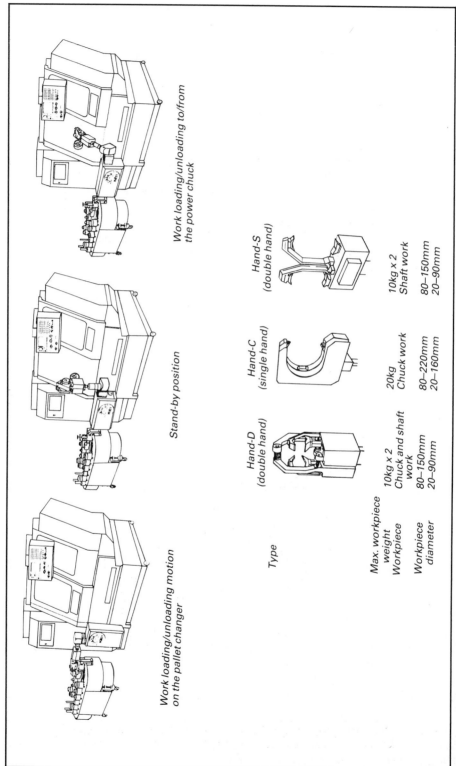

Work loading/unloading to/from the power chuck

Stand-by position

Work loading/unloading motion on the pallet changer

Type	Hand-D (double hand)	Hand-C (single hand)	Hand-S (double hand)
Max. workpiece weight	10kg x 2	20kg	10kg x 2
Workpiece	Chuck and shaft work	Chuck work	Shaft work
Workpiece diameter	80–150mm 20–90mm	80–220mm 20–160mm	80–150mm 20–90mm

Fig. 12.19   Example of part loading sequence and hand specifications (courtesy Yamazaki Machinery Works, Nagoya, Japan)

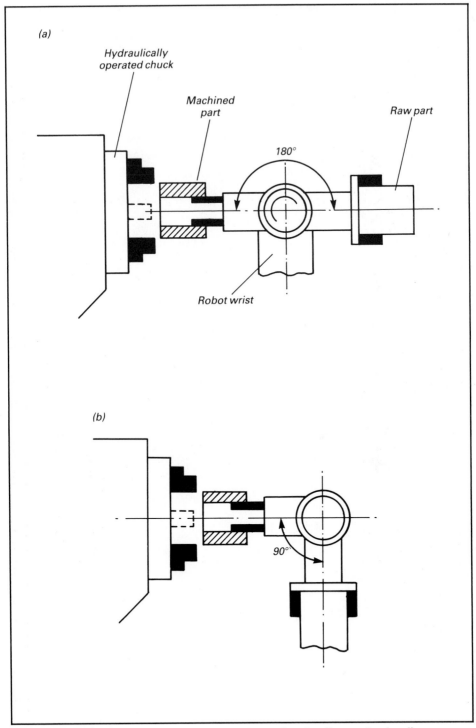

Fig. 12.20 Dual action gripper application ideas when loading and unloading machine tools with integrated manipulators: (a) mounted 180° apart and (b) mounted 90° apart

Fig. 12.21 Part changing sequence utilising a Fanuc machine loading/unloading manipulator

12.6 Welding robot tools

The majority of today's robot population is used in welding. Welding robots perform best those jobs requiring flexible automation. Plants dealing with sheet metal and structural metal products are perfect applications for welding robots.

Since no outside material is used in spot welding, the basic metallurgical composition of the joint remains unchanged, and thus represents a wide application area. In most cases an assembly can be more economically joined by spot welding than by mechanical fasteners.

The major advantages of the robotised welding process are that it is fast and easily adaptable for almost random manufacture, and that the current, the timing, the electrode force and the positioning are controlled by the robot's computer. The process ensures a high level of consistency and high production rates, and since this application area often lacks in skilled labour it also represents an economic solution.

The structure of the arm and the weight of the welding attachment are two important areas for consideration. Resistance spot welding equipment, despite miniaturisation, consists of heavy and rigid devices with load capacities of over 20kg (see Fig. 12.22).

When selecting welding guns for industrial robots, the following data should be considered:

- Total weight of the gun.
- Total mounted length.
- Power supply requirements.
- Throat size.
- Weld stroke opening.
- Total opening (retract stroke).
- Bore size.
- Weld force at 80 psi.

When developing a system, the part orientation and part handling is as important as the examination of the loaded part. Out-of-specification workpieces need additional fixturing and adaptive tracking facilities for the robot, which can cause major problems in unmanned welding operations.

Arc welding is more complicated, since the metal in the molten weld pool would react chemically with oxygen and nitrogen in the air, thus affecting the solidified joint. A protective shield of gas, vapour or slag must therefore be applied. Most arc welding robots accomodate gas welding facilities, with a bar continuous consumable electrode and a separately applied insert protective gas (Fig. 12.23).

From the trajectory control point of view the continuous path control with circular interpolation are important features because the shape of the parts dictate the robot motions. In the case of arc welding, voltage, amperage, wire feed rates and travel have to be programmed. Adaptive contact and non-contact sensors are often used for seam tracking, arc sensing and positioning, because unlike repetitive spot welding arc welding is typically an area of flexible small batch manufacture where hard fixturing would be expensive.

Fig. 12.22 Spot welding gun for industrial robots (courtesy of Goodrich Division, Courac Corporation, Michigan)

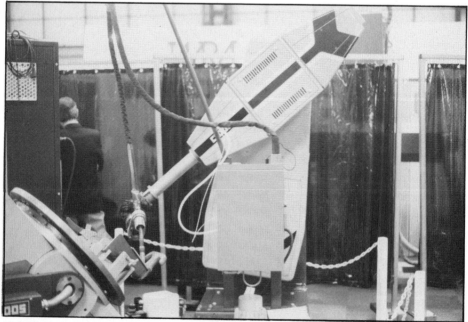

Fig. 12.23 Arc welding industrial robot (Note the programmable part holding workstation)

12.7 Machining robots and tools

Machining robots are relatively recently developed applications of the rapidly expanding robot population. With their ± 0.08–1.0mm usual positioning accuracy they are less accurate than, for example, five axes milling centres (typically offering a repeatibility value within ± 0.01–0.005mm). Nevertheless they are applied for drilling, deburring, grinding, routing, plasma cutting, laser cutting and welding, where they are more cost effective than dedicated machinery or manual methods.

In the aircraft industry, where wing drilling is a time consuming operation if done manually, robotic systems can perform the task in approximately half the time taken by dedicated machinery and approximately one-sixth of the time compared to manual operations. If 3D surfaces have to be drilled then normality sensing devices or off-line robot programming methods are used.

With changeable drilling heads, different hole sizes can be drilled and the tool changing time is a few seconds or less. Fig. 12.24 illustrates the KUKA Series IR 200 industrial robot as used for drilling three-dimensional surfaces utilising the Desoutter drilling attachment.

By developing different robot tools the same robotised cell can be used not only for drilling but also for other operations, such as routing and deburring (Fig. 12.25). Cost evaluation for robot routing and drilling as compared to manual and dedicated machinery has shown that robots perform the job more economically.

In the case of most milled or turned components made from solid aluminium, steel, cast iron, titanium, etc., the tool leaves burrs around the edges. These burrs must be removed and the webb corners radiused by machines rather than manually, due to the need for consistency, the environmental and operator safety aspects and the long time involved in manual operation.

Typical components suitable for robotised deburring include cast iron crankshafts, transmission cases and aluminium cylinder heads, steel and cast iron housings, rear wheel spindles, gears and gearbox housings, stainless steel and various plastic components. In the case of machine tool components, or structural airframe components such as wing parts and stringers, robotised deburring is an important possibility for flexible automation.

The direct benefits achieved by robotised as opposed to manual deburring and routing can be summarised as follows:

- Uniform quality.
- Shorter processing (i.e. throughput) time.
- Lower supervisory cost and reduction in direct labour.
- Lower cost of environmental facilities and safety devices (i.e robots can work in hostile environments, but humans should not or could not).
- Easily adaptable random manufacturing facility.
- Robots can be programmed for a wide variety of jobs.
- Work experience can be duplicated by means of program transfer, rather than expensive and time consuming training.

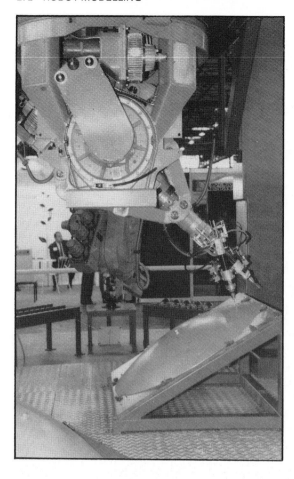

Fig. 12.24 Drilling holes into a three-dimensional sheet metal surface using a KUKA robot and a Desotter drilling head

Experiments have shown that deburring can be carried out by slow running rotary brushes made of stainless steel or abrasive flap wheels mounted on a quick release mechanism. In the case of aluminium components, brushes are usually made of a nylon monofilament that contains either aluminium oxide or silicon carbide to provide a flexible but wear resistant deburring tool.

In machining, deburring, grinding and routing operations it is important to analyse whether the robot should handle the part or the tool. Since each alternative can have its advantages in each individual case, some aspects for consideration are summarised below.

If the robot handles the tool rather than the parts (Fig. 12.25(c)), then:

● The full trajectory and tool orientation facilities of the robot can be applied when programming the tool path.
● Real-time feedback processing actions can be taken relating to different tool paths.
● There is no limitation in the part weight.
● Smaller robots can be used, assuming that the tools applied are lighter than the parts to be processed.

- If the parts are heavy, the throughput time will be shorter since the robot arm can be moved faster than the heavy workpieces.
- The tool holder must have some level of compliance to avoid tool breakage (often adaptive control systems must be implemented).

As can be seen from this list, if the parts are relatively heavy in most cases the overall cost is smaller if the robot holds the tool rather than handles the part.

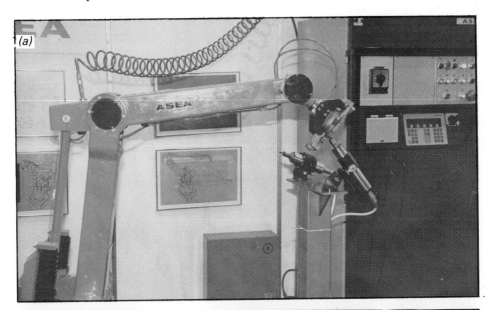

Fig. 12.25(a,b) *Routing with an ASEA robot ('part in the hand' arrangement). (Note the different tools used in the routing/deburring heads)*

Fig. 12.25(c) Engine block deburring operation with a Fanuc robot ('tool in the hand' arrangement). (See also Fig. 12.57 (a and b) for the Automated Robot Hand Changer)

If the robot holds the part (Fig. 12.25(a,b)) in front of the tool:

● Different tools can be applied without the need for tool changing (this is in most cases faster and less expensive than the tool changing operation).
● There is no limitation in the power required for the process.
● Many in-process machining, deburring and inspection activities can be integrated in one cell.

When designing deburring tools for industrial robots, one of the most important tasks is to incorporate a passive or active sensor into the tool and a real-time sensory feedback processing (or adaptive control) system into the control system of the robot to avoid overload and tool breakage. Because of the importance of this topic Fig. 12.26 summarises possible measuring principles and the measured variables.

Deburring tools include carbide rotary files, reciprocating files, rotary brushes (nylon bristles containing aluminium oxide and steel-wire brushes), and countersink tools (for holes and round edges) with flexible guidepins. Regardless of the tool being used the surface speed and the major deburring (i.e. cutting) force should be kept constant. To make sure these conditions apply, electric motors are often used since they are easy to control.

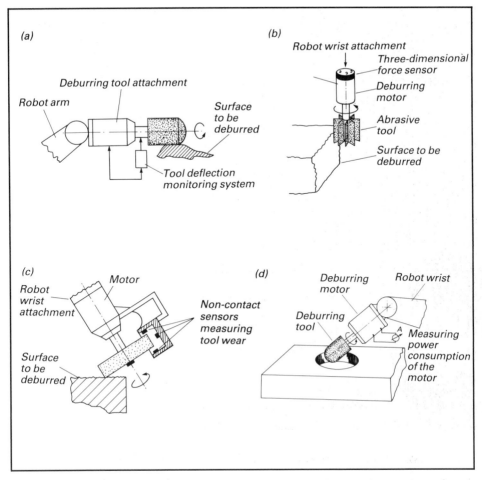

Fig. 12.26 *Possible measuring principles and the measured variables when designing robotised deburring tools: (a) measuring tool deflection, (b) measuring force, (c) measuring tool wear in real-time, and (d) measuring power consumption*

When used frequently, the direction of rotation of some tools (e.g. steel-wire tools) must be altered to maintain sharpness. In many other cases tools must be changed after a certain number of operations, as is done on machine tools.

Since some tools are sensitive to vibration, unnecessary vibration should be avoided. This can be achieved by:

- Selecting the correct part material and tool shape/material combination.
- Utilising rigid fixtures and 'soft' tools (i.e. tool holders incorporating compliance devices), or compliance fixtures and rigid tools.
- Decreasing the feedrate of the deburring tool.
- Changing (usually decreasing) the deburring tool's speed of rotation.

In addition to drilling, deburring and routing, other manufacturing jobs are also performed by industrial robots (Fig. 12.27).

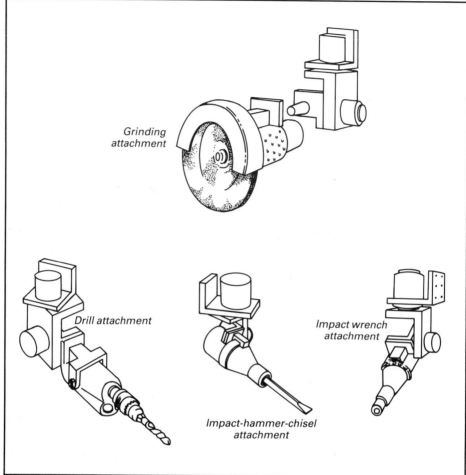

Fig. 12.27 Examples of different machining tools to be mounted on robots (courtesy of Air Technical Industries)

12.8 Non-contact sensing tools

Optical scanning, laser scanning and other non-contact measuring and surface inspection methods are in use on machine tools and coordinate measuring machines (CMMs), and so their adaptation for industrial robots is relatively simple. Obviously, since robots compared to machine tools and CMMs have larger positioning uncertainty, the methods used are different but the basic principles remain the same.

Non-contact sensors in robotics are used for a variety of applications in which the sensors can be situated remotely from the measuring instrument, enabling many parameters to be monitored in locations and orientations difficult for other machines having a lesser degree of freedom, or for human operators. Their use in automated assembly is significantly increasing and is considered to be one of the most important aspects of robot systems carrying out assembly work.

Non-contact sensory robot tool applications include:

● Assembly.
● Inspection (surface and shape).
● Condition monitoring.
● Thickness measurement.
● Distance measurement.
● Eccentricity measurement.
● Film coating measurement.
● Heat measurement.
● Colour sensing.
● Electromagnetic field measurement.
● Gas leakage (smell) measurement.

By way of example, the automated dry leak test method utilising industrial robots developed by Wickman Automation Ltd is considered. Instead of the messy and time consuming fluid method by which cars are sprayed with water to test where doors or other fittings leak, Wickman has developed the dry leak test method using a small measured quantity of helium gas mixed with air and introduced into the car body at low pressure. Robots fitted with detector heads (Fig. 12.28(a)) designed and patented by BL Technology scan the body along a pre-programmed path, determined by quality control requirements. The system can work as fast as 300mm per second and the computer analysis of the measured data offers 400 separate pieces of information.

Besides measured values, the output (Fig. 12.28(b)) indicates the positions of the 'sniffing heads' where leakage was detected, allowing test engineers to quickly find the faults on the car body. The system monitor, as part of the computer system (Fig. 12.28(c)), prints out all functions of the system and compiles a trend analysis to alert management if more than the predefined (i.e. acceptable) number of faults occur in any area of the body within a defined period of time.

Since in robotic applications non-contact distance measurement methods have a wide application possibility, this area is discussed in more detail. The most important physical principles involved in non-contact distance sensing include:

● Mechanical effects.
● Electromechanical effects.
● Capacitive effects.
● Electrooptical effects.
● Radioactive effects.
● Magnetic effects.

A classification and an overview of non-contact distance sensors and some of their application areas is shown in Fig. 12.29.

To give some examples, mechanical non-contact distance sensors are discussed first.

Pressure or vacuum based pneumatic sensors, or air gauges, can be used to detect the presence of any object by blocking a roughly positioned vacuum nozzle in the end-effector of the robot.

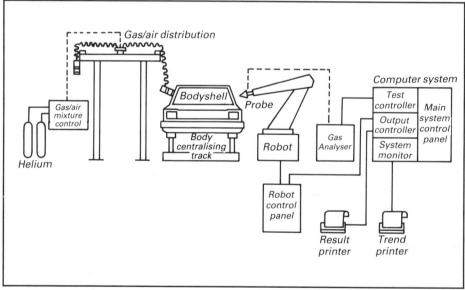

Fig. 12.28 (a) The robot and the gas-leak sensory tool in (b) Wickman's automated dry leak test system (See also facing page)

Acoustic sensors use the time it takes for a sound wave to return to source from the target through a conducting medium, such as air, water or gas. End-effectors used in underwater applications often use such sensors. The major drawback of such currently available sensors is that they tend to be inaccurate under a distance of about 300mm.

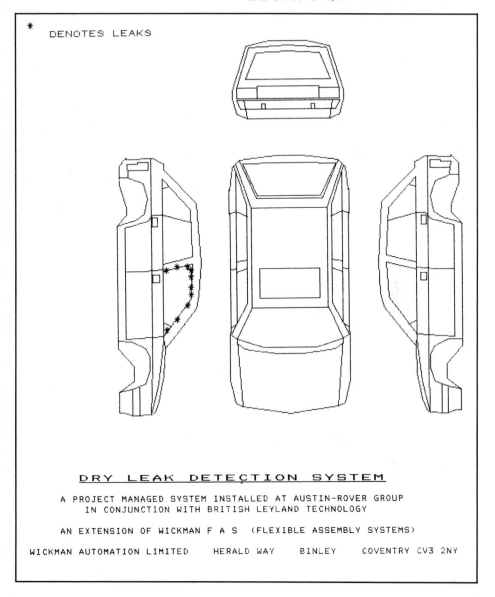

```
*
   DENOTES LEAKS
```

DRY LEAK DETECTION SYSTEM

A PROJECT MANAGED SYSTEM INSTALLED AT AUSTIN-ROVER GROUP
IN CONJUNCTION WITH BRITISH LEYLAND TECHNOLOGY

AN EXTENSION OF WICKMAN F A S (FLEXIBLE ASSEMBLY SYSTEMS)

WICKMAN AUTOMATION LIMITED   HERALD WAY   BINLEY   COVENTRY CV3 2NY

The operation of a large number of electronic gauging systems are based on the electrical capacitance or inductivity between the sensor and the test surface. This principle is used in applications such as object presence monitoring, roundness and clearance check, and thickness measurement, film coating measurement, and eccentricity measurement. Fig. 12.30 illustrates some possible tasks such sensors can solve.

With infrared sensors, where a beam of infrared light is emitted from a source, any light that is reflected is picked up by a receiver (usually phototransistor) and a voltage is produced. This voltage is not however a reliable indicator of distance and is very difficult to use, other than to indicate whether or not an object is present (Fig. 12.31).

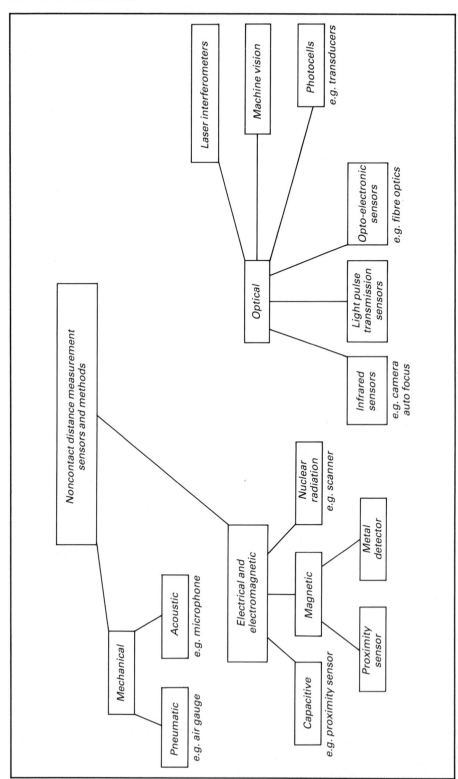

Fig. 12.29 A classification and overview of non-contact distance sensors and some of their possible application areas

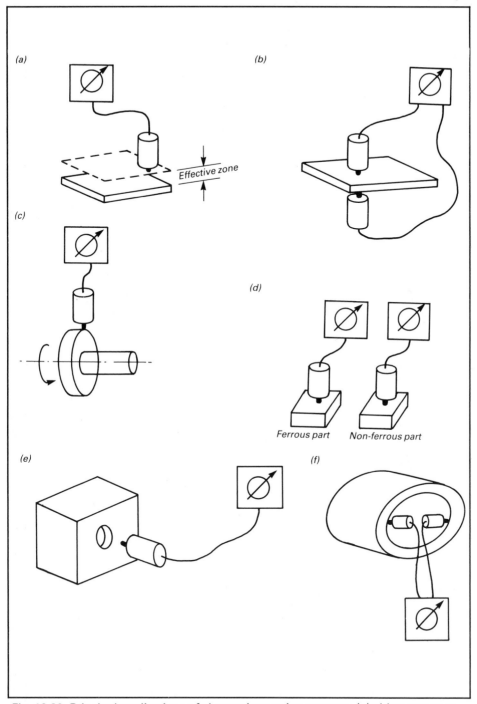

Fig. 12.30 Principal applications of electronic gauging sensors: (a) object presence measurement (part detection), (b) object thickness measurement, (c) roundness measurement with non-contact sensor, (d) selection of ferrous/non-ferrous materials, (e) non-contact sensors can be used to find objects, shapes, etc., and (f) roundness measurement with non-contact sensors

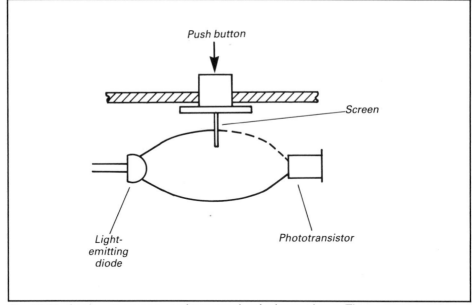

Fig. 12.31 Analogue sensor – a photo-mechanical transducer. The pressure on the button and its position are measured by the variation in the light received and detected by the phototransistor

The optical fibre sensor in its simplest form consists of a light source, a photodetector and two adjacent optical fibres. Light is transmitted throught the first optical fibre and projected onto an object, while the reflected light is collected by the second fibre and transmitted to a photodetector. With proper electronic compensation to account for variations in reflectivity, the intensity of received light can be as small as 0.4 – 1mm. It also requires a small stand-off distance for proper operation. Beyond the upper limit of the sensing range the output voltage falls inversely as the square of the distance.

To illustrate a possible use of photoelectric sensors a case study on the recognition of simple shapes using the area measurement method is now presented. This was carried out by one of the authors (P.G. Ranky) and M.A. Walker at Trent Polytechnic, Nottingham, UK.

The purpose of the exercise was to investigate different low cost sensors working in robotised assembly and test operations. For this study a Unimation PUMA 560 robot was used (Fig. 12.32) with the 'Ranky-type' automated robot hand changer.

Shape recognition is normally done by expensive vision systems. In several cases, however, mainly when the variety of possible different simple two-dimensional shapes are known and the robot's job is to select from these known shapes, low cost sensors can solve the task adequately. Possible approaches solving the shape recognition task in such cases include:

● The algorithmic approach, when a broad scan is done by the robot of the object present and a recording made of a few positions (e.g. 10 to 12 locations) where the sensor passes an edge and consequently where the

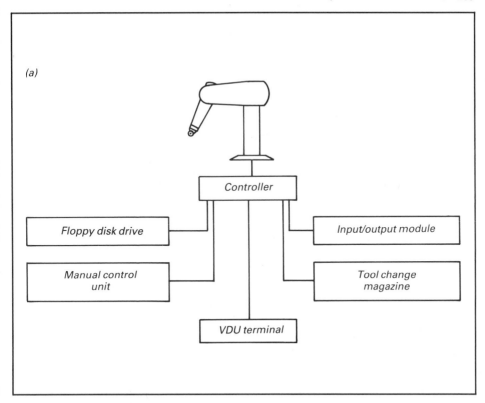

(a)

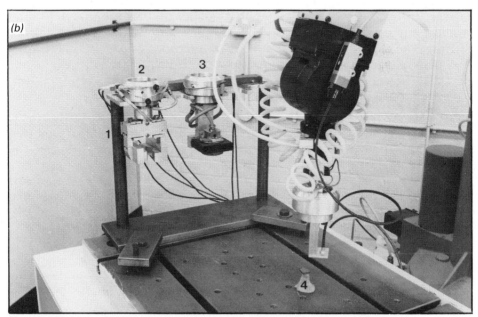

(b)

Fig. 12.32(a) The PUMA robot configuration used in the study, and (b) the 'Ránky-type' tool changer and the sensor tools used in different studies: 1, tool magazine of the AHRC; 2, parallel gripper; 3, tool with capacitive sensor; 4, throw-away carbide tool tip

signal changes from low to high or vice versa, depending how it was set. In many cases using only a few measured locations it is possible to calculate which one of the several known objects has been found.

● Another possible solution is to write an algorithm and program that make the sensor follow the edge of the object (edge following method). From the results of such measurement it should be possible to calculate the number of corners, the length of selected sides and approximate values for the angles at the corners. This information is normally enough for the shape to be matched to one of the known shapes.

● Since both of the above methods need computing tasks which proved to be impossible in VAL, because of the integer based arithmetic and because it was impossible to extract the calculated transformations from the VAL controller and transfer the values to any VAL routine, we have followed the area measurement method. This involves the following simple algorithm: by first finding the object and then scanning a small area on which the object is known to lie, the area of the object can be calculated. This involves passing over the object on several evenly spaced parallel paths. By measuring the distance travelled over the object on each pass, the area of the object can be calculated by mutliplying the sum of the distances measured by the distance between each pass. This method is relatively slow on any robot, but can be programmed even in VAL.

The sensor used is an OMRON E32-DA50, E3S-XE1, made by Omron Tatsisi Electronics Co., Japan (Fig. 12.33). This is a diffuse reflective type, infrared proximity detector. It consists of an amplifier unit (E3S-XE1) which contains a photoelectric switch and an infrared light emitting diode (LED). The other part of the sensor is a fibre unit (E32-DA50) made up of two parallel fibre-optic cables: one to transmit light from the LED to the probe at the end of the cable, and the other to transmit any reflected light to the photoelectric switch (Fig. 12.34).

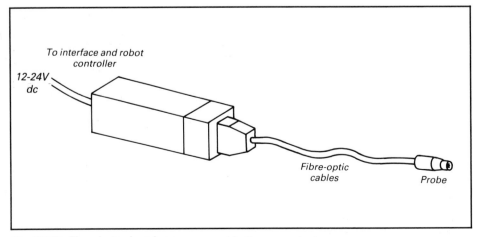

Fig. 12.33 The photoelectric sensor used in the case study

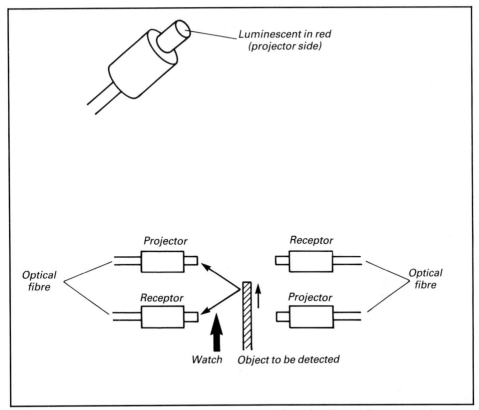

Fig. 12.34 Object detection principle (courtesy of OMRON Tatsisi Electronics Co., Kyoto, Japan)

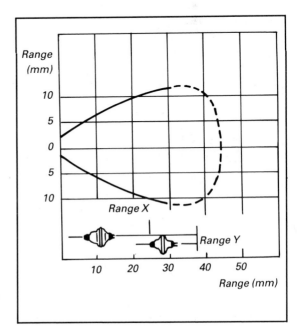

Fig. 12.35 The graph of operation territory (typification), (courtesy of OMRON Tatsisi Electronics Co.)

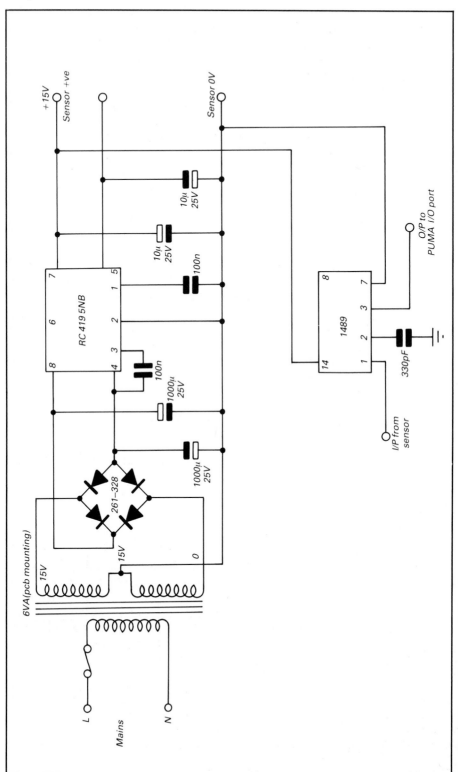

Fig. 12.36 Circuit diagram of the signal reversing unit

The sensor requires an input voltage of 12–24V(dc) at a current of up to 50mA. It gives an output signal (interfaced with the input ports of the PUMA) of either 0V (when an object is detected), or slightly less than the input voltage when no object is detected. Both opaque and transparent surfaces can be detected at distances up to 8mm, although this is lower for some surfaces (Fig. 12.35).

Because the REACTI instruction in the VAL language reacts to high signals, but unfortunately the sensor offers low signal when an object is detected, a signal reversing circuit had to be incorporated (Fig. 12.36). Two ICs were used, the first turning the 0 and +15V input signals to +15 and −15V output signals, and the second turning the output signals from the first to 15V and to 0V output signals, suitable for VAL. The circuit also incorporates a transformer, as the VAL controller does not offer any additional independent power supply for sensors.

The entire sensory input data handling sofware was written in VAL (Figs. 12.37 and 12.38).

The first major task is to find the object. This is done by the 'FIND' routine (Fig. 12.37). The sensor is first moved to a location 'START' at the corner of the work area, the arm is then moved back and forth in the X direction, moving 5mm increments along the Y axis after each 300mm motion along the X axis. (The area of search has been defined as an area of 300×300mm). During these motions the REACTI instruction is used to monitor the signal from the sensor and when a high signal is received the arm immediately stops moving and program control is transferred to a routine called AREA.

AREA (Fig. 12.38) narrows down the area on which the object is known to lie. This is done by the following operations:

● Moving 35mm along the X axis so that the sensor is clear of the left-hand edge of the object. (The extra 5mm allows for the distance the arm travels before the signal from the sensor is received by the controller). It also steps back 4mm on the Y axis so that the sensor is at least level with the top edge of the shape.

● Using the IFSIG instruction in VAL whether or not an object is present at a particular point is checked, by moving the arm along the Y axis in short steps until it reaches the bottom of the sector on which the object is known to lie, or until the object is detected. If the bottom of the sector is reached, the arm repeats this second process starting from a point 5mm further along the X axis. If the object is found, the arm moves to the point at which the previous scan started. This point is then known to be within 5mm of both the top left-hand edges, and therefore defines the location of the object to within known limits.

The next process is to measure the area of the shape in order to select different shapes by comparing their sizes. Finally a routine called 'GRAPHICS' is used to display the shape of the found object on the CRT using simple alphanumeric characters, since VAL is not capable of handling graphics.

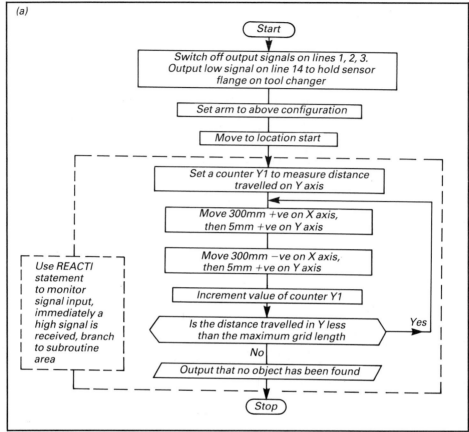

Fig. 12.37(a) The 'FIND' routine flowchart

Several other experiments have been done within this project, which also relate to robot testing (see also Chapter Eleven). One such test was to find out the effect of the arm speed on the accuracy of the detection process, as well as on the repeatability errors of the arm. (The results of these tests were used to determine the values to be used in the shape identification program.) The speed test was important, since the slow speed of the process seemed to be the biggest drawback of using such sensors for shape identification or recognition.

This test involved varying the arm speed of the robot, while other parameters were kept constant. The results given in Fig. 12.39(a) show what was expected, i.e. the faster the arm is moving, the further it travels during this response period. Below a robot motion speed of 100, the speed is negligible. But above this speed the area measured by the system drops alarmingly, due to the fact that the robot moves so far during the response time that the 35mm step size used to move the arm past the left-hand edge of the object is not sufficient, and therefore the object is not fully scanned, giving reduced (i.e. false) area measurement.

The accuracy and repeatability of the area measurements were also tested by measuring a disc of known area (i.e. $324.3mm^2$) at arm speed 70.

```
.PROGRAM FIND
   1.         SIGNAL -1,-2,-3,-14,,,,
   2.         TYPE ******************************
   3.         TYPE * FIND ROUTINE IS RUNNING... *
   4.         TYPE ******************************
   5.         ABOVE
   6.         REMARK *** USE 'ABOVE' ARM ORIENTATION
   7.         MOVES START
   8.         REMARK *** MOVE THE SENSOR TO LOCATION 'START'
   9.         REACTI 10, AREA ALWAYS
  10.         REMARK *** WHEN A POSITIVE SIGNAL IS RECEIVED FROM CHANNEL 10
  11.         REMARK *** EXECUTE THE 'AREA' SUBROUTINE IMMEDIATELLY
  12.         SETI Y1=0
  13.         REMARK *** SET THE VALUE OF A COUNTER (Y1) WHICH IS USED TO
  14.         REMARK *** CHECK HOW FAR THE SENSOR HAS BEEN MOVED IN THE
  15.         REMARK *** 'Y' AXIS TO ZERO
  16.     10  DRAW 300.00, 0.00, 0.00
  17.         REMARK *** MOVE 300 MM ALONG THE 'X' AXIS
  18.         DRAW 0.00, 5.00, 0.00
  19.         REMARK *** MOVE 5 MM ALONG THE 'Y' AXIS
  20.         DRAW -300.00, 0.00, 0.00
  21.         REMARK *** MOVE -300 MM ALONG THE 'X' AXIS
  22.         DRAW 0.00, 5.00, 0.00
  23.         REMARK *** MOVE 5 MM ALONG THE 'Y' AXIS
  24.         SETI Y1 = Y1 + 1
  25.         REMARK *** INDICATE THAT A FURTHER 10 MM HAS BEEN TRAVELLED
  26.         REMARK *** IN THE 'Y' DIRECTION
  27.         IF Y1 LT 30 THEN 10
  28.         REMARK *** REPEAT UNTIL THE SENSOR HAS NOT MOVED 300 MM YET
  29.         REMARK *** IN THE 'Y' DIRECTION
  30.         TYPE ******************************
  31.         TYPE * WARNING. NO OBJECT FOUND. *
  32.         TYPE ******************************
  33.         REMARK *** IF THIS STAGE OF THE PROGRAM HAS BEEN REACHED
  34.         REMARK *** NOTHING HAS BEEN DETECTED. SORRY...
  35.         HALT
  36.         TYPE ============= STOPPED WITH HALT =================
.END
```

Fig. 12.37(b) VAL code of the 'FIND' routine

The results of ten measurements at different points of the workspace (given in Fig. 12.39(b)) show that in 99% of the measurements taken, the area recorded will be within 15mm^2 of the mean area measured for this particular object. (Note that as a comparative measurement the results are usually acceptable since the robot was capable of selecting the required shapes without any difficulty; see also Fig. 12.39(c)).

Finally, the results of successfully selecting six different but simple shapes using this sensor are shown. The accurate areas of these shapes were not important as the identification is based on comparison rather than on the absolute values of areas. The test involved taking the area measurements at five separate locations for each of the shapes. The results (shown in Fig. 12.40) illustrate that although the method is slow and inaccurate, it is of very low cost and could be used, for example, for identifying throw away carbide tips of machining tools, or in assembly work if a self-recovery feature is required after losing a component.

Vision systems are of extreme importance in planning the task of the industrial robot as well as in increasing its level of 'intelligence'. Since there is much literature dealing with this topic only a brief summary is given of some possible applications mainly relating to robotised assembly and end-of-arm tooling.

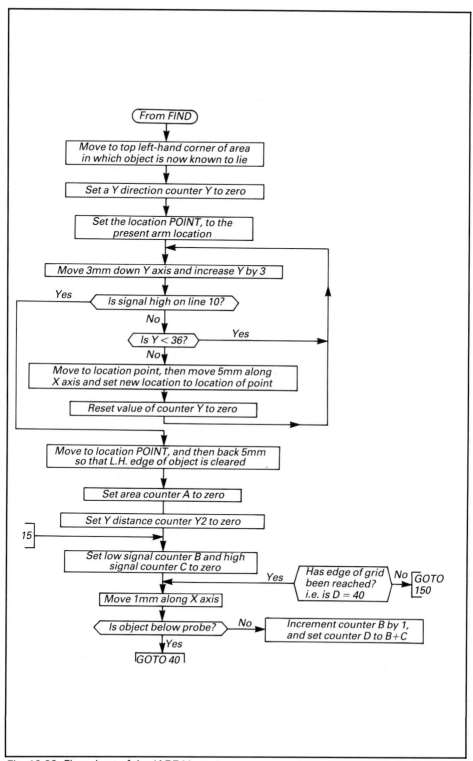

Fig. 12.38 Flowchart of the 'AREA' routine

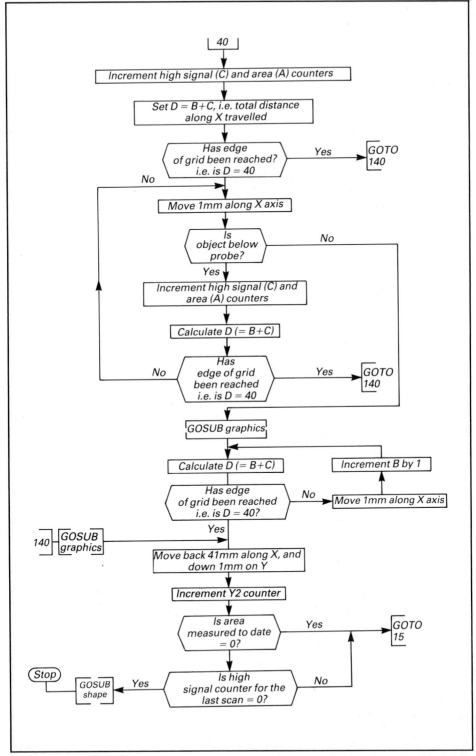

Fig. 12.38 continued

(a)

Arm speed (% of default)	Execution time (min: s)	Area measured (mm²)
5	13:57	376
10	9:00	377
20	7:02	378
30	6:37	375
40	6:19	380
50	5:50	378
60	5:39	380
70	5:33	380
80	5:14	377
90	5:10	379
100	5:08	379
110	5:06	328
120	4:57	247

(b)

Test number	Area recorded
1	346
2	334
3	338
4	338
5	331
6	333
7	326
8	338
9	337
10	343

(c)

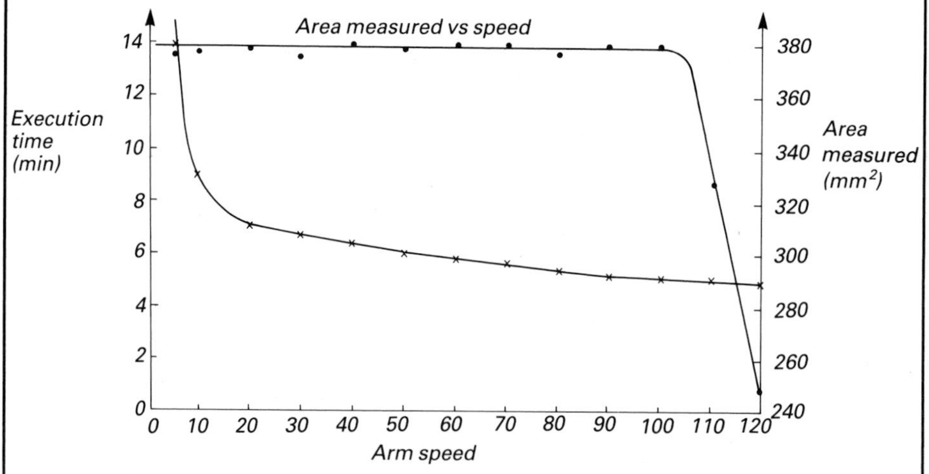

Fig. 12.39 (a) The effect of the arm speed on the area measured by the sensors; (b) at arm speed 70% the recorded area is within 15mm² of the mean area of the measured object; (c) graph showing relationship between arm speed, execution time and area measured

Shape	Area measured 1	2	3	4	5	Mean area	Lower limiting area value
Circle	778	768	760	770	772	769.6	700
Square	520	536	538	528	532	530.4	515
Wide rectangle	498	502	491	495	503	497.8	470
Thin rectangle	445	439	444	441	448	443.4	410
Diamond	369	375	372	367	370	370.6	345
Triangle	321	323	319	327	322	322.4	280

Fig. 12.40 Successfully selected objects with the robot

Vision systems use cameras of various types to look at the work area, and a control system incorporating a vast amount of software for analysing the image (expert systems). Images may be binary (i.e. high and low values, or '1's and '0's) indicating whether or not anything is present in a particular segment of the field of view, or pixel, or they may be greyscale images giving a value for the 'greyness' of the image in a particular pixel. For the acquired images a pattern recognition process must be carried out, which is in most cases complex and time consuming even in computing terms (Fig. 12.41).

Various methods have been developed for using vision systems with robots (Fig. 12.42). These methods include:

● A single overhead camera used to locate the object and guide the control of the robot to find it. This method is relatively fast since the camera can see a relatively large area of the work table. However, it is inaccurate, as once the robot arm is near the object the camera's view is obscured.

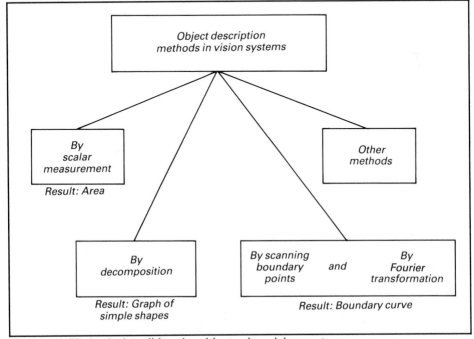

Fig. 12.41 Methods describing the object using vision systems

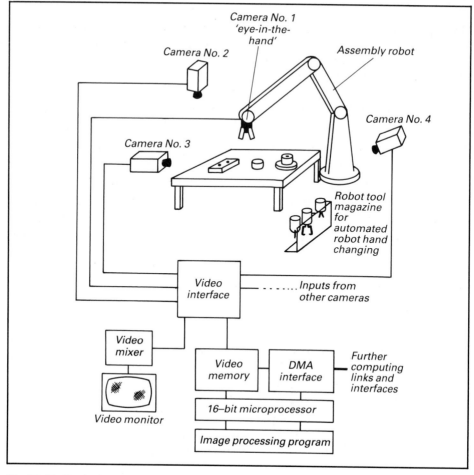

Fig. 12.42 Typical vision system architecture, using many cameras, one of them being an 'eye-in-the-hand' (i.e. robot tool integrated camera) for flexible assembly

● Two or more cameras produce a three-dimensional image of the work-space. This method involves using even more complicated algorithms and is far from being perfected. However, it has the advantage that it can overcome the distance measurement and parallax errors which occur when only one camera is used.

● The 'eye in hand' method is probably the most important from the tooling point of view. This involves placing one or more miniature cameras inside the robot tool. The advantage of this is that the camera can 'see' the object at all times during operation, thus enabling the robot to make positional and orientational adjustments when necessary. The best results can be achieved if the 'eye in hand camera' input, as well as the images offered by outside cameras, are processed in real-time. Currently such systems are expensive and have limited use. However, it is expected that they will become increasingly popular in the future particularly on automated assembly.

12.9 Soft-finger grippers

The idea of designing so called 'soft finger' grippers, or hands for industrial robots has emerged from the need to manipulate hard, soft and fragile objects of various shapes. (Obviously the other solution is automated robot hand changing, discussed later in this chapter.)

The first example of a versatile robot hand was developed by Hirose and Umetani at the Tokyo Institute of Technology [3]. In their interpretation, the soft gripping action must incorporate the following two functions:

- The finger should actively conform to the periphery of objects of any shape, including concave.
- The finger should produce uniform pressure after gripping the object.

The mechanism used in the gripper, shown in Fig. 12.43, uses adjacent links and pulleys connected by a spindle, and are free to rotate around it. The mechanism is actuated by a pair of wires, the 'grip wire' and the 'release wire'. These wires pass from the base to the tip of fingers, as shown in Fig. 12.44.

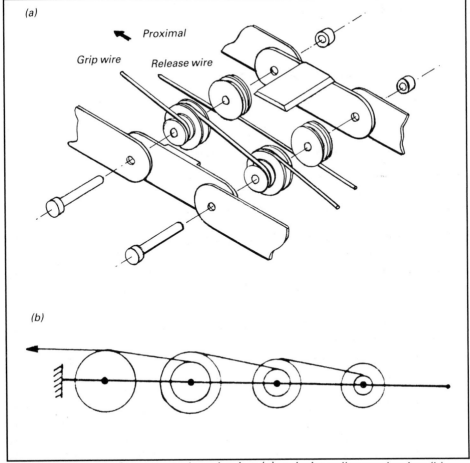

Fig. 12.43 Soft gripper segmental mechanism (a) and wire pulley mechanism (b)

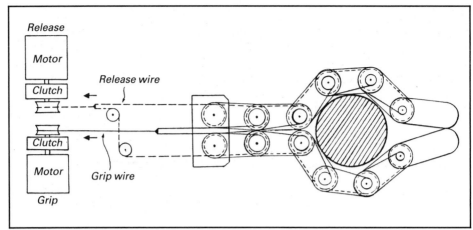

Fig. 12.44 Total mechanism of the soft gripper

This gripper can conform to objects of almost any shape and size, provides uniform gripping pressure along the whole segmented finger, and has a relatively simple mechanism to control. Fig. 12.45 illustrates some application possibilities.

12.10 Compliance hands and robotised assembly tools

The increasing intelligence of industrial robots through their sensory feedback processing devices, the risk of the high capital investment in dedicated assembly machines, and the need for flexible, programmable assembly systems, have all boosted the development of assembly robots (Figs. 12.46 and 12.47).

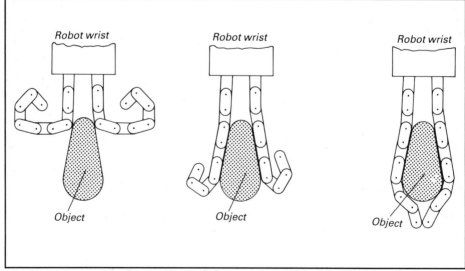

Fig. 12.45 Examples of using the 'soft-finger gripper' (as shown in Fig. 12.44)

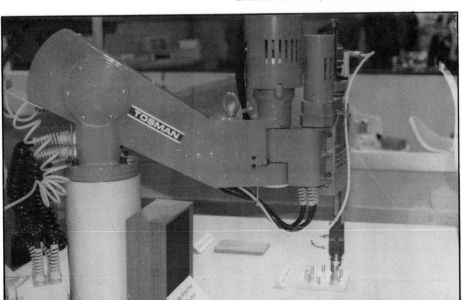

Fig. 12.46 Tosman assembly robot

Fig. 12.47 KUKA robot utilised in car component assembly

The major problem was to design accurate, fast and sensory feedback controlled devices, capable of being interfaced with the 'outside world' via their high-level controllers. Because of these requirements robotised assembly may be regarded as the most advanced use of computer controlled manipulators at present.

The main features of robots most capable of doing assembly work include:

● High positioning and orientation accuracy (i.e. under ±0.1mm orientation errors and approximately ±0.1mm or smaller positioning errors along each axis, and max. ±0.3mm in 3D).
● Relatively large number of input/output signal handling capable of monitoring each arrival and departure, and the gripper status, etc. from the robot program.
● Capacity for real-time sensory feedback processing (i.e. as a minimum requirement force sensing in the gripper, if necessary extendable with vision, slip and torque sensing).
● Fast, vibration-free movement.
● Preferably continuous path control (although for several jobs point-to-point controlled robots can be adequate).
● Preferably a powerful, modular high-level programming language, with the possibility of writing parameterised subroutines, rather than only teach box programming.
● Interfacing facility with other devices, such as other robots, computers, sensors, programmable part positioning and orientating devices, etc. via standard communication interfaces.

When carrying out complex assembly tasks, several elements and subassemblies of the component have to be manipulated and joined. Besides the arm itself, the importance of the robot tools and grippers must also be emphasised. These often have to be tailored for each different assembly application and their design demands a considerable amount of experience. The design of robot tools has to be adapted to the geometrical irregularities of the components, as well as to their softness or rigidity. In other words, grippers must often sense force and torque and also be equipped with slip sensing.

The Model AST-100 robot tool (Astek Engineering, Inc., Massachusetts) is a device for removing misalignment between parts during automated assembly operations. It allows the robot to compensate for positioning errors due to part variation, arm positioning errors, or fixturing tolerances, by minimising the assembly forces and the risk of parts jamming or breaking. The tool is designed around a set of six elastomeric shear pads, which make it stiff in compression, yet relatively soft in shear. Forces generated by mating surfaces cause elastomeric pads to shear. This allows parts to translate and/or rotate, removing misalignment. It can also be precluded with a set of anti-rotation pins, so that torque can be transmitted (Fig. 12.48).

The Astek GTP-45 robot tool can hold parts rigidly during automated assembly operations. It is single acting with a spring return, and is equipped

Fig. 12.48 Compliance gripper
utilising elastomeric pads
(courtesy of Astek Engineering
Inc., Watertown,
Massachusetts)

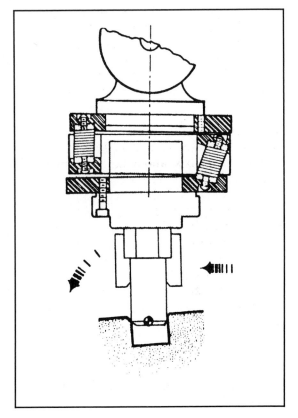

with two or three fingers. The optional finger position sensor consists of a
Hall effect switch and an activating magnet. When the fingers grasp a part,
the piston is stopped mid-stroke and the sensor remains inactive. If the
fingers do not grasp a part then the piston moves to the bottom of its stroke
and activates the sensor. This output signal can then be used to warn the
controller that nothing has been gripped (Fig. 12.49).

Artificial hands are intended to copy the capabilities of the human
hand. Unfortunately such robot tools are often limited in application in
automated assembly, because they are often slow, too large, and do not
satisfy the requirements of strength and reliability.

Tokuji Okada at The Electrotechnical Laboratory in Tokyo and Seiji
Tsuchiya in Ciba, Japan have designed a mechanical hand with high mani-
pulative capacity and flexibility. The object was to design a mechanical
hand to relieve human operators from their mechanical labour. The structure contains an arm very similar in size to the human arm and a three-
fingered tool capable of grasping objects (Fig. 12.50). The system incorporates 11 controlled axes, since the fingers can be controlled separately as
well as the joints of the arm. The system does not incorporate force
feedback, but does work smoothly and accurately.

Because robot tools are often one-off products, they represent a relatively large proportion of the total cost. This fact should not be overlooked
when planning the implementation of robotised assembly cells. Fig. 12.51

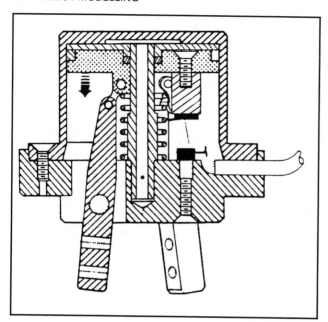

Fig. 12.49 The ASTEK GTP-45 robot tool with finger position sensor (courtesy of Astek Engineering Inc.)

illustrates such an expensive robot tool used by a KUKA robot for assembling car components. Because of the need for increased reliability this tool also incorporates force and torque sensing.

To illustrate how force sensing can be achieved two methods are now introduced. Firstly the pressure sensitive pad, developed by P. G. Ránky at Trent Polytechnic, Nottingham, has almost linear characteristics, within a limited range of pressure. The interesting feature of the material is that it is a low-cost foam commonly used for packaging integrated circuits; its pressure versus resistance sensitivity having been accidentally found by the author. The major advantage of this sensor is that it can also be used for

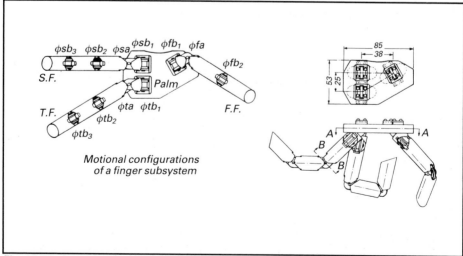

Fig. 12.50 A mechanical hand incorporating 11 controlled axis, offering high manipulativity and flexibility

Fig. 12.51 Sensory feedback robot tool for car component assembly work

slip sensing. The problem with such sensors is their limited force sensing capabilities as well as their sensitivity to humidity and heat. Thus, before use, they have to be calibrated. The principle of this sensor is illustrated in Fig. 12.52.

The second method uses the Astek six-axis force sensor with onboard microprocessor control. This is a sophisticated robot tool for assembly. It incorporates strain gauges and the necessary electronics to measure the six-axis of force which may be applied to the sensor body by an arbitrary load. The tool incorporates a microprocessor driven circuit, which can

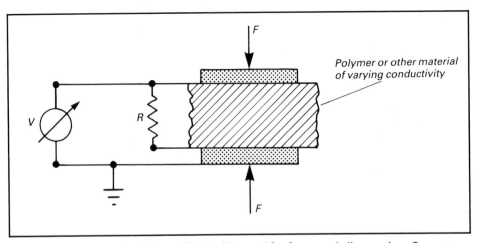

Fig. 12.52 An example of the artificial skin used for force and slip sensing. Current can be measured when force is applied to the upper and lower electrode. (The polymer is compressed, the resistance (R) decreases and the current increases)

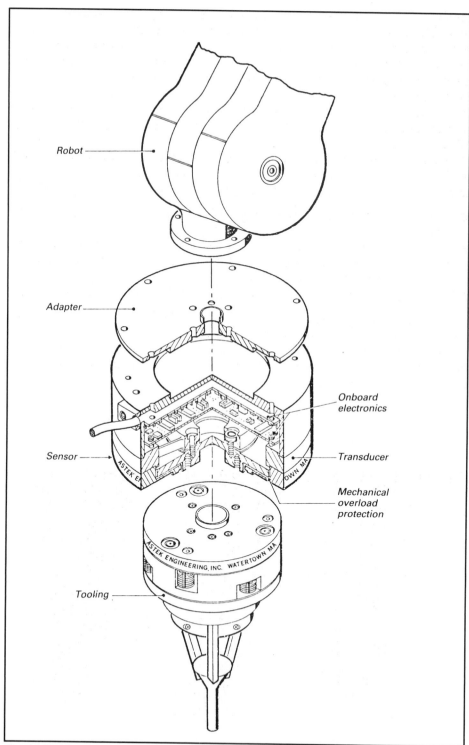

Fig. 12.53 The ASTEK six-axis force sensor tool with on-board microprocessor (courtesy of Astek Engineering Inc.)

resolve six force components in real-time in any specified coordinate system. The tool provides a standard RS-232C compatible serial bit stream output containing measurement data from the instantaneous forces acting on the robot tool body. The operator can then handle this information from his robot program and decide the necessary operating parameters according to the particular job. The data sampling rate of this gripper may range from 1 to 480Hz, determined by the operating baud rate and the selected output data format. The body of this assembly tool is illustrated in Fig. 12.53.

Since automated screwfeeders and screwdrivers are often used in robotised assembly, the Desoutter automatic screwfeeder and screwdriver system is now introduced. The operating principle is as follows (Fig. 12.54):

● The screw is clamped in position and is ready to be used by the screwdriver.
● The gate opens only when the screwdriver is ready, i.e. when the screw detection sensor at this gate is actuated, thus opening it and releasing the next screw for fastening by the screwdriver.
● The screwdriver withdraws to allow a screw to be 'blasted' into position.

Finally, again it must be emphasised, that before designing tools for robotised assembly one should start with the analysis of the components to be assembled. Often it is more cost effective to redesign the component for mechanised assembly, rather than to adjust the assembly system and the expensive tooling to an over-engineered design.

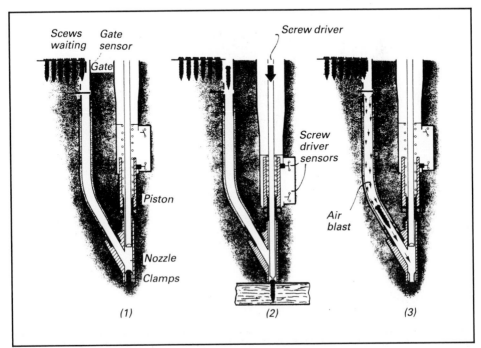

Fig. 12.54 The Desoutter automatic screwfeeder and screwdriver system

12.11 Automated robot hand changers (ARHCs)

If the industrial robot could change its tools and grippers like a human operator when using different tools, then it would become more versatile and allow automated assembly of small batch sizes, where dedicated machinery is uneconomic. It would also use few robots, since robots would rather change tools than distribute the workload on many different devices.

On the negative side, changing grippers and robot tools takes time, so by increasing the versatility of the robotised assembly cell the cycle time is also increased. On average, the time taken to drive the arm to the gripper magazine is a few seconds. The robot arm has then to move to the part feeding station, which can take a few seconds more. Having finished the job with one tool, it has to be changed again, which can cause another 2 – 4 seconds delay in the cycle time.

The simplest method of changing robot tools automatically is shown in Fig. 12.55. Here, the self-powered tools are mounted onto a tool rack or magazine and have a standard mechanical interface, so that the robot can easily pick up in any order. The activation of the power supply is performed simply by high setting of the appropriate output channel. This method is often used if the tools and their power supply requirements are very different from each other, and/or when their power supply requirement is relatively high.

Tools are often arranged around a pneumatically indexing disk (Fig. 12.56) or if there are only two or three tools, at 30° – 90° from each other (Fig. 12.24). These solutions are usually very productive, since the tool or finger changing time can be as low as 0.5 – 1 second. The major problem is

Fig. 12.55 Simple automated robot hand changing method and tool magazine

that they often become too large, thus preventing certain operations from being carried out by the robot. In such cases fingers or tools must be changed individually.

There are a number of ARHC systems offered by manufacturers. These are used mainly in robotised assembly, but also find use in deburring (Fig. 12.57), welding (Fig. 12.58) and other operations such as jet-washing, drilling, rivetting, glueing, etc.

When selecting and/or designing ARHC systems, the following import- ant aspects should be considered:

- Weight (should be light enough not to decrease loading capacity).
- Longitudinal and axial sizes (should be small enough not to limit oper- ations which it would otherwise be feasible to carry out).
- Longitudinal and axial accuracy (should be as good or better than the robot can offer without the ARHC system mounted on it).
- Torque transfer capacity (job dependent, but generally should be high).
- Coupling stiffness (job dependent; must be adequate to prevent the need for compensating the overall deflection of the arm).
- The number of possible pneumatic connection lines.
- The number of possible electronic connection lines (crucial if a large number of sensors are utilised in the hand).
- Easy adaptation to different robot wrists and end-of-arm flanges to provide interchangeability, and thus low-cost tooling for many different robots.

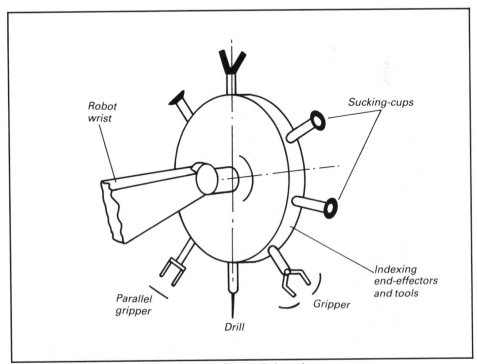

Fig. 12.56 Indexing robot tool for very fast tool changing

Fig. 12.57 Fanuc hand changing system (ARHC) utilised for deburring

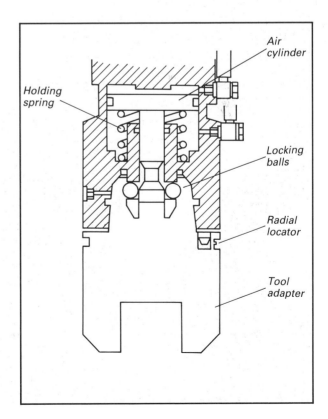

Fig. 12.58 Fanuc ARHC system and principle of operation – the arm pushes down and the balls lock a new hand

Air cylinder

Holding spring

Locking balls

Radial locator

Tool adapter

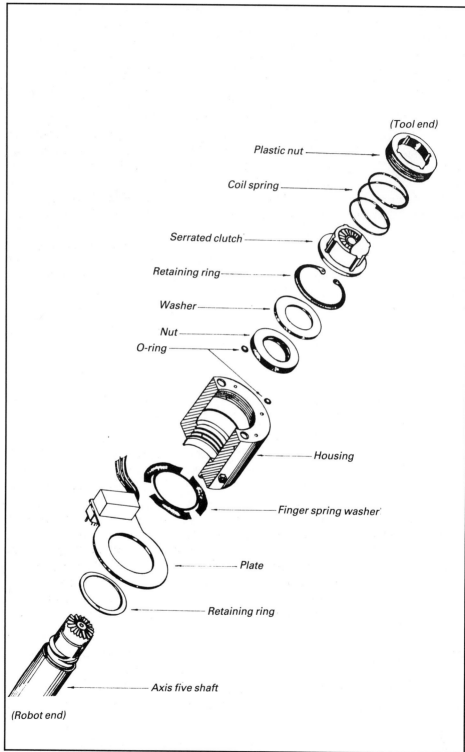

(Tool end)

Plastic nut

Coil spring

Serrated clutch

Retaining ring

Washer

Nut

O-ring

Housing

Finger spring washer

Plate

Retaining ring

Axis five shaft

(Robot end)

Fig. 12.59 The Intelledex ARHC system (courtesy of Intelledex Inc., Corvallis, Oregon)

As an example, the operation of the Intelledex ARHC system (Fig. 12.59) is now introduced. The tool interface engages and locks itself to the robot by means of a shaft at the end of axis 5. The shaft is extended with a short section of centralising thread and a serrated clutch face at the end. The serrated clutch is connected to a one way roller clutch assembly which allows rotation in one fixed direction only. The interface itself has a mating thread and a fixed serrated clutch which is spring loaded. When axis 5 thread is screwed into the interface the serrated clutches are engaged and the roller clutch prevents counter-clockwise rotation so that the interface cannot unscrew off the shaft. A piston located internally onto the shaft can be extended to separate the serrated clutches and allow the unit to be removed by rotating in the counter-clockwise direction.

The 'Ranky-type' ARHC system (Fig. 12.60, see also 12.32(b)) offers automatic reconnection of the tool, not only mechanically but also pneumatically and electronically. The major strength of the design is its simplicity, providing a low cost and light standard tool changer for assembly automation. As over 100 subsequent tool loading/unloading tests have shown, the tool changing time on the Unimation PUMA 560 robot is in the region of 4 seconds, offering 100 out of 100 reliable tool change operations. With careful tuning the tool changing cycle time can be decreased to about 3 seconds, but risk is taken if the robot drifts, or does not consistently maintain its positioning and orientation errors.

This ARHC system consists of the following main components:

● The standard flange, or interface, mounted on each hand capable of reconnecting the male part of the robot hand mechanically, pneumatically and electronically.

● The male part of the robot hand, mounted on the robot arm itself. This is a parallel gripper mechanism providing two moving fingers, each of them containing the necessary pneumatic and electronic connections which are capable of mating the necessary female counter parts in the standard flange, or interface of the tools. (Fig. 12.60 (a-d) illustrates the working prototype design of this ARHC, Fig. 12.60(e) shows the pneumatic circuits, and Fig. 12.60(f) shows the input/output signal handling used for PUMA 560 robot for loading/unloading a tool, and for gripper in/out operations.) The final design contains minor changes in the shape of the male part (i.e. the collecting fingers) and in the way the interfaces are calibrated (i.e. adjusted in terms of robot hand orientation error correction), (Fig. 12.61).

● The robot hand magazine is the device where the tools are held and is interfaced with the input port of the industrial robot. It can contain several different tools, depending on its length (Fig. 12.62).

● The last part of this ARHC system is the robot software, helping in loading and unloading tools, testing tool magazine contents, etc. (Sample programs written in VAL are shown in Figs. 12.63 and 12.64 for tool loading/unloading operations.)

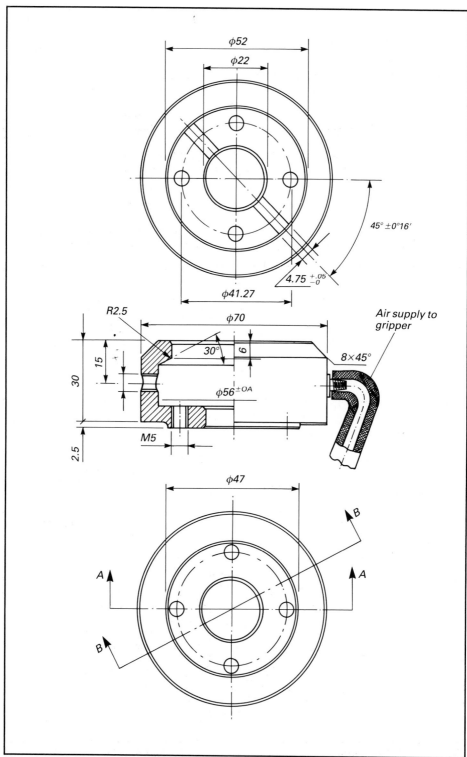

Fig. 12.60(a) The standard interface of the 'Ránky-type' ARHC system

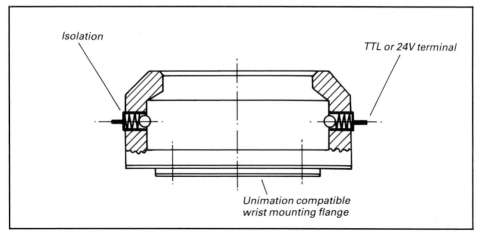

Fig. 12.60(b) The Electronic connections in the 'Ránky-type' ARHC system

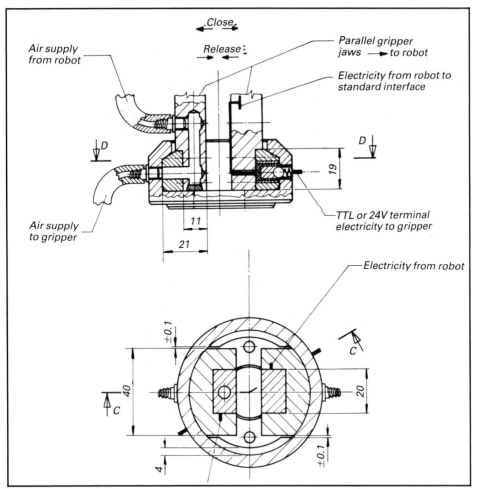

Fig. 12.60(c) The operating principle of the 'Ránky-type' automated robot hand changing (ARHC) system

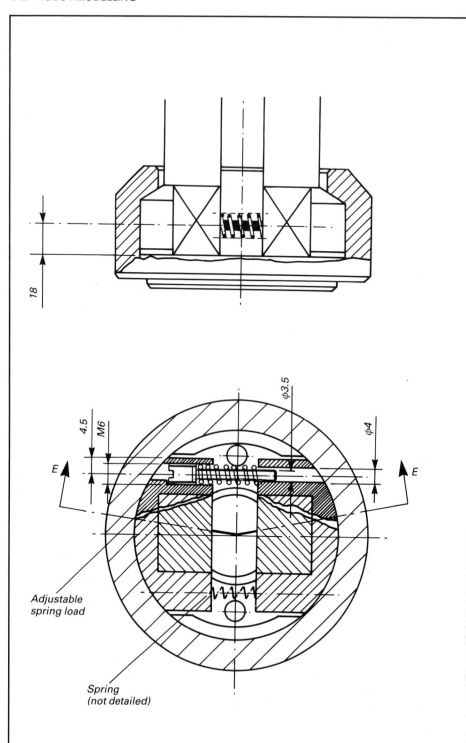

Fig. 12.60(d) The collecting fingers (i.e. the male part) of the 'Ránky-type' ARHC system

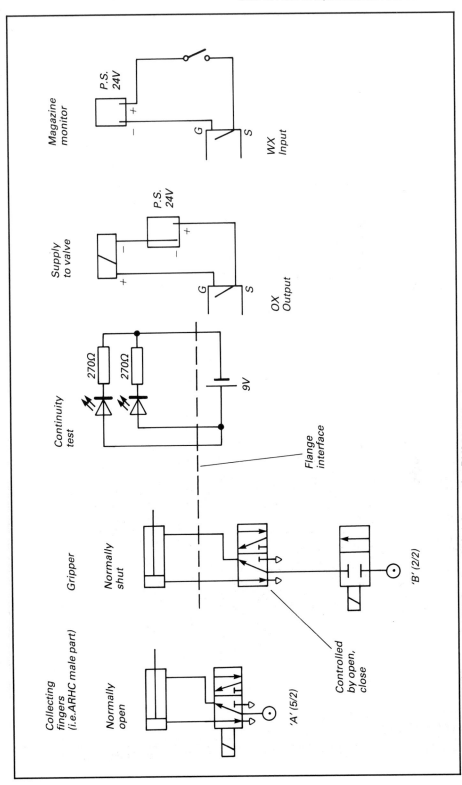

Fig. 12.60(e) 'Ránky-type' ARHC control circuits on the PUMA 560 robot

I/O	No of port	Function	Polarity	Effect
I	11	Monitor station 1	−11 +11	No tool here Tool here
I	12	Monitor station 2	−12 +12	No tool here Tool here
I	13	Monitor station 3	−13 +13	No tool here Tool here
O	14	Switch valve A	−14 +14	Collecting fingers open Collecting fingers shut
O	15	Switch valve B	−15 +15	Air supply off Air supply on

Fig. 60(f) I/O table – signal lines

This ARHC system can be utilised in a variety of cases when the tools differ widely and where sensing is an important requirement.

To summarise, one must realise that since the robot arm itself is incapable of doing much alone, the peripheral devices, such as the end-effectors and special purpose tooling, part orientation, part storage and part feeding devices, safety and guarding equipment and their communication with the robotised cell controller, are of crucial importance and must be considered carefully when designing and/or purchasing robots.

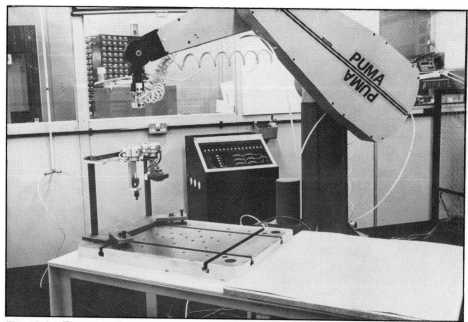

Fig. 12.61 The overall arrangement of the ARHC tool magazine

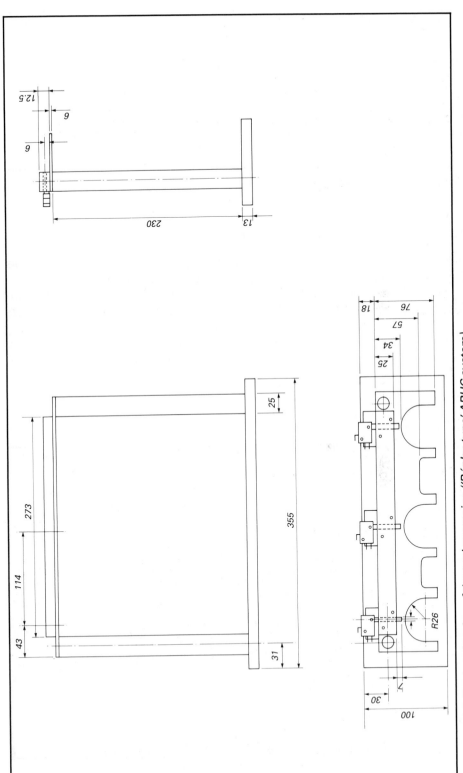

Fig. 12.62(a) Design drawing of the tool magazine ('Ránky-type' ARHC system)

Fig. 12.62(b) Photograph of the tool magazine with a parallel gripper, a pneumatic sucking tool and a sensory tool

References

[1] Fan Yu Chen, 1982. Gripping mechanisms for industrial robots. An overview. *Mechanisms and Machine Theory,* 17(5): 299-311.
[2] Neubauer, G., 1982. Pneumatic grippers for a vice-like grip or a gentle squeeze. *Machine Design,* November 25, 1982.
[3] Hirose, S. and Ometani, Y., 1978. The development of soft gripper for the versatile robot hand. *Mechanisms and Machine Theory,* 13:351-359.

```
.PROGRAM GETTOOL
   1.        TYPE ***************************************************
   2.        TYPE * GETTOOL (SINGLE HAND CHANGE) ROUTINE RUNNING...*
   3.        TYPE *---------------------------------------------------*
   4.        TYPE * BEFORE USING THIS SUBROUTINE THE '#AWAY' AND     *
   5.        TYPE * THE 'TOOL' LOCATIONS MUST HAVE BEEN TAUGHT.      *
   6.        TYPE ***************************************************
   7.        PAUSE TYPE 'PRO' TO CONTINUE...
   8.        SIGNAL 14
   9.        REMARK *** CLOSE MALE PART OF THE ARHC SYSTEM
  10.        ABOVE
  11.        REMARK *** 'ABOVE' CONFIGURATION IS USED
  12.        APPROS TOOL, 100.00
  13.        REMARK *** MOVE 100 MM ABOVE THE LOCATION 'TOOL', WHERE
  14.        REMARK *** THE NEW HAND IS HELD IN THE ARHC SYSTEM
  15.        SPEED 20.00
  16.        REMARK *** BE CAREFUL...
  17.        MOVES TOOL
  18.        REMARK *** STRAIGHT LINE MOTION INTO THE TOOL FLANGE
  19.        SIGNAL -14, , , , , , ,
  20.        DELAY 1.00
  21.        REMARK *** SIGNAL -14 OPENS THE MALE PART OF THE ARHC
  22.        REMARK *** UNIT, THUS PICKING UP THE NEW HAND.
  23.        REMARK *** DELAY 1 SEC IS NOT NECESSARY, BUT ADVISABLE
  24.        REMARK *** BECAUSE OF THE PNEUMATIC CIRCUITS ON THE PUMA
  25.        MOVEST #AWAY1, 0.00
  26.        MOVEST #AWAY2, 0.00
  27.        MOVEST #AWAY3, 0.00
  28.        MOVEST #AWAY4, 0.00
  29.        REMARK *** THESE PRECISION POINTS DEFINE THE SAFE PATH
  30.        REMARK *** TO AND FROM THE TOOL HOLDER. THE MOTION HERE
  31.        REMARK *** AWAY.
  32.        MOVET HOME, 0.00
  33.        REMARK *** THIS IS A TAUGHT SAFE LOCATION. NOW YOU CAN
  34.        REMARK *** USE THE NEW TOOL.
  35.        RETURN 0
.END


.PROGRAM PUTTOOL
   1.        MOVE HOME
   2.        MOVE #AWAY4
   3.        SPEED 20.00
   4.        MOVES #AWAY3
   5.        MOVES #AWAY2
   6.        MOVES #AWAY1
   7.        MOVES TOOL
   8.        SIGNAL 14, , , , , , ,
   9.        TYPE *** OLD HAND RELEASED ***
  10.        RETURN 0
```

Fig. 12.63 Single tool load/unload routines in VAL as utilised in the 'Ránky-type' ARHC system

Fig. 12.64 Tool loading sequence ('Ránky-type' ARHC system): (a) step 1, (b) step 2, and (c) step 3

Chapter Thirteen

ROBOTISED MANUFACTURING, ASSEMBLY SYSTEM DESIGN AND SIMULATION SOFTWARE

THIS chapter is concerned with some important software tools to be utilised when designing and/or simulating flexible manufacturing systems (FMS) and computer integrated manufacturing systems (CIM) with robots. The discussed principles and the demonstrated programs can also be utilised in mechanised assembly or flexible assembly systems (FAS) underlying how similar flexible machining and assembly systems are from the system analysis point of view. As already emphasised, the computer integrated manufacturing technology has a common base of knowledge and principles which can be applied to a wide range of processes.

Flexible manufacturing systems deal with high-level distributed data processing and automated material flow using computer-controlled machine tools, robots, coordinate measuring machines (CMM), assembly robots and automated material handling and storage devices, etc., with the aim of combining the benefits of a highly productive but inflexible transfer line, and a highly flexible but inefficient job shop.

In the case of scheduling flexible assembly systems the modular systems approach identifies robotised assembly cells, or processing stations, capable of loading and unloading fixtured components following the established optimal sequence, changing grippers and tools automatically, sending and receiving data, and generally acting as intelligent nodes in the computer network.

Processing cells can have different physical characteristics and a variety of interconnections with different material handling systems (e.g. industrial robots, automated guided vehicles, conveyors, buffer stores, etc.) and they have to process data and communicate via a communications network within the distributed data processing system.

To illustrate some of the basic features of FMS and FAS systems, Fig. 13.1 shows a flexible machining cell, Fig. 13.2 a robotised assembly and inspection cell, and Fig. 13.3 a flexible assembly line.

Fig. 13.1 Flexible machining cell utilising integrated part loading and unloading robot

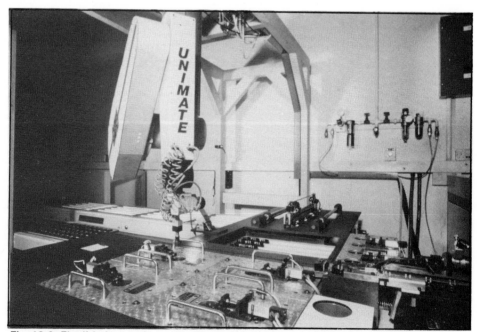

Fig. 13.2 Flexible inspection cell with conveyor part transfer system testing transducers and palettising the components (Courtesy of Plessey Office Systems Ltd, Nottingham, UK)

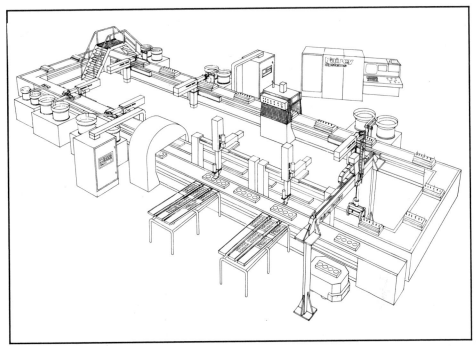

Fig. 13.3 Flexible assembly line incorporating industrial robots and other programmable part handling and transfer devices (Courtesy of Fairey Automation Ltd, Swindon, UK)

The discussed packages and methods include scheduling, capacity planning and mechanised/robotised assembly line balancing techniques, and sample runs of these programs representing some of the most important system design and simulation tools of FAS and FMS.

Scheduling is a process that relates specific events to specific times or to a specific span of time. It is of extreme importance in any complex manufacturing or assembly activity because scheduling faults can create delays in the system, or low utilisation levels of resources (i.e. robots, machines, etc.) and/or disorder if sequencing faults generate long queues.

Capacity planning means the calculation of the measure of capacity to enable FMS and robot cell and system requirements to be identified for each component, batch, or batches for all selected production routes. Capacity planning is generally applied during the factory, or manufacturing system planning and design phase, as well as when the capacity requirements of a certain schedule need to be checked.

Assembly line balancing in most cases must start with a list of operations to be performed on assembly machines, or by assembly robots in a given order to complete a subassembly or a component. Specifying the time given for each operation as well as the order in which the operations must be completed, the task of the assembly line balancing program is to assign operations to workstations or assembly robots in the best way.

Some mathematical models are now introduced and how such programs can assist the FAS/FMS system designer are illustrated.

13.1 Scheduling FMS and FAS cells

Scheduling in the manufacturing industry means the allocation of jobs to be processed on specific machines, robots or cells in the case of FMS and FAS, in a specified time span. (A job is a product to be manufactured on a machine, or assembled on a robot, or on a series of machines or robots by performing different operations on it in a programmed order.)

Scheduling is usually more effective if applied at different levels and if certain real-time data feedback is provided between the different organisational levels to be able to consider certain changes and variations in the production plan. During the fine tuning phase, scheduling deals with individual sequencing of jobs on robots, or in general on machines, in order to achieve certain management objectives.

Assumptions in the '*n*' jobs single machine scheduling task

The '*n*' jobs single processor (i.e. machine, or robot, etc.) problem can be analysed using the following assumptions:

● There are '*n*' independent jobs available for processing at time zero ($t=0$).
● Jobs are not divided or interrupted when being processed. In other words, if processing has already started it must be finished.
● Setup times are independent of the job sequence and can be included in processing times.
● All CNC part programs, or robot programs are available in the controller, or via a DNC link before processing the component.
● The FMS cell is continuously available for processing during the analysed period of time (i.e. shift).
● Parts can wait in a buffer store or in a pallet magazine, but the processing station (i.e. the assembly cell or machine) is not permitted to be idle.

Important data and terminology

Before discussing and demonstrating the SCHEDULE program the most important data and terminology relating to the scheduling problem are first summarised:

● *Processing time (ti)* is defined as the calculated time or the forecasted estimate of how long it will take to complete job *i*.

● *Flow time (Fi)*, or manufacturing interval, is defined as the time span between the point at which job *i* is available for processing and the point at which it is completed. In other words, *Fi* is the total time job *i* spends in the system, and is equal to the sum of the processing time (*ti*) and the waiting time (*wi*) of job *i*, i.e. $Fi = ti + wi$.

● *The makespan,* or the maximum flow time of the schedule, is always constant for a given number of jobs or batches of jobs.

● *Due date (di)*, or delivery date for job *i*, is the deadline at which the processing of job *i* is due to be completed, and beyond which it would be considered as tardy.

● *Lateness (Li)* is the deviation between the actual completion time (*Ci*) and the due date (*di*) of operation *i*, i.e. $Li = Ci - di$. Lateness can be positive or negative. Positive lateness means that the operation has been completed after its due date (known as tardiness). Negative lateness means that the operation has been completed earlier than its due date.

The SCHEDULE program

The SCHEDULE program uses a variety of rules to achieve different job sequence optimisation goals:

● Shortest processing time (SPT) rule.
● Weighted shortest processing time (WSPT) rule.
● Earliest due date (EDD) rule.

Following the SPT rule the mean flow-time is minimised by sequencing the SPT task first.

The algorithm based on the WSPT rule minimises the weighted mean flow-time (w_i) by sequencing the jobs in the following order:

$$\frac{t_1}{w_1} \leqslant \frac{t_2}{w_2} \leqslant \frac{t_3}{w_3} \ldots \leqslant \frac{t_n}{w_n}$$

When applying the EDD rule maximum job lateness and maximum job tardiness are minimised.

Before running the program a list of jobs and their processing times on each machine or robot should be prepared. If the WSPT and/or the EDD rules are going to be used, then the weight factors and/or the due dates must be specified for each job.

The program offers the possibility of testing different solutions, in which the input data are only estimates, thus helping to provide design guidelines for FMS cell designers and shop floor managers. (Since all Library packages are aimed at international use and as it was a design concept to ensure the maximum user-flexibility, none of the programs use any fixed time or length dimensions unless unavoidable because of input data safety.)

The program offers the possibility of selecting different job sequence optimisation rules:

M (ean flow times minimised . . .
L (ate jobs, lateness minimised . . .
T (ardy jobs, mean tardiness minimised . . .

The program also allows the use of the standard commands of the Library, i.e. PRINT and/or DISPLAY the input data and the results, or the input only, SAVE the output in a user specified disk file, make FORMATTED PRINT files and modify the processing time, the weight factors and/or the due date constraints with the WHAT IF? COMMANDS FOR ANY JOB.

The example shown in Fig. 13.4 (a and b) demonstrate the SCHEDULE program when utilised for job sequencing. (Note the way the job sequence alters depending on the optimisation rule applied.)

```
          *** MEAN FLOW TIME MINIMISED ***

      ****************************************
      *        CALCULATED JOB SEQUENCE       *
      ****************************************
          SCHEDULED POSITION   JOB NUMBER
      ------------------------+---------------
                      1           JOB   5
                      2           JOB   8
                      3           JOB   3
                      4           JOB   1
                      5           JOB   6
                      6           JOB   7
                      7           JOB   2
                      8           JOB   4
                      9           JOB  10
                     10           JOB   9
      ------------------------+---------------
```

TOTAL NUMBER OF JOBS SCHEDULED FOR THE CELL IN THIS RUN = 10

```
          EVALUATION OF THE CALCULATED SEQUENCE
          *************************************
```

MAKESPAN	=	93.00
NUMBER OF TARDY JOBS	=	5
LATENESS	=	22.00
MEAN LATENESS	=	2.20

```
********************************************************************
*                        TABLE OF RESULTS                         *
********************************************************************
```

JOB NUMBER	PROCESSING TIME	STARTING...FINISHING TIME		DUE TIME	LATENESS
1	5	18.00	23.00	12	11.00 *
2	9	44.00	53.00	23	30.00 *
3	6	12.00	18.00	22	-4.00
4	10	53.00	63.00	43	20.00 *
5	7	0.00	7.00	56	-49.00
6	13	23.00	36.00	78	-42.00
7	8	36.00	44.00	25	19.00 *
8	5	7.00	12.00	34	-22.00
9	18	75.00	93.00	21	72.00 *
10	12	63.00	75.00	88	-13.00

**)>> PLEASE NOTE that jobs marked with * are TARDY

```
**********************************************
*       INPUT DATA ECHO AND JOB RECORD       *
**********************************************

JOB No.  1
*********
```

PROCESSING TIME	=	5
IMPORTANCE WEIGHTING	=	1
DUE TIME	=	12

Fig. 13.4(a) and (b) Sample runs of the SCHEDULE program

```
JOB No.  2
*********

      PROCESSING TIME        =      9
      IMPORTANCE WEIGHTING    =      1
      DUE TIME               =     23

JOB No.  3
*********

      PROCESSING TIME        =      6
      IMPORTANCE WEIGHTING    =      2
      DUE TIME               =     22

JOB No.  4
*********

      PROCESSING TIME        =     10
      IMPORTANCE WEIGHTING    =      1
      DUE TIME               =     43

JOB No.  5
*********

      PROCESSING TIME        =      7
      IMPORTANCE WEIGHTING    =      3
      DUE TIME               =     56

JOB No.  6
*********

      PROCESSING TIME        =     13
      IMPORTANCE WEIGHTING    =      2
      DUE TIME               =     78

JOB No.  7
*********

      PROCESSING TIME        =      8
      IMPORTANCE WEIGHTING    =      1
      DUE TIME               =     25

JOB No.  8
*********

      PROCESSING TIME        =      5
      IMPORTANCE WEIGHTING    =      2
      DUE TIME               =     34

JOB No.  9
*********

      PROCESSING TIME        =     18
      IMPORTANCE WEIGHTING    =      1
      DUE TIME               =     21

JOB No. 10
*********

      PROCESSING TIME        =     12
      IMPORTANCE WEIGHTING    =      1
      DUE TIME               =     88
```

Fig. 13.4(a) continued

```
                    *** MAXIMUM LATENESS MINIMISED ***

          ***************************************
          *       CALCULATED JOB SEQUENCE        *
          ***************************************
                 SCHEDULED POSITION    JOB NUMBER
          -----------------------+------------------
                        1             JOB    1
                        2             JOB    9
                        3             JOB    3
                        4             JOB    2
                        5             JOB    7
                        6             JOB    8
                        7             JOB    4
                        8             JOB    5
                        9             JOB    6
                       10             JOB   10
          -----------------------+------------------

    TOTAL NUMBER OF JOBS SCHEDULED FOR THE CELL IN THIS RUN = 10


                 EVALUATION OF THE CALCULATED SEQUENCE
                 *************************************


        MAKESPAN                =      93.00

        NUMBER OF TARDY JOBS    =       9

            LATENESS            =      93.00
        MEAN LATENESS           =       9.30
```

```
**********************************************************************
*                         TABLE OF RESULTS                           *
**********************************************************************
JOB NUMBER  PROCESSING TIME   STARTING...FINISHING TIME  DUE TIME  LATENESS
----------+----------------+---------+----------------+---------+----------
     1            5            0.00         5.00           12       -7.00
     2            9           29.00        38.00           23       15.00 *
     3            6           23.00        29.00           22        7.00 *
     4           10           51.00        61.00           43       18.00 *
     5            7           61.00        68.00           56       12.00 *
     6           13           68.00        81.00           78        3.00 *
     7            8           38.00        46.00           25       21.00 *
     8            5           46.00        51.00           34       17.00 *
     9           18            5.00        23.00           21        2.00 *
    10           12           81.00        93.00           88        5.00 *
----------+----------------+---------+----------------+---------+----------
 **)>> PLEASE NOTE that jobs marked with   *   are TARDY
```

Fig. 13.4(b)

13.2 Capacity planning software

The purpose of capacity planning software is to calculate a measure of capacity to enable FMS cell and system requirements to be identified for each component, batch, or batches for all selected production routes in the robotised manufacturing or assembly system. Capacity planning is generally applied during the factory or manufacturing system planning and design phase, as well as when the capacity requirements of a certain schedule need to be checked.

The CAPACITY program demonstrated in this section can be used as a simulation tool for FMS and FAS cell requirements and cell efficiency planning. This is useful if the system is manufacturing and/or assembling a series of batches of components or mixed type of components in any order during a user-specified production period. When analysing the results, the bottleneck cells of each different production route can be identified, and the number of components produced by the FMS or FAS can be determined, for each production period.

By utilising this program as a system architecture design tool, the capacity of the each cell as well as the total system can be optimised.

The program follows the concept of the 'variable route' FMS part programming method. This concept can be implemented not only in machining but also in assembly or other flexible manufacturing shops.

Capacity planning in an assembly shop

Capacity planning in a robotised assembly shop is demonstrated by means of an example. In the given sample run the determination of robot and/or assembly head requirements is done by taking care of losses of defective machines, components and various differences in the operation of the system.

Having selected one of the possible and/or prescheduled assembly routes, what the CAPACITY program can do in performing the job using existing resources is examined by running the program (Fig. 13.5).

Mathematical model of the CAPACITY program

The mathematical model considers several production cells used in any user-defined order for the batches or single components. The determination of FMS cell (i.e. machining, assembly, inspection, part washing, etc.) requirements is performed by the successive use of the following formula:

$$\text{Requirement} = \frac{\text{Time} \times \text{Output rate}}{60 \times \text{Shift} \times \text{Efficiency}}$$

where 'Requirement' is the FMS cell requirement (i.e. the number of machines required); 'Time' is the processing time per component per hour per manufacturing cell (i.e. machine); 'Output rate' is the production rate per cell, i.e. manufactured (good) components per production period (i.e. *S*); 'Shift' is the duration of an operating period in hours; and 'Efficiency' is

```
FMS CELL REQUIREMENTS FOR THE SPECIFIED ROUTE IN THE FMS
***********************************************************


FINISHED PARTS required per production period =   1250.00 <<((**


PRODUCTION      RAW         CELL       DEFECT    UNADJUSTED    ADJUSTED
  CELL     PARTS REQUIRED  EFF. [%]   RATE [%]   REQUIREMENT  REQUIREMENT
_____I_____I_____I_____I_____I_____

    1        1888.00       97.08       4.80        8.10         8.00
    2        1797.00       98.33       3.40        8.76         9.00
    3        1736.00       95.28       3.10        6.26         7.00
    4        1682.00       96.67       2.40        2.61         3.00
    5        1642.00       95.00       4.00       10.80        11.00
    6        1576.00       92.00       4.50        7.07         7.00
    7        1505.00       98.00       2.50        6.40         7.00
    8        1467.00       96.00       3.40        6.69         7.00
    9        1417.00       93.00       2.00        3.56         4.00
   10        1389.00       97.00       3.00        9.55        10.00
   11        1347.00       96.00       2.30        5.85         6.00
   12        1316.00       89.00       5.00        3.70         4.00


------------------------------------------------------------------------


    ****************
      FMS CELL 1
    ****************

    INPUT DATA ECHO
    ---------------

    LENGTH OF PRODUCTION PERIOD   =    8.00 HOURS

    AVERAGE DOWNTIME            =   30.00 MINS
    AVERAGE SETUP TIME          =   12.00 MINS
    ANALYSED PERIOD OF TIME     =   24.00 HOURS

    PROCESSING TIME PER COMPONENT =    2.00 MINS
    DEFECT RATE OF THE CELL       =    4.80 %
    (Assuming that bad components cannot be reworked)

    CALCULATED RESULTS
    ------------------

    FMS CELL EFFICIENCY             =     97.08 %
    RAW PARTS REQUIRED PER PERIOD   =   1888.00
    UNADJUSTED EQUIPMENT REQUIREMENT =     8.10 FMS CELLS
    ADJUSTED EQUIPMENT REQUIREMENT   =     8.00 FMS CELLS
```

Fig. 13.5 Sample run of the CAPACITY program as used for analysing an assembly system

```
*****************
   FMS CELL 2
*****************

INPUT DATA ECHO
---------------

LENGTH OF PRODUCTION PERIOD   =     8.00 HOURS

AVERAGE DOWNTIME              =    11.00 MINS
AVERAGE SETUP TIME            =    13.00 MINS
ANALYSED PERIOD OF TIME       =    24.00 HOURS

PROCESSING TIME PER COMPONENT =     2.30 MINS
DEFECT RATE OF THE CELL       =     3.40 %
(Assuming that bad components cannot be reworked)

CALCULATED RESULTS
------------------

FMS CELL EFFICIENCY             =      98.33 %
RAW PARTS REQUIRED PER PERIOD   =    1797.00
UNADJUSTED EQUIPMENT REQUIREMENT =      8.76 FMS CELLS
ADJUSTED EQUIPMENT REQUIREMENT  =      9.00 FMS CELLS



*****************
   FMS CELL 3
*****************

INPUT DATA ECHO
---------------

LENGTH OF PRODUCTION PERIOD   =     8.00 HOURS

AVERAGE DOWNTIME              =    45.00 MINS
AVERAGE SETUP TIME            =    23.00 MINS
ANALYSED PERIOD OF TIME       =    24.00 HOURS

PROCESSING TIME PER COMPONENT =     1.65 MINS
DEFECT RATE OF THE CELL       =     3.10 %
(Assuming that bad components cannot be reworked)

CALCULATED RESULTS
------------------

FMS CELL EFFICIENCY             =      95.28 %
RAW PARTS REQUIRED PER PERIOD   =    1736.00
UNADJUSTED EQUIPMENT REQUIREMENT =      6.26 FMS CELLS
ADJUSTED EQUIPMENT REQUIREMENT  =      7.00 FMS CELLS



*****************
   FMS CELL 4
*****************

INPUT DATA ECHO
---------------

LENGTH OF PRODUCTION PERIOD   =     8.00 HOURS
```

Fig. 13.5 continued

```
          AVERAGE DOWNTIME              =     12.00 MINS
          AVERAGE SETUP TIME           =     56.00 MINS
          ANALYSED PERIOD OF TIME      =     34.00 HOURS

          PROCESSING TIME PER COMPONENT =     0.72 MINS
          DEFECT RATE OF THE CELL       =     2.40 %
          (Assuming that bad components cannot be reworked)

          CALCULATED RESULTS
          ------------------

          FMS CELL EFFICIENCY                =      96.67 %
          RAW PARTS REQUIRED PER PERIOD      =    1682.00
          UNADJUSTED EQUIPMENT REQUIREMENT   =      2.61 FMS CELLS
          ADJUSTED EQUIPMENT REQUIREMENT     =      3.00 FMS CELLS




          ****************
            FMS CELL 5
          ****************

          INPUT DATA ECHO
          ---------------

          LENGTH OF PRODUCTION PERIOD    =     8.00 HOURS
          FMS CELL EFFICIENCY            =    95.00 %
          PROCESSING TIME PER COMPONENT  =     3.00 MINS
          DEFECT RATE OF THE CELL        =     4.00 %
          (Assuming that bad components cannot be reworked)

          CALCULATED RESULTS
          ------------------

          RAW PARTS REQUIRED PER PERIOD     =    1642.00
          UNADJUSTED EQUIPMENT REQUIREMENT  =      10.80 FMS CELLS
          ADJUSTED EQUIPMENT REQUIREMENT    =      11.00 FMS CELLS




          ****************
            FMS CELL 6
          ****************

          INPUT DATA ECHO
          ---------------

          LENGTH OF PRODUCTION PERIOD    =     8.00 HOURS
          FMS CELL EFFICIENCY            =    92.00 %
          PROCESSING TIME PER COMPONENT  =     1.98 MINS
          DEFECT RATE OF THE CELL        =     4.50 %
          (Assuming that bad components cannot be reworked)

          CALCULATED RESULTS
          ------------------

          RAW PARTS REQUIRED PER PERIOD     =    1576.00
          UNADJUSTED EQUIPMENT REQUIREMENT  =      7.07 FMS CELLS
          ADJUSTED EQUIPMENT REQUIREMENT    =      7.00 FMS CELLS
```

Fig. 13.5 continued

```
****************
  FMS CELL 7
****************

INPUT DATA ECHO
---------------

LENGTH OF PRODUCTION PERIOD    =     8.00 HOURS
FMS CELL EFFICIENCY            =    98.00 %
PROCESSING TIME PER COMPONENT  =     2.00 MINS
DEFECT RATE OF THE CELL        =     2.50 %
(Assuming that bad components cannot be reworked)

CALCULATED RESULTS
------------------

RAW PARTS REQUIRED PER PERIOD       =    1505.00
UNADJUSTED EQUIPMENT REQUIREMENT =       6.40 FMS CELLS
ADJUSTED EQUIPMENT REQUIREMENT   =       7.00 FMS CELLS




****************
  FMS CELL 8
****************

INPUT DATA ECHO
---------------

LENGTH OF PRODUCTION PERIOD    =     8.00 HOURS
FMS CELL EFFICIENCY            =    96.00 %
PROCESSING TIME PER COMPONENT  =     2.10 MINS
DEFECT RATE OF THE CELL        =     3.40 %
(Assuming that bad components cannot be reworked)

CALCULATED RESULTS
------------------

RAW PARTS REQUIRED PER PERIOD       =    1467.00
UNADJUSTED EQUIPMENT REQUIREMENT =       6.69 FMS CELLS
ADJUSTED EQUIPMENT REQUIREMENT   =       7.00 FMS CELLS




****************
  FMS CELL 9
****************

INPUT DATA ECHO
---------------

LENGTH OF PRODUCTION PERIOD    =     8.00 HOURS
FMS CELL EFFICIENCY            =    93.00 %
PROCESSING TIME PER COMPONENT  =     1.12 MINS
DEFECT RATE OF THE CELL        =     2.00 %
(Assuming that bad components cannot be reworked)

CALCULATED RESULTS
------------------

RAW PARTS REQUIRED PER PERIOD       =    1417.00
UNADJUSTED EQUIPMENT REQUIREMENT =       3.56 FMS CELLS
ADJUSTED EQUIPMENT REQUIREMENT   =       4.00 FMS CELLS
```

Fig. 13.5 continued

```
****************
   FMS CELL 10
****************

INPUT DATA ECHO
---------------

LENGTH OF PRODUCTION PERIOD   =     8.00 HOURS
FMS CELL EFFICIENCY           =    97.00 %
PROCESSING TIME PER COMPONENT =     3.20 MINS
DEFECT RATE OF THE CELL       =     3.00 %
(Assuming that bad components cannot be reworked)

CALCULATED RESULTS
------------------

RAW PARTS REQUIRED PER PERIOD    =    1389.00
UNADJUSTED EQUIPMENT REQUIREMENT =       9.55 FMS CELLS
ADJUSTED EQUIPMENT REQUIREMENT   =      10.00 FMS CELLS


****************
   FMS CELL 11
****************

INPUT DATA ECHO
---------------

LENGTH OF PRODUCTION PERIOD   =     8.00 HOURS
FMS CELL EFFICIENCY           =    96.00 %
PROCESSING TIME PER COMPONENT =     2.00 MINS
DEFECT RATE OF THE CELL       =     2.30 %
(Assuming that bad components cannot be reworked)

CALCULATED RESULTS
------------------

RAW PARTS REQUIRED PER PERIOD    =    1347.00
UNADJUSTED EQUIPMENT REQUIREMENT =       5.85 FMS CELLS
ADJUSTED EQUIPMENT REQUIREMENT   =       6.00 FMS CELLS


****************
   FMS CELL 12
****************

INPUT DATA ECHO
---------------

LENGTH OF PRODUCTION PERIOD   =     8.00 HOURS
FMS CELL EFFICIENCY           =    89.00 %
PROCESSING TIME PER COMPONENT =     1.20 MINS
DEFECT RATE OF THE CELL       =     5.00 %
(Assuming that bad components cannot be reworked)

CALCULATED RESULTS
------------------

RAW PARTS REQUIRED PER PERIOD    =    1316.00
UNADJUSTED EQUIPMENT REQUIREMENT =       3.70 FMS CELLS
ADJUSTED EQUIPMENT REQUIREMENT   =       4.00 FMS CELLS
```

Fig. 13.5 (continued)

the FMS cell efficiency (percent), as defined later. So that at each cell (i.e. machine) the various operating conditions of the particular machine and the previous cells involved in the given route are accounted for.

To estimate the output rate, we must be aware of the fact that often some defective components cannot be eliminated from the process. They often must be transferred from one cell to another, thus the total number of components produced on the FMS must be calculated for each cell:

$$\text{Total part requirement} = \text{Good} + \text{Defective components}$$

This calculation must also consider the fact that because of the defective components there are gradually less and less good components transferred to the next (as programmed) cell, although the system must cope with the part handling tasks of all (i.e. good and defective) components at the same time.

Prior to the determination of the desired amount of components to be manufactured, the execution of a lot-size analysis program can be very useful. This will enable the production engineer to determine economic batch sizes and cycle times considering the actual product mix. As an example of typical results, Fig. 13.6 presents a sample display of the LOTSIZE program.

```
     ***  TABLE OF RESULTS  >>> NOTE: values are not rounded

  **>> The OVERALL ECONOMIC BATCH SIZE is based on the optimal production
       cycle time for for all components of =  14.28 DAYS
..............................................................................
   PRODUCT    MIN. COST    TOTAL PERIOD    OVERALL ECON.    ECON. PROD.
    TYPE      BATCH SIZE    BATCH SIZE      BATCH SIZE      CYCLE TIME
              [UNIT]        [UNIT]        **>> [UNIT]        [DAY]
       I              I              I              I
  ____I_____I_____I_____I_____

      1          28.28          60.00          14.28          28.28
      2          46.92         620.00          42.85          15.64
      3       77706.21     1095000.00       42833.97          25.91
      4         479.17        6944.00         714.14           9.58
      5          72.31         360.00         114.26           9.04
      6        1094.53        5280.00        1428.27          10.95
      7         336.29        1056.00         328.50          14.62
      8         142.68         930.00         171.39          11.89
      9        1764.53        6432.00        1571.10          16.04
     10         151.61         620.00         171.39          12.63


                 ****)))>> PRESS RETURN TO CONTINUE...
```

Fig. 13.6 Sample results of the LOTSIZE program, offering the optimisation of batch sizes and their cycle times in which they should be manufactured to minimise variable cost. Values are calculated for each batch separately as well as for the mix of batches. (Note that one could argue that the true FAS or FMS is flexible enough to manufacture different batch sizes in random order at the minimum or the same cost. This is true but unfortunately there are very few, if any, 'true' FAS or FMS systems implemented in industry and until then batch size analysis is a useful tool to cut the cost)

Machine, or robot cell efficiency is also taken into account and is calculated for each robot as follows:

$$\text{Efficiency} = 1 - \frac{\text{Setup time} + \text{Downtime}}{\text{Analysed period}}$$

where 'Setup time' is given in minutes per cell per analysed period of time; 'Downtime' is given in minutes per cell per analysed period of time; and 'Analysed period' is the time utilised for the evaluation of the average setup and downtime values.

In FMS or FAS, machine and robot setup times are relatively small, if for example palletised part transfer or robotised part loading/unloading is employed. In conventional machine and assembly shops setup time and downtime are usually several fold greater than in FMS or FAS. (A sample screen input asking for such data is shown in Fig. 13.7.)

When providing these data, one can experiment with different data combinations and see their effect on the cell requirement analysis. (This can be obtained by using the W(hat if? command). Using the program this way enables the optimum, the realistic and the pessimistic machine and robot cell requirement values to be designed, simulated and/or checked. This point is illustrated in Fig. 13.8. Of the displayed results one can see the way the system designer can influence whether the line should be under-utilised or overloaded for a given component following a certain production route.

If more batches of components are machined or assembled on the same system, during the same time period in a mixed, random or any scheduled (i.e. optimised) order, then the maximum cell requirement for the given

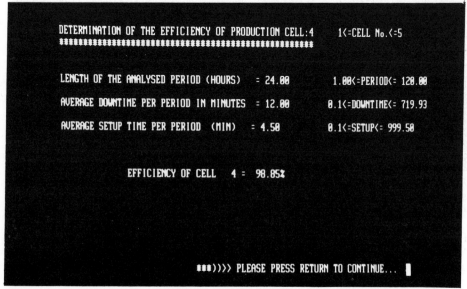

Fig. 13.7 The cell efficiency calculation method in the CAPACITY program can be applied for many different processes, including machining and assembly

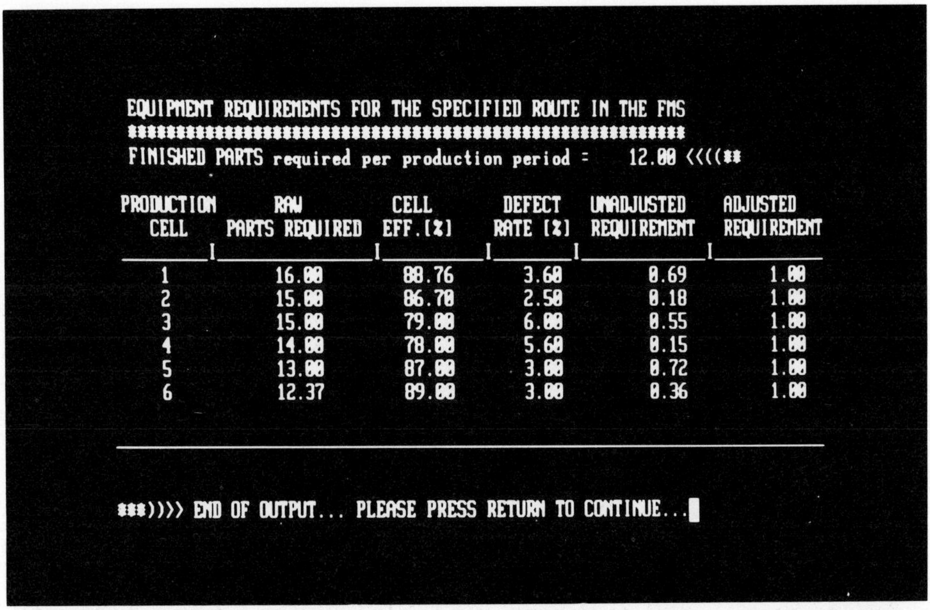

EQUIPMENT REQUIREMENTS FOR THE SPECIFIED ROUTE IN THE FMS
****
FINISHED PARTS required per production period = 12.00 <<<(**

PRODUCTION CELL	RAW PARTS REQUIRED	CELL EFF.[%]	DEFECT RATE [%]	UNADJUSTED REQUIREMENT	ADJUSTED REQUIREMENT
1	16.00	88.76	3.60	0.69	1.00
2	15.00	86.70	2.50	0.18	1.00
3	15.00	79.00	6.00	0.55	1.00
4	14.00	78.00	5.60	0.15	1.00
5	13.00	87.00	3.00	0.72	1.00
6	12.37	89.00	3.00	0.36	1.00

***))>> END OF OUTPUT... PLEASE PRESS RETURN TO CONTINUE...█

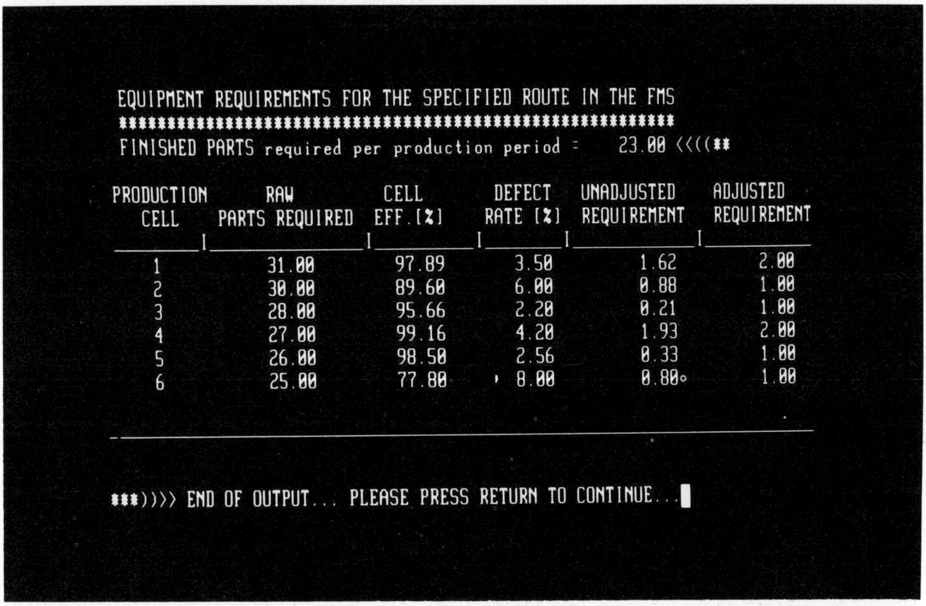

EQUIPMENT REQUIREMENTS FOR THE SPECIFIED ROUTE IN THE FMS
****
FINISHED PARTS required per production period = 23.00 <<<(**

PRODUCTION CELL	RAW PARTS REQUIRED	CELL EFF.[%]	DEFECT RATE [%]	UNADJUSTED REQUIREMENT	ADJUSTED REQUIREMENT
1	31.00	97.89	3.50	1.62	2.00
2	30.00	89.60	6.00	0.88	1.00
3	28.00	95.66	2.20	0.21	1.00
4	27.00	99.16	4.20	1.93	2.00
5	26.00	98.50	2.56	0.33	1.00
6	25.00	77.80	8.00	0.80°	1.00

***))>> END OF OUTPUT... PLEASE PRESS RETURN TO CONTINUE...█

Fig. 13.8 Two photographs taken from the computer's screen illustrate the way the simulation facility of the CAPACITY program can be used for determining cell (i.e. machine or robot) utilisation levels

period of time can be obtained by adding the unadjusted cell requirement values together for each batch (or individual component if the batch size is 1) in the selected route.

Logically, the sum of this data will provide the total requirement for the given time period. In order to make a perfect analysis for selected intervals of the production period, scheduling programs as discussed above should also be used. These programs can simulate the time distribution of the peak loads with different scheduling methods. This should also apply if random (or dynamic) scheduling is used.

13.3 Robotised assembly line balancing

Assembly line balancing in most cases must start with a list of operations to be performed on assembly machines, or by assembly robots in a given order to complete a subassembly or a component. Often this order is expressed in terms of a precedence graph (Fig. 13.9). Specifying the time given for each operation as well as the order in which the operations must be completed, the task of the assembly line balancing program is to assign operations to workstations or assembly robots in the best way. (Some assembly systems requiring balancing are shown in Figs. 13.10–13.12).

The optimisation rule might be to maximise the production rate for a given product, or more generally to minimise the product unit cost. Whatever the method used, the 'rule of thumb' is that each workstation or robot should take the work in such a way that the system is neither under utilised nor overloaded, but working in a 'balanced' way.

Balancing is of extreme importance if manual and mechanised assembly, or other operations, are mixed in the same system, since imperfectly balanced assembly lines can put unnecessary pressure on the workforce.

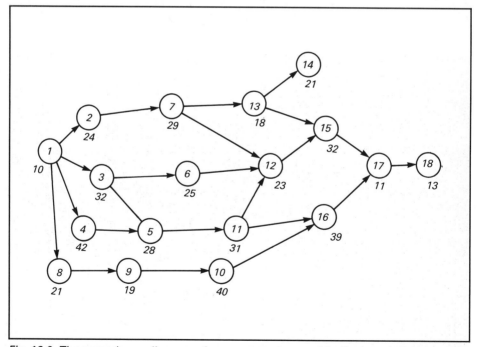

Fig. 13.9 The precedence diagram of a component assembly. Each node in the graph represents an operation or task. The number below each node indicates the length of time the operation takes place. The arrows indicate the order of operations, i.e. the precedence rules

Fig. 13.10 CAM shaft driven mechanised assembly systems require careful balancing (Courtesy of Plessey Office Systems Ltd, Nottingham, UK)

Fig. 13.11 Mechanised assembly machine with rotary indexing table; another example where the work load of the assembly heads must be balanced at system design stage

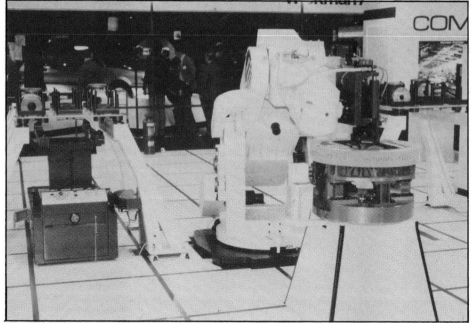

Fig. 13.12 Flexible robot cell making use of the Comau Smart robot and a pallet carrier AGV system

Main features of the BALANCE program

It must be emphasised that although mathematical models are never as perfect as the 'real thing', a vast amount of experience can be gained by simulation. Thus the BALANCE program is used in this text to offer some experience in this area.

Some general features of this program include:

- Capability of evaluating all possible solutions for minimising the number of robots (or assembly heads) required in an automated assembly system using a heuristic calculation model.
- User-friendly and run-time-checked screen management system incorporated to ensure a robust user interface.
- The program can be used as a simulation tool by using the 'What if?' input data edit command.
- Results can be saved in a user-specified text file on disk. (This file can also be used in user-written programs for further processing.)
- The input data collection and the calculated results can be downloaded via any valid port into any piece of equipment capable of receiving this data (in ASCII format) for further processing, etc.

Use of the BALANCE program

Before running the program the logically correct precedence diagram of the part or parts to be assembled on the assembly line in question should be prepared. In most cases the sequence of assembly operations is restricted in terms of the order in which the operations can be carried out. Sometimes alternatives are availbale or can be generated using simulation methods.

In any case it is important to analyse the assembly system carefully from the engineering point of view:

- What are the necessary operations?
- What are the alternative operations?
- What is the order of the selected operations?
- How many robots are required to perform the selected operations in the designed order?
- What happens if the cycle times change?
- Should the production rate be higher or should one employ less robots? Is there an optimum cycle time, if so what is it?

It is important to remember that in a 'real' FAS robots can accomodate a large variety of jobs in random order. Because they have an automated hand changing facility, the parts are transferred automatically between the robotised cells, and the necessary sensing, control and data communication tasks are carried out by the robot cell controller.

In this sense the robot cell is an assembly station capable of working unmanned, receiving and transferring parts, tools (i.e. robot hands) and data automatically. A robot in this sense is really a 'computer with hands'.

Fig. 13.13 illustrates the way this program can be used. (Note the way different cycle time values affect the number of robots required to perform the necessary operations.)

Main menu of the program

The main menu of the program includes the following commands:

P (rint input and calculated results

This will put all input data and the calculated data for each cell to the PRINTER: port.

R (erun program with the same data

This program uses a heuristic method for calculating the minimum number of robots required to perform the specified number of operations in their given order. The algorithm developed generates a number of feasible solutions in a random order and selects the answer which has the minimum number of robots (or assembly heads). Because of the random number generation, different solutions can occur every time the program is rerun, although the alterations are usually minimal.

N (ew data input and new run . . .

This command will clear all the memory used in the previous run.

W (hat if? Experiment by changing the maximum cycle time.

```
INPUT DATA ECHO
***************

BATCH CODE    :Sample batch A-010
PRODUCT NAME :Disk drive assembly
ANALYST       :Paul Ranky
DATE          :23 August 1984

COMMENT       :SAMPLE RUN

------------------------------------------------------------------

NUMBER OF OPERATIONS           =    18
MAXIMUM ASSEMBLY CYCLE TIME    =    55.000
NUMBER OF TRIALS IN THIS RUN   =    20


            OPERATION      DURATION        PRECEDENCE CONSTRAINTS
------------------------------------------------------------------

                1           10.000
                2           24.000          1,
                3           32.000          1,
                4           42.000          1,
                5           28.000          3,   4,
                6           25.000          3,
                7           29.000          2,
                8           21.000          1,
                9           19.000          8,
               10           40.000          9,
               11           31.000          5,
               12           23.000          6,   7, 11,
               13           18.000          7,
               14           21.000         13,
               15           32.000         13,  12,
               16           39.000         10,  11,
               17           11.000         15,  16,
               18           13.000         17,




CALCULATED RESULTS FOR THE OPTIMUM SOLUTION
********************************************


NUMBER OF ROBOT STATIONS REQUIRED          =    10
TOTAL CYCLE TIME REQUIREMENT               =    53.000

CYCLE TIME DIFFERENCE (target-calculated best) =    2.000
```

Fig. 13.13(a)–(d) Sample runs of the BALANCE program with different cycle time values (Note the way different cycle time values affect the number of robots or assembly heads required for the same job)

```
******    TABLE OF RESULTS    ******
```

ROBOT STATION	OPERATION	DURATION
1	1	10.000
	4	42.000
	SUB-TOTAL...	52.000
2	8	21.000
	2	24.000
	SUB-TOTAL...	45.000
3	3	32.000
	9	19.000
	SUB-TOTAL...	51.000
4	5	28.000
	6	25.000
	SUB-TOTAL...	53.000
5	11	31.000
	SUB-TOTAL...	31.000
6	7	29.000
	12	23.000
	SUB-TOTAL...	52.000
7	10	40.000
	SUB-TOTAL...	40.000
8	16	39.000
	SUB-TOTAL...	39.000
9	13	18.000
	15	32.000
	SUB-TOTAL...	50.000
10	17	11.000
	14	21.000
	18	13.000
	SUB-TOTAL...	45.000

Fig. 13.13(a) continued

```
INPUT DATA ECHO
**************

BATCH CODE    :Sample batch A-010
PRODUCT NAME  :Disk drive assembly
ANALYST       :Paul Ranky
DATE          :23 August 1984

COMMENT       :SAMPLE RUN

------------------------------------------------------------


NUMBER OF OPERATIONS              =     18
MAXIMUM ASSEMBLY CYCLE TIME       =     42.000
NUMBER OF TRIALS IN THIS RUN      =     20


            OPERATION      DURATION          PRECEDENCE CONSTRAINTS
      ------------------------------------------------------------------

                1           10.000
                2           24.000           1,
                3           32.000           1,
                4           42.000           1,
                5           28.000           3,    4,
                6           25.000           3,
                7           29.000           2,
                8           21.000           1,
                9           19.000           8,
               10           40.000           9,
               11           31.000           5,
               12           23.000           6,    7, 11,
               13           18.000           7,
               14           21.000          13,
               15           32.000          13, 12,
               16           39.000          10, 11,
               17           11.000          15, 16,
               18           13.000          17,


CALCULATED RESULTS FOR THE OPTIMUM SOLUTION
********************************************


NUMBER OF ROBOT STATIONS REQUIRED              =      14
TOTAL CYCLE TIME REQUIREMENT                   =      42.000

CYCLE TIME DIFFERENCE (target-calculated best) =       0.000


            ******    TABLE OF RESULTS    ******


   ROBOT STATION        OPERATION              DURATION
  -------------------------------------------------------------

          1                 1                  10.000
                            3                  32.000
                      SUB-TOTAL...             42.000

          2                 8                  21.000
                            9                  19.000
                      SUB-TOTAL...             40.000
```

Fig. 13.13(b)

3	10	40.000
	SUB-TOTAL...	40.000
4	6	25.000
	SUB-TOTAL...	25.000
5	2	24.000
	SUB-TOTAL...	24.000
6	4	42.000
	SUB-TOTAL...	42.000
7	7	29.000
	SUB-TOTAL...	29.000
8	13	18.000
	14	21.000
	SUB-TOTAL...	39.000
9	5	28.000
	SUB-TOTAL...	28.000
10	11	31.000
	SUB-TOTAL...	31.000
11	12	23.000
	SUB-TOTAL...	23.000
12	16	39.000
	SUB-TOTAL...	39.000
13	15	32.000
	SUB-TOTAL...	32.000
14	17	11.000
	18	13.000
	SUB-TOTAL...	24.000

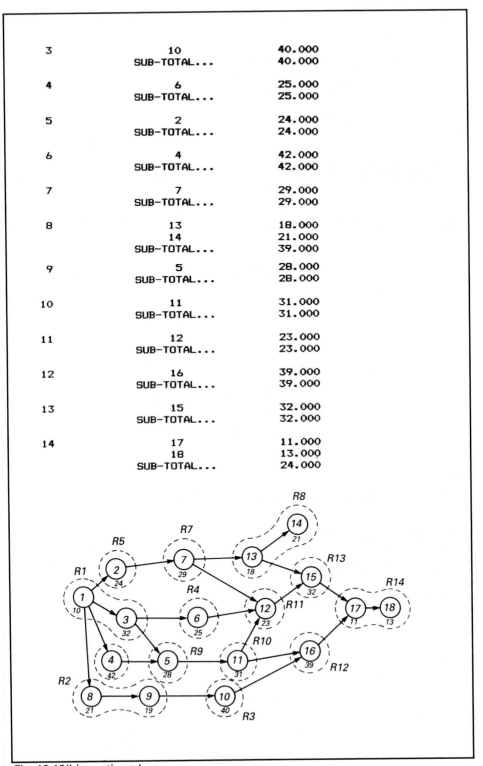

Fig. 13.13(b) continued

```
INPUT DATA ECHO
***************

BATCH CODE    :Sample batch A-010
PRODUCT NAME :Disk drive assembly
ANALYST       :Paul Ranky
DATE          :23 August 1984

COMMENT       :SAMPLE RUN

-------------------------------------------------------------------

NUMBER OF OPERATIONS                =    18
MAXIMUM ASSEMBLY CYCLE TIME         =    90.000
NUMBER OF TRIALS IN THIS RUN        =    20


            OPERATION      DURATION           PRECEDENCE CONSTRAINTS
-------------------------------------------------------------------

                1          10.000
                2          24.000            1,
                3          32.000            1,
                4          42.000            1,
                5          28.000            3,   4,
                6          25.000            3,
                7          29.000            2,
                8          21.000            1,
                9          19.000            8,
               10          40.000            9,
               11          31.000            5,
               12          23.000            6,   7,  11,
               13          18.000            7,
               14          21.000           13,
               15          32.000           13,  12,
               16          39.000           10,  11,
               17          11.000           15,  16,
               18          13.000           17,




CALCULATED RESULTS FOR THE OPTIMUM SOLUTION
*******************************************


NUMBER OF ROBOT STATIONS REQUIRED              =      6
TOTAL CYCLE TIME REQUIREMENT                   =      80.000

CYCLE TIME DIFFERENCE (target-calculated best) =      10.000
```

Fig. 13.13(c)

```
******   TABLE OF RESULTS   ******
```

ROBOT STATION	OPERATION	DURATION
1	1	10.000
	8	21.000
	4	42.000
	SUB-TOTAL...	73.000
2	3	32.000
	5	28.000
	9	19.000
	SUB-TOTAL...	79.000
3	11	31.000
	10	40.000
	SUB-TOTAL...	71.000
4	6	25.000
	2	24.000
	7	29.000
	SUB-TOTAL...	78.000
5	12	23.000
	16	39.000
	13	18.000
	SUB-TOTAL...	80.000
6	15	32.000
	17	11.000
	14	21.000
	18	13.000
	SUB-TOTAL...	77.000

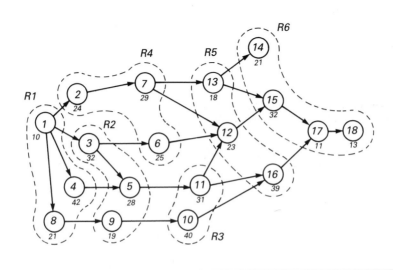

Fig. 13.13(c) continued

```
INPUT DATA ECHO
***************

BATCH CODE    :Sample batch A-010
PRODUCT NAME  :Disk drive assembly
ANALYST       :Paul Ranky
DATE          :23 August 1984

COMMENT       :SAMPLE RUN

------------------------------------------------------------

NUMBER OF OPERATIONS              =    18
MAXIMUM ASSEMBLY CYCLE TIME       =    200.000
NUMBER OF TRIALS IN THIS RUN      =    20


              OPERATION      DURATION          PRECEDENCE CONSTRAINTS
         -----------------------------------------------------------------

                  1          10.000
                  2          24.000            1,
                  3          32.000            1,
                  4          42.000            1,
                  5          28.000            3,    4,
                  6          25.000            3,
                  7          29.000            2,
                  8          21.000            1,
                  9          19.000            8,
                 10          40.000            9,
                 11          31.000            5,
                 12          23.000            6,    7, 11,
                 13          18.000            7,
                 14          21.000            13,
                 15          32.000            13, 12,
                 16          39.000            10, 11,
                 17          11.000            15, 16,
                 18          13.000            17,




CALCULATED RESULTS FOR THE OPTIMUM SOLUTION
*******************************************


NUMBER OF ROBOT STATIONS REQUIRED            =      3
TOTAL CYCLE TIME REQUIREMENT                 =      182.000

CYCLE TIME DIFFERENCE (target-calculated best) =    18.000
```

Fig. 13.13(d)

```
******   TABLE OF RESULTS   ******
```

ROBOT STATION	OPERATION	DURATION
1	1	10.000
	4	42.000
	3	32.000
	2	24.000
	6	25.000
	8	21.000
	5	28.000
	SUB-TOTAL...	182.000
2	9	19.000
	11	31.000
	7	29.000
	10	40.000
	12	23.000
	13	18.000
	14	21.000
	SUB-TOTAL...	181.000
3	15	32.000
	16	39.000
	17	11.000
	18	13.000
	SUB-TOTAL...	95.000

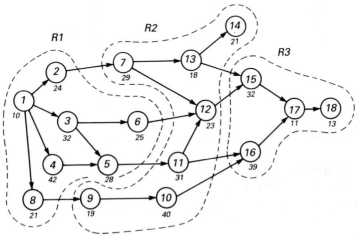

Fig. 13.13(d) continued

This is a useful feature if one has already at least one valid data collection and wishes to experiment with it by changing the maximum cycle time of the assembly system. As can be seen in the sample runs (Fig. 13.13), its value greatly influences the number of robots required.

S (ave results in a text file specified by yourself

This command allows the transfer of information from the memory to a disk file specified by the user, and data storage.

Mathematical model of the program

The core of the mathematical model of the program is based on an extended version of the COMSOAL algorithm. The algorithm of the BALANCE program employs a method of generating sequences of operations that can be described as follows:

Step 1.   For each operation on the assembly system form a list of the number of pre-tasks (i.e. operations).

Step 2.   Form a new list of operations from those which have no immediate pre-tasks and whose duration times are no greater than the remaining time of the robot station.

Step 3.   Form a scan list of this new list in a random order. (i.e. select operations randomly and form a new list).

Step 4.   Eliminate selected operation from the list and deduct 1 from the operations immediately following the selected operation from the list.

Step 5.   Reduce remaining time in the robot cell by the duration of the operation.

Step 6.   Repeat Steps 2 – 5 until all tasks have been allocated. Here, procedure evaluation is progressive, i.e. the available time is diminished as each operation is generated, and the operation too large for the remaining operation becomes the first at the next assembly station.

It is assumed in the algorithm that each operation in the assembly system could be performed by any of the robot cells, as long as the precedence diagram is observed. (By using flexible robot cells equipped with automated hand changers this provides no real problem even in a mixed product assembly situation.)

It is also assumed that at each robot a number of operations can be performed and that each robot is allowed the same maximum length of time to complete the assigned operations. The equal intervals an assembled component leaves the assembly system is the cycle time of the line. This is a sum of the time spent at a single robot plus the time needed to move the component or subassembly between robot cells. It is assumed that the time spent for material transportation between robot cells is the same between each robotised assembly cell. This is no real restriction if one employs a conveyor, or buffer stores, or the same type of robots working at the same speed when loading each other.

The BALANCE program can be used for balancing mixed product assembly lines. In this case compared to the single product balancing problem some operations are no longer performed on all assembly robots (i.e. assembly heads or assembly cells). To perform this task in a simple way, probably the best solution is to first design the precedence diagrams of the different products separately, and then analyse them together by simply adding the relevant processing times at the shared robot stations.

APPENDIX

THIS Appendix illustrates some theoretical results relating to what has been discussed in the first part of the book, as well as to provide some easy-to-follow, self-documented software procedures written in UCSD Pascal on the Apple II computer. The listed procedures can be integrated into a program written by the reader.

The purpose of the OPT_BASE program to be discussed at a procedure level is to enable the reader to simulate a two-link robot arm. The user must only input the X, Y coordinates of the points that are required for the robot to reach. These points are relative to a user-defined zero reference coordinate system. For the purpose of this program it is suggested that the zero reference point is no more than 2000mm from the furthest point.

Given this information the program will return the optimum base position, as illustrated, relative to the zero reference, as well as the two-arm length and their angular ranges of movement. The solution will be optimal, in other words the position of the robot base is such that the accessible region of the robot arm is the minimum area.

Following this concept the 'best-fit' robot arm can be designed for certain applications. Probably the most important area where this program could be applied is when designing machine integrated manipulators and robot arms for loading and unloading components (Fig. 1 illustrates a typical application of this kind).

1. The algorithm used in the OPT_BASE program

The program contains some modifications, but it must be mentioned that the algorithms adapted were inspired by two papers[1,2]. (The modifications allow the OPT_BASE program to offer in many cases a 5–10% smaller area compared to the results described in the given references.)

To summarise the calculation method (Fig. 2) the X coordinate of the end of the robot arm can be calculated as follows:

$$X = L_1\sin\theta_1 + L_2\sin(\theta_1 + \theta_2) \tag{1}$$

and similarly

$$Y = L_1\cos\theta_1 + L_2\cos(\theta_1 + \theta_2) \tag{2}$$

where Y is the Y coordinate of the end of the robot arm.

The angle of the robot link No. 1 ($\theta_{1,i}$) required to reach point i is equal to:

$$\cos^{-1}\frac{Y_i - Y_0}{\sqrt{(X_i - X_0)^2 + (Y_i - Y_0)^2}}$$

$$\cos^{-1}\frac{(X_i - X_0)^2 + (Y_i - Y_0)^2 + L_1^2 - L_2^2}{2L_1\sqrt{(X_i - X_0)^2 + (Y_i - Y_0)^2}} \tag{3}$$

and respectively for the second arm:

$$\theta_{2,i} = \cos^{-1}\frac{(X_i - X_0)^2 + (Y_i - Y_0)^2 - (L_1^2 + L_2^2)}{2L_1L_2} \tag{4}$$

The distance L_i is the distance from point (X_0, Y_0) to point (X_i, Y_i), and can be calculated as follows:

$$L_i = \sqrt{(X_i - X_0)^2 + (Y_i - Y_0)^2} \tag{5}$$

Fig. 1 Machine integrated loading/unloading manipulator in a CNC turning cell

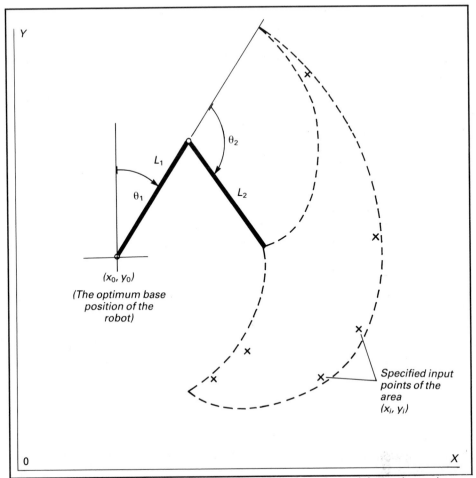

Fig. 2 Illustration to the calculation of the 'best-fit' robot arm and the optimum base position

The area of the accessible region (A) is:

$$A = F (\theta_{1max} - \theta_{1min}) (L_1 + L_2)^2 \qquad (6)$$

where

$$F = \frac{(\cos\theta_{2min} - \cos\theta_{2max})}{1 + (L_2/L_1)^2} \qquad (7)$$

The methodology used in the OPT_BASE program is a grid search method. This involves choosing points over a given area, performing a set of calculations at the chosen point and then comparing the results with the results obtained at previous points.

Basically the algorithm used is as follows:

- The first position of the grid search is chosen (X_0, Y_0).
- The length of (X_i, Y_i) to each (X, Y) is computed, using Eqn. (5).
- The maximum value of L is computed and the length L_1 and L_2 are determined.

- $\theta_{1,i}$ and $\theta_{2,i}$ are calculated using Eqns. (3) and (4). The θ_{1min}, θ_{1max}, θ_{2min} and θ_{2max} values are established also from these equations.
- The accessible region area is calculated from Eqn. (6).
- The last value and the currently calculated value of the area are compared and the lower area value provides the basis for the synthesis of the two link arm, i.e. for calculating L_1,L_2 and X_0,X_0, θ_{1min}, θ_{1max}. θ_{2min}, θ_{2max}.

In order to search every possible point for the robot base position it would require a very large number of calculations. To avoid this the search is broken down into grids as seen from the illustrated procedures.

2. Some important procedures used in the OPT_BASE program

Probably the most important procedures used in the program are those which are searching for the optimum X and Y base coordinate pair. (Note that in each case to minimise computation, time integer variables have been used during the rough search procedure. Even so, some of the results shown took the Apple II computer over 45 minutes to work on some of the graphs. This relatively long time could be reduced in a number of different ways, including further optimisation of the algorithm. However, to provide an easy-to-follow code this has been avoided as much as is feasible.)

The MAX_MIN procedure (Fig. 3) determines the maximum and minimum values of both the X and Y values from the input data collection. These are used for determining the size of the grid search area.

```
PROCEDURE MAX_MIN;   (* FINDS MAX/MIN INPUT VALUES *)

VAR COUNTER: INTEGER;           (* COUNTER *)

BEGIN

  X_MAX:= CX[1];                     (* INPUT X CO-ORDINATES *)

  MIN_X:= CX[1];


  MAX_Y:= CY[1];                     (* INPUT Y CO-ORDINATES *)

  Y_MIN:= CY[1];


  FOR COUNTER:= 2 TO SPEC_INPUT_POS DO
                        (* NUMBER OF SPECIFIED INPUT POSITIONS *)

    BEGIN

      IF CX[COUNTER] > X_MAX THEN X_MAX:= CX[COUNTER];

      IF CX[COUNTER] > MIN_X THEN MIN_X:= CX[COUNTER];

      IF CX[COUNTER] > MAX_Y THEN MAX_Y:= CY[COUNTER];

      IF CX[COUNTER] > Y_MIN THEN Y_MIN:= CY[COUNTER];

    END;

END; (* MAX_MIN PROCEDURE *)
```

Fig. 3 The UCSD-Pascal listing of the MAX_MIN procedure

```
PROCEDURE ROUGH_SEARCH;    (* SEARCHES EVERY 100 MM FOR OPTIMUM *)

BEGIN

    AREA_MIN_X:= 0;

    AREA_MIN_Y:= TRUNC((Y_MIN+20)/100);

    SCALE:= 100;                    (* GRID SEARCH SCALE FACTOR *)

    AREA_MAX_Y:= ROUND(MAX_Y/100);

    AREA_MAX_X:= ROUND((MAX_X-300)/100);

    GRID_SEARCH;
```

Fig. 4 The ROUGH_SEARCH procedure

The ROUGH_SEARCH procedure (Fig. 4) sets up the grid search parameters. This procedure searches the grid area at 100mm intervals in both the X and Y direction. It also calls the GRID_SEARCH procedure, discussed later. The results of the optimum position are used in other procedures, although the calculated position is only the first approximation towards the optimum solution.

The NARROW procedure (Fig. 5) uses the previously determined point and searches an area of 200×200mm with the previous best point as the centre of this area. The area is grid search at 10mm intervals again, calling the procedure GRID_SEARCH discussed later.

Procedure FINE (Fig. 6) uses the previously determined point and searches an area of 20×20mm centering on that point. The area is grid searched at 1mm intervals using the GRID_SEARCH procedure with the optimum result returned being the final position for the base X, Y coordinate pair.

```
PROCEDURE NARROW;    (* SEARCHES EVERY 10 MM AROUND PREVIOUS POINT *)

(* VALUE... RESULTS STORED FOR COMPARISON IN A REAL ARRAY *)

BEGIN

    AREA_MIN_X:= ROUND((VALUE[2]-100)/10);

    AREA_MIN_Y:= ROUND((VALUE[3]-100)/10);

    SCALE:= 10;

    AREA_MAX_Y:= ROUND((VALUE[3]+100)/10);

    AREA_MAX_X:= ROUND((VALUE[2]+100)/10);

    GRID_SEARCH;

END; (* PROCEDURE NARROW *)
```

Fig. 5 The NARROW procedure

```
PROCEDURE FINE; (* SEARCHES EVERY 1 MM AROUND PREVIOUS BEST POINT *)

BEGIN

    AREA_MIN_X:= ROUND(VALUE[2]-10);

    AREA_MIN_Y:= ROUND(VALUE[3]-10);

    SCALE:= 1;

    AREA_MAX_X:= AREA_MIN_X + 20;

    AREA_MAX_Y:= AREA_MIN_Y + 20;

    GRID_SEARCH;

END; (* PROCEDURE FINE *)
```

Fig. 6 The FINE procedure

The GRID_SEARCH procedure consists of two nested loops, one of them incrementing the X coordinate and the other one for incrementing the Y coordinate during the grid search. Within the loops, Eqn. (5) is performed in a procedure called LENGTH, whiles Eqns. (3) and (4) are performed in the procedure named THETA. Both procedures return maximum and minimum values as required.

The value of 'A' is computed and compared with the value returned by the last time through the loop (or a constant if this is the first loop). If the value of 'A' is lower then the values of A, all are stored for future use.

This procedure returns the optimum values for each time it is called, and thus allows the final optimal position to be established for the robot arm base in relationship to the specified area points.

3. Sample runs of the OPT_BASE program

To illustrate the algorithm discussed, some sample runs are shown of the OPT_BASE program with their input data, calculated results and the produced graphs (Figs. 8–11).

References

[1] Tsai, Y. C. and Soni, A. H., 1981. Accessible region and synthesis of robot arms. *J. Mechanical Design,* 103: 803-811.
[2] Tsai, Y. C. and Soni, A. H., 1983. An algorithm of the workspace of a general *n-R* robot. *Trans. ASME,* 105 (March): 52-56.

```
PROCEDURE GRID_SEARCH;   (* GRID SEARCH FOR OPTIMAL BASE POSITION *)

VAR X_SEARCH_COUNT, Y_SEARCH_COUNT : INTEGER;
                                  (* X AND Y GRID SEARCH COUNTERS *)

    A, F : REAL;     (* COMPUTATIONAL VALUES *)

BEGIN
  FOR Y_SEARCH_COUNT:= AREA_MIN_Y TO AREA_MAX_Y DO

    BEGIN

      ACTUAL_BASE_Y:= Y_SEARCH_COUNT * SCALE;

      FOR X_SEARCH_COUNT:=AREA_MIN_X TO AREA_MAX_X DO

          BEGIN

            ACTUAL_BASE_X:=X_SEARCH_COUNT * SCALE;

            LENGTH;    (* PROCEDURE CALCULATES LENGTH FROM BASE
                           TO EACH POINT *)

            THETA;     (* PROCEDURE CALCULATES ARM ANGLES TO EACH
                           SPECIFIED INPUT POSITION *)

            F:= (COS(MIN_THETA2) - COS(MAX_THETA2) / 2;

            A:= F * (MAX_THETA1 - MIN_THETA1) * L * L;

            IF   A < VALUE[1] THEN

                    BEGIN

                      VALUE[1]:= A;

                      VALUE[2]:= ACTUAL_BASE_X;

                      VALUE[3]:= ACTUAL_BASE_Y;

                      VALUE[4]:= L1;

                      VALUE[5]:= L2;

                      VALUE[6]:= MIN_THETA1;

                      VALUE[7]:= MAX_THETA1;

                      VALUE[8]:= MIN_THETA2;

                      VALUE[9]:= MAX_THETA2;

                      VALUE[10]:= (TRUNC(VALUE[6] * 100 * 57.2958)) / 100;

                      VALUE[11]:= (TRUNC(VALUE[7] * 100 * 57.2958)) / 100;

                      VALUE[12]:= (TRUNC(VALUE[8] * 100 * 57.2958)) / 100;

                      VALUE[13]:= (TRUNC(VALUE[9] * 100 * 57.2958)) / 100;

                    END

               ELSE
          END;

      END;

  END; (* PROCEDURE GRID_SEARCH *)
```

Fig. 7 The GRID_SEARCH procedure

```
                        INPUT DATA ECHO
                        ***************

POINT No.        X CO-ORDINATES        Y CO-ORDINATES
-----------------------------------------------------------

    1              1000.00               150.00

    2              1100.00               150.00

    3              1094.00               106.00

    4              1050.00               150.00

    5              1050.00               300.00

    6              1094.00               556.00

    7              1300.00               600.00

    8              1400.00               600.00



                 CALCULATED OPTIMAL RESULTS
                 **************************


        X CO-ORDINATE OF THE BASE = 426.00

        Y CO-ORDINATE OF THE BASE = 540.00

        ARM LENGTH 1              = 487.92

        ARM LENGTH 2              = 487.92

        MIN. VALUE OF THETA 1     =  41.57

        MAX. VALUE OF THETA 1     =  86.49

        MIN. VALUE OF THETA 2     =   0.00

        MAX. VALUE OF THETA 2     =  93.56
```

Fig. 8 Sample run No. 1

```
                        INPUT DATA ECHO
                        ***************

     POINT No.       X CO-ORDINATES       Y CO-ORDINATES
     ----------------------------------------------------------

         1              800.00              500.00

         2              900.00              500.00

         3             1000.00              500.00



                    CALCULATED OPTIMAL RESULTS
                    **************************


         X CO-ORDINATE OF THE BASE = 610.00

         Y CO-ORDINATE OF THE BASE = 610.00


         ARM LENGTH 1              = 202.61

         ARM LENGTH 2              = 202.61


         MIN. VALUE OF THETA 1     =  62.81

         MAX. VALUE OF THETA 1     = 105.75


         MIN. VALUE OF THETA 2     =   0.00

         MAX. VALUE OF THETA 2     = 114.22
```

Fig. 9 Sample run No. 2

```
                        INPUT DATA ECHO
                        ***************

    POINT No.       X CO-ORDINATES        Y CO-ORDINATES
    ------------------------------------------------------

        1              800.00              500.00

        2              900.00              500.00

        3             1000.00              500.00

        4             1100.00              500.00

        5             1000.00              750.00


                  CALCULATED OPTIMAL RESULTS
                  **************************


        X CO-ORDINATE OF THE BASE = 610.00

        Y CO-ORDINATE OF THE BASE = 532.00


        ARM LENGTH 1              = 245.52

        ARM LENGTH 2              = 245.52


        MIN. VALUE OF THETA 1     =  32.71

        MAX. VALUE OF THETA 1     =  93.73


        MIN. VALUE OF THETA 2     =   0.00

        MAX. VALUE OF THETA 2     = 133.25
```

Fig. 10 Sample run No. 3

```
                    INPUT DATA ECHO
                    ***************

POINT No.      X CO-ORDINATES       Y CO-ORDINATES
-------------------------------------------------------

   1              1000.00           150.00

   2              1904.00           106.00

   3              1094.00           556.00

   4              1400.00           600.00



                CALCULATED OPTIMAL RESULTS
                ***************************


     X CO-ORDINATE OF THE BASE = 513.00

     Y CO-ORDINATE OF THE BASE = 477.00


     ARM LENGTH 1              = 447.75

     ARM LENGTH 2              = 447.75


     MIN. VALUE OF THETA 1     =  32.97

     MAX. VALUE OF THETA 1     =  82.14


     MIN. VALUE OF THETA 2     =   0.00

     MAX. VALUE OF THETA 2     =  98.17
```

Fig. 11 Sample run No. 4

```
                          INPUT DATA ECHO
                          ***************

    POINT No.        X CO-ORDINATES        Y CO-ORDINATES
    ---------------------------------------------------------

         1              1000.00              150.00

         2              1904.00              106.00

         3              1094.00              556.00

         4              1400.00              600.00


                     CALCULATED OPTIMAL RESULTS
                     **************************


         X CO-ORDINATE OF THE BASE = 513.00

         Y CO-ORDINATE OF THE BASE = 477.00


         ARM LENGTH 1               = 447.75

         ARM LENGTH 2               = 447.75


         MIN. VALUE OF THETA 1      =  32.97

         MAX. VALUE OF THETA 1      =  82.14


         MIN. VALUE OF THETA 2      =   0.00

         MAX. VALUE OF THETA 2      =  98.17
```

Fig. 11 Sample run No. 4

AUTHORS' ADDRESSES

Dr. Paul G. Ránky
Associate Professor
ITI, The University of Michigan
PO Box 1485
Ann Arbor, Michigan 48106
USA
Tel. (313) 764-6775

Dr. C. Y. (Peter) Ho
Professor
University of Missouri – Rolla
Computer Science Department
Rolla, Missouri 65401
USA
Tel. (314) 341-4991

FMS Software Library

For further information on the 'FMS Software Library' referred to in the text, please contact either Paul Ránky at the above address, or the UK distributor:

John G. Crouchley
MALVA Ltd
Real-time Software Specialists
70 Aschurch Drive
Wollaton
Nottingham NG8 2RA
Tel. (0602) 284593